Protein Bioinformatics

Protein Bioinformatics: An Algorithmic Approach to Sequence and Structure Analysis

Ingvar Eidhammer and Inge Jonassen

Department of Informatics, University of Bergen, Norway

William R. Taylor

Division of Mathematical Biology, National Institute for Medical Research, London, UK

John Wiley & Sons, Ltd

Other Wiley Editorial Offices

John Wiley & Sons Inc., 111 River Street, Hoboken, NJ 07030, USA

Jossey-Bass, 989 Market Street, San Francisco, CA 94103-1741, USA

Wiley-VCH Verlag GmbH, Pappellaee 3, D-69469 Weinheim, Germany

John Wiley & Sons Australia Ltd, 33 Park Road, Milton, Queensland 4064, Australia

John Wiley & Sons (Asia) Pte Ltd, 2 Clementi Loop #02-01, Jin Xing Distripark, Singapore 129809

John Wiley & Sons Canada Ltd, 22 Worcester Road, Etobicoke, Ontario, Canada M9W 1L1

Wiley also publishes its books in a variety of electronic formats. Some content that appears in print may
not be available in electronic books.

British Library Cataloguing in Publication Data

A catalogue record for this book is available from the British Library

ISBN 0-470-84839-1

Produced from LATEX files supplied by the authors, typeset by T&T Productions Ltd, London.

This book is printed on acid-free paper responsibly manufactured from sustainable forestry
in which at least two trees are planted for each one used for paper production.

Contents

5   Scoring Matrices   101

6   Profiles   123

Preface

Aim of the book

Our aim has been to write a book that can serve as a textbook as well as a reference book covering the central algorithms for protein bioinformatics, focusing on methods for sequence and structure comparison and pattern discovery.

The book is for both researchers and students in molecular biology who want to understand more fully the programs or methods they are using and also for computer scientists who wants to learn how algorithms are used to attack the central problems in molecular biology. The first group of readers should, by reading the book, become more aware of the possibilities and the limitations of the available methods and programs, gain a better basis for choosing the right program and the right parameters and options, and finally become more competent to understand and evaluate the results. The computer scientist (or computer science student), on the other hand, may find interesting problems, and he or she may gain the understanding necessary to come up with improved or new algorithms in bioinformatics. The book will also provide a common basis for collaborative projects in the area by providing both communities with a common reference.

A challenge in writing this book has been to find the right balance making the material interesting and understandable for both groups of readers. We attempt to do this by focusing on the ideas of the methods/programs while giving only a general idea about their areas of application. The methods are also described formally as algorithms, avoiding all details that are not central to the overall understanding of the method. In order to help the explanation we include a large number of examples that illustrate the basic concepts and ideas. In addition there are a number of exercises in each chapter. Finally, each chapter has a bibliographic section providing pointers to the original research papers as well as reviews and books for the reader who wants to learn about the subject in more depth.

A large number of bioinformatics books have appeared over the last few years, several of which are aimed at practitioners of bioinformatics, and many of which give excellent descriptions of the available databases for DNA and protein sequences, protein structures, and descriptions of how to use common programs and on-line tools. These books typically go in very little detail on the internal working of the algorithms used in the programs or tools. There are also books with a more theoretical focus on the algorithmic or statistical aspects of bioinformatics. Typically, these are aimed at students or researchers with an extensive background in computer science and/or statistics. We found that there is a shortage of books that aimed at both bioinformatics practitioners and developers.

Prerequisites

We assume that the reader has a very basic knowledge of mathematics. No knowledge of computer science or programming is assumed. Readers with very little background in mathematics and computer science should be able to follow most of the book, but may find some sections hard to follow (which can be skipped). In a similar manner we assume no prior knowledge of molecular biology, but some background in biology will help to understand in what context the algorithms are used and the reasoning behind the assumptions made and the choice of heuristic procedures. In several of the chapters examples use the insulin protein, and this protein is therefore described in the appendix.

We include in the appendix short introductions to the central concepts of molecular biology and computer science, as well as some very basic mathematics. In the appendix we define the central concepts and introduce notations and terminology used in the book, and we recommend all readers to first read or take a look at this. These introductions are of course very brief and readers may want to consult additional books or other material.

For a fuller understanding of the topics we provide biological motivation and the context for each of the computational problems and algorithms covered in the book.

What is bioinformatics anyway?

During the twentieth century, science and technology advanced tremendously. Molecular biology and computer science and technology are areas where really revolutionary events and developments took place during the century. Since the discovery of the structure of DNA by Watson and Crick in 1953, molecular biology has taken off and gained many new insights. The development of digital computers went through major breakthroughs in the 1940s and 1950s. It is the developments in these areas that have made possible the newer discipline of bioinformatics that is now attracting many students and researchers from both the computational and the biological communities.

One reason why computer scientists are attracted to molecular biology is that the way information is encoded in DNA is in some way similar to the way it is coded in computers. While computers on a basic level deal with zeros and ones (bits), DNA carries information as chains of molecules (nucleotides) that come in four different types.

Biological and biomedical research has been revolutionized by a number of technological advances, including automatic methods for DNA sequencing, advances in methods for the determination of three-dimensional structures of proteins, and the development of high-throughput methods for measurements of mRNA and protein abundances. All these data carry information of value for understanding the underlying biology. To a large extent it requires the application of computers to uncover this information. Biologists are becoming more and more dependent on computers for storing and analysing data, both data that they are producing in their own laboratories and data produced elsewhere. Public repositories of different sources of data are valuable resources, but one needs computer programs to extract information from these, for example, to make inferences about DNA sequences emerging from sequencing projects.

Bioinformatics has grown into a large topic, but still one of the most widely used tools in bioinformatics is that for searching a sequence database for all sequences similar to a given query sequence. The query can be, for example, a new DNA sequence potentially coding for a protein. If the query is found to be significantly (beyond what can be expected to happen by chance) similar to a database protein sequence, one can hypothesize that the new gene codes for a protein homologous (evolutionarily related) to the database sequence and that the query sequence may code for a protein with structural and functional properties similar to those of the database protein. Algorithms for performing this type of analysis are discussed in depth in Chapters 1–3.

Another widely used approach is to collect a set of related protein sequences, align them, and study the alignment to gain information about the relationships between the sequences under study and the functional and structural properties common to the proteins encoded by them. For example, if one finds conserved positions (columns in the alignment where all sequences have the same amino acid), the amino acid in this position may be crucial to the function or the structure of the protein. Also, one may find patterns of hydrophobicity/hydrophilicity (amino acids that 'dislike' and 'like' water, respectively) that may be hints about the existence of secondary structure elements, e.g. alpha helices with one side on the protein surface and one side buried. Multiple alignments also form the starting point for phylogenetic studies, i.e. the estimation of trees reflecting the evolutionary relationship between a set of proteins (or genes). Both for protein functional and structural studies and for phylogenetic analyses, it is crucial that the multiple alignment is as accurate as possible. Methods for the automatic analysis of multiple sequences are treated in Chapters 4–7.

Proteins are formed by chains of amino acids and proteins are both building blocks for cellular structures and the major working horses (performing, for example, metabolism and signalling) in living cells. In order to perform their functions, proteins fold into three-dimensional structures. It is interesting to study the structures both to understand the functional mechanisms at a molecular level (that may be involved in disease), to better understand the evolution of proteins (since the structure of a protein changes more slowly in evolution than does its sequence), and to identify the common building blocks of protein structures. It is also interesting and important to form a classification of the 'universe of protein structures'. A common approach to doing this is to combine some manual classification with methods for automatic comparison of protein structures. In Part II (Chapters 8–14) we describe methods for comparing structures and we discuss structure classification systems.

A major challenge in bioinformatics is to predict a protein's three-dimensional structure given its sequence (the order of amino acids along its chain(s)). No methods exist that produce accurate models for arbitrary proteins. The most successful methods produce models of structures by using existing known protein structures as a starting point. In one approach called threading one takes one protein sequence and one protein structure and evaluates whether the sequence 'fits into' the structure. This can be done using algorithms similar to those used for comparing protein structures. We give a brief introduction to protein structure prediction and threading in Part III (Chapter 15).

What about DNA?

The book mainly focuses on methods for the analysis of proteins, i.e. protein sequences and protein structures. Most of the algorithms for protein sequences are also readily applicable to nucleotide (DNA or RNA) sequences. There are, however, a number of algorithms for the analysis of nucleotide sequences that are not covered in this book. These includes gene prediction, prediction and analysis of RNA secondary (and tertiary) structure, sequence assembly, and EST clustering.

Website

The book's website is http://www.ii.uib.no/proteinbioinformatics/.

The website will contain solutions to exercises, new literature, pointers to programs and biological databases, etc.

Notation

The following notation predominates.

- $a, b, c, \ldots$ denotes unspecified amino acids.

- A, C, D, $\ldots$ (the one-letter code) is used for special amino acids.

- C_α and C^α are both used for the backbone α-carbon atom.

- $\mathcal{C}$ is a general alphabet, mostly used for the set of amino acids.

- $\mathcal{M}$ is used for the set of amino acids.

- s is used for a general sequence, and $\{s^1, s^2, \ldots, s^n\}$ for a set of sequences.

- $S^{1,n}$ is used for the set $\{s^1, s^2, \ldots, s^n\}$.

- q is used for a (query) sequence, and d for a database sequence.

- q_i is the ith residue in q.

- $q_{i \ldots j}$ is the subsequence (substring) of q from q_i to q_j.

- r is used for a general residue.

- S and S' are used for scoring, mostly between segments, sequences or structures.

- R is used for a scoring matrix, R_{ab} is the scoring between a and b.

- $\mathcal{A}$ denotes an alignment.

- $\mathcal{E}$ denotes an equivalence.

- $A, B, C, \ldots$ are used for structures.

- $a_i, a_k, a_r, b_j, b_l, b_s, \ldots$ are used for residues when comparing structures.

- $A_i, A_k, A_r, B_j, B_l, B_s, \ldots$ are used for components of structures when comparing structures.

- P is used as a path in dynamic programming.

- $(S, P) = \mathcal{A}_R(A, B, \dots)$ denotes an alignment done by dynamic programming on the structures $A, B, \dots$ using the scoring matrix R. S is the scoring, and P the path.

- T denotes a transformation (for superposition).

Acknowledgements

This book is built on lecture notes for courses in bioinformatics algorithms taught at the University of Bergen. We acknowledge all students who have followed the courses for valuable comments and suggestions, and especially we would like to thank Øystein Tenfjord Engelsen and Trond Hellem Bø. We also appreciate many informative and inspiring conversations with Rein Aasland and Kjell Petersen. Some of the material covered builds on joint work with Alvis Brazma and David Gilbert. We have also been in contact with several researchers in the field during the work on this book, and we would like to thank all and especially Ruth Nussinov, and Oranit Dror for valuable help. Finally, thanks to our families for keeping up with us during busy writing periods.

Part I

SEQUENCE ANALYSIS

.

1

Pairwise Global Alignment of Sequences

Comparing sequences, structures (and sequences with structures) is the most fundamental operation in protein sequence and structure analysis. When a comparison indicates a similarity between two proteins, it can immediately suggest relationships involving structure, function and the evolution of the two proteins from a common ancestor protein. When one of the proteins is well characterized (in terms of structure and function), the connection with a novel sequence allows all the hard-earned biological data to be transferred to the new protein. The degree of certainty with which this transfer can be made depends on how similar the two sequences are, but even for distant relationships it is likely that the overall structure of the two proteins (their fold) will have remained the same and even tentative suggestions of function can be used as a basis to suggest further experiments on the novel protein.

The comparison of two proteins is mostly made by trying to align the sequences (structures or sequences/structures). In making an alignment, a 1:1 correspondence is set up between the residues of the two proteins. This has the evolutionary implication that at one time the paired residues were the same in an ancestral protein and have diverged through the accumulation of point mutations (in their DNA). Point mutation is not the only process at work and extra residues may have been inserted or deleted giving rise to breaks or gaps in the alignment. These are referred to as insertions and deletions or, jointly, as indels. The simplest operation to explain is the global alignment of two sequences, in which the two proteins have maintained a correspondence over their entire length. An alternative is to align only the most similar part of the proteins, which is called local alignment and which will be considered in the next chapter.

In this chapter, we describe the basic algorithm for making an alignment (called dynamic programming) before considering more specialized comparison methods in later chapters. The basic dynamic programming algorithm will recur throughout these and other chapters and is perhaps the most widely used and important algorithm in bioinformatics. Variations of it are used for local alignment and it can be extended to align more that two sequences (multiple alignment). In later chapters we will also

Protein bioinformatics: an algorithmic approach to sequence and structure analysis
I. Eidhammer, I. Jonassen and W. R. Taylor © 2004 John Wiley & Sons, Ltd   ISBN: 0-470-84839-1

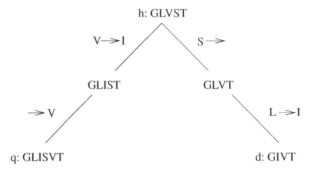

Figure 1.1   An evolution from h to q and d.

describe how it has been adapted to compare two structures and for the hybrid task of comparing a sequence with a structure.

1.1   Alignment and Evolution

An evolutionary perspective is important for getting an understanding of the function of proteins. That means, given two proteins, one often wants to find the evolutionary relationship between them. When only the sequences of the proteins are known, one attempts to reveal the relationship by *aligning* the sequences. The alignment should therefore show the mutations that have happened in the evolution of the two sequences.

Example

Let $h = $ GLVST be the ancestor of two sequences $q = $ GLISVT and $d = $ GIVT. Assume that the evolution is as shown in Figure 1.1, where $a \to b$ means substitution from a to b, $a \to$ means deletion of a, and $\to a$ means insertion of a.

An alignment should show the corresponding positions of q and d, and where insertions and deletions have occurred. Thus, the 'true' alignment can be found by using h as a template,

$$h: \quad \text{GLVS T}$$

$$q': \quad \text{GLISVT}$$
$$d': \quad \text{GIV--T}$$

where '–' (denoted by *blank*) means deletion or insertion (*indels*). One or several contiguous blanks are called a *gap*. d' is d with possible insertions of gaps.   △

When the evolutionary history is not known (and h is not known), a given alignment can be interpreted in different ways. If we assume that only single mutations have happened (only one residue change in each mutation), the alignment between q and d in the example can be interpreted as two substitutions, and either two insertions, two deletions, or one deletion and one insertion. That means, even by knowing the correct alignment, we are not able to reconstruct for certain the evolutionary history, the true one is only one of several possibilities.

When trying to reconstruct the evolution, one needs to have a *model*, telling how to construct the tree from an alignment. One such model can be to not introduce more mutations than necessary, resulting in the following relation between q and d:

$$q = \text{GLISVT}: \quad \text{I} \leftrightarrow \text{L}; \text{V} \leftrightarrow \text{I}; \leftarrow \text{S} \rightarrow; \leftarrow \text{V} \rightarrow : d = \text{GIVT}.$$

The new symbols are introduced to show that we do not know the direction of the substitutions, and for each blank we do not know whether an insertion or a deletion has happened. Several histories can be constructed from this relation (where the true history is one of those), for example,

$$q = \text{GLISVT}: \quad \text{L} \rightarrow \text{I}; \text{I} \rightarrow \text{V}; \text{S} \rightarrow; \text{V} \rightarrow : d = \text{GIVT}.$$

meaning that q is an ancestor of d.

When only the sequences are known, it is even more difficult to reconstruct the true evolutionary history. First, one can try to align them, and then construct the history from the constructed alignment. For constructing an alignment we again need a model, and the same simple model can be used: try to minimize the number of mutations. An alignment of q and d in accordance with this would be

$$
\begin{aligned}
q': &\quad \text{GLISVT} \\
d': &\quad \text{G-I-VT}
\end{aligned}
$$

showing two indels. One history could be

```
        h*:GLIVT
          / \
     ->S/   \ L->
       /     \
  q:GLISVT  d:GIVT
```

Since $h*$ is not the same as the true h, using our model with this example does not give us the true evolutionary history from the alignment only. Despite this drawback, and in the absence of a better alternative that is not too complicated, we will often use this model, since it is so simple. It should be mentioned that constructing alignments for predicting evolution is only meaningful for sequences of *homologous* proteins, i.e. proteins with a common ancestor. But whether the sequences are homologous or not is not often known, and in this context we see an important aspect of making alignments: to assess if homology exists. Being able to construct a 'good' alignment can indicate homology. Homology can then be used to predict the structure and/or the function of proteins for which those are not known, since two homologous proteins often have similar structures. This is one motivation for database searching: given a query sequence q, find the sequences in a database D which make 'good' alignments to q. This is treated in Chapter 2.

1.2   What is an Alignment?

An alignment of two sequences q and d must satisfy the following constraints

- All symbols (residues) in q and d have to be in the alignment, and in the same order as they appear in q and d.

- We can align one symbol from q with one from d.

- A symbol can be aligned with a *blank*, written as '−'.

- Two blanks cannot be aligned.

Example

A possible alignment of the insulin proteins from sheep and zebrafish is

```
Fish:  MAVWLQAGALLVLLVV-SSVSTNPGTPQHLCGSHLVDALYLVCGPTGFFYNPK--R
Sheep: MALWTRLVPLLALLALWAPAPAHAFVNQHLCGSHLVEALYLVCGERGFFYTPKARR

Fish:  DVE-PLLGFLPPKSAQETEVADFAFKDHAELIRKRGIVEQCCHKPCSIFELQNYCN
Sheep: EVEGPQVGAL--ELAGGPG-AG-GL-EGPP-Q-KRGIVEQCCAGVCSLYQLENYCN
```

△

1.3   A Scoring Scheme for the Model

From our simple model for constructing alignments, we can define a scoring scheme for scoring the alignments.

- Each column can be given a score, independent of the other columns, meaning that we think of all mutations as single mutations:

  the score of a column with two amino acids a, b is denoted by R_{ab};

  the score of a column with blank can be $-g$, where g is called the *penalty* of a blank.

- The score of the alignment can be found as the sum of the score of all columns (additive scoring scheme).

Note the correspondence between score and penalty of a column with blank, the score is the negative of the penalty.

Example

Let a scoring scheme be

- $R_{ab} = 1$ for $a = b$, 0 for $a \neq b$;

- $g = 1$.

Then the score of some different alignments of the same sequences are

```
ALIGN1:
q': V  -  E  I  T  G  E  I  S  T
d': P  R  E  -  T  E  R  I  -  T
    0 -1  1 -1  1  0  0  1 -1  1          Score 1


ALIGN2:
q': V  E  I  T  G  E  I  S  T
d': P  R  E  T  -  E  R  I  T
    0  0  0  1 -1  1  0  0  1             Score 2


ALIGN3:
q': -  V  E  I  T  G  E  -  I  S  T
d': P  R  E  -  T  -  E  R  I  -  T
   -1  0  1 -1  1 -1  1 -1  1 -1  1       Score 0
```

△

Note that which alignment will score highest depends on the scoring scheme used. Hence, finding the highest-scoring alignment does not necessarily means the 'best alignment', if the scoring scheme is bad. Therefore, choosing which scoring scheme to use is an important and difficult task. Also note that there may be more than one alignment with the maximum score.

1.4 Finding Highest-Scoring Alignments with Dynamic Programming

We now realize that, even for small sequences, there exists a large number of possible alignments, and it is impractical to generate all of them and calculate their scores in order to find the best. Fortunately, there exists a method which in an efficient way can be used to find the best alignment, *for a given scoring scheme*.

This method is based on a general programming paradigm, called *dynamic programming*. The main idea is that results found early in the solution procedure are used in later calculations. This paradigm was first used for biosequences by Needleman and Wunsch (1970). The task of finding the highest-scoring alignment(s) is done in two steps.

1. Using dynamic programming, find the highest possible score.

2. Find (one, several or all) alignments achieving the highest score by using the intermediate results from the first step.

To explain the method we introduce some notation.

- We have sequence q of length m, and sequence d of length n. For example, $q = $ VEITGEIST $(m = 9)$, $d = $ PRETERIT $(n = 8)$.

- q_i is the ith symbol of q, d_j is jth symbol of d.

- $q_{1...i}$ is the sequence of the first i symbols of q. For example, $q_{1...0} = \varepsilon$ (the empty sequence), $q_{1...1} = \text{V}$, $q_{1...4} = \text{VEIT}$, $q_{1...m} = q$.

- $d_{1...j}$ is the sequence of the first j symbols of d.

- R_{ab} is the scoring between a, b.

- $H_{i,j}$ is the highest score which can be achieved by aligning $q_{1...i}, d_{1...j}$.

- g is the penalty for a blank.

Note that $H_{m,n}$ will be the highest score which can be achieved by aligning q and d.

Use of the dynamic programming paradigm here implies that we can determine $H_{i,j}$ by using one or more of $H_{k,l}$, $0 \leqslant k \leqslant i$, $0 \leqslant l \leqslant j$. This means that $H_{m,n}$ can be found by first finding some $H_{i,j}$ for $i \leqslant m$, $j \leqslant n$. This calculation can be done in a systematic way, as will be described in the following subsections.

1.4.1   Determine $H_{i,j}$

The alignment for $(q_{1...i}, d_{1...j})$ can only end with one of three different columns:

$$
\begin{array}{c|c|c}
q_i & - & q_i \\
\hline
- & d_j & d_j
\end{array}
$$

We will find an expression for $H_{i,j}$ by regarding each of these cases, and from that determine the correct value for $H_{i,j}$.

We use $i = 3$ and $j = 4$ as an example in the explanation, $(q_{1...i} = \text{VEI}, d_{1...j} = \text{PRET})$. Assume we know $H_{i-1,j}$, $H_{i,j-1}$, $H_{i-1,j-1}$.

1. The alignment ends with $(q_i, -)$. For the example it is $(\text{I}, -)$. The alignment is then

$$
\begin{array}{c|c}
q'_{1...i-1} & q_i \\
d'_{1...j} & -
\end{array}
\qquad \text{for the example it may be} \qquad
\begin{array}{c|c}
\text{V-E-} & \text{I} \\
\text{PRET} & -
\end{array}
$$

   Since we have additive scoring, we see that we must add the penalty for blank to the score of aligning $q_{1...i-1}, d_{1...j}$, which is $H_{i-1,j}$, hence, $H_{i,j}^{(1)} = H_{i-1,j} - g$ ($g = 1$ in the example).

2. The alignment ends with $(-, d_j)$ $(-, \text{T})$. By using the same explanation as above we find the alignment to be

$$
\begin{array}{c|c}
q'_{1...i} & - \\
d'_{1...j-1} & d_j
\end{array}
\qquad \text{for the example it may be} \qquad
\begin{array}{c|c}
\text{VEI} & - \\
\text{PRE} & \text{T}
\end{array}
$$

$H_{i,j}^{(2)} = H_{i,j-1} - g$.

3. The alignment ends with (q_i, d_j). The alignment is then

$$
\begin{array}{|c|}
\hline
q'_{1\ldots i-1} \quad \boxed{q_i} \\
d'_{1\ldots j-1} \quad \boxed{d_j} \\
\hline
\end{array}
\qquad \text{for the example it may be} \qquad
\begin{array}{c}
\text{V--E} \quad \boxed{\text{I}} \\
\text{PRE} \quad \boxed{\text{T}}
\end{array}
$$

$H_{i,j}^{(3)} = H_{i-1,j-1} + R_{q_i d_j}$ ($R_{\text{IT}} = 0$ in the example).

We then have three alternatives for aligning $q_{1\ldots i}, d_{1\ldots j}$, depending on the last column. We choose one with highest score, such that the value for $H_{i,j}$ becomes $H_{i,j} = \max[H_{i,j-1} - g, H_{i-1,j} - g, H_{i-1,j-1} + R_{q_i d_j}]$.

Note that for this to be correct the scoring scheme must be additive, and that each blank must have the same score (linear scoring). Then $H_{m,n}$ will get the score of the best alignment of (q, d).

1.4.2 Use of matrices

To help in the aligning process, it is appropriate to arrange the scores $H_{i,j}$ in a two-dimensional matrix of size $(m+1) \cdot (n+1)$ as shown in Figure 1.2. The arrows show which earlier filled cells are used for calculating the value of a cell, $H_{i,j}$ ($i = 3$, $j = 4$ in the example).

We see that the matrix can be filled in row by row from the upper left corner down to the bottom right corner. However, we must have start values, otherwise, for example, $H_{1,1}$ cannot be calculated. Therefore, we have to initialize the values in row and column 0. $H_{0,j}$ is highest score for aligning $q_{1\ldots 0}, d_{1\ldots j}$, which means aligning the empty sequence (ε) to $d_{1\ldots j}$. This is done by expanding ε with j blanks, such that, for example, the alignment of $q_{1\ldots 0}, d_{1\ldots 3}$ becomes

```
- - -
PRE
```

Each blank has a score $-g$, meaning that $H_{0,j} = -jg$. Figure 1.2 shows the initialized matrix when $g = 1$.

Now the rest of the values can be filled in, to $H_{m,n}$, as shown in Figure 1.3(a), where our simple scoring scheme is used. The arrows show which neighbour cells are used for getting the maximum score of a cell. Note that in some cases, two of the neighbour cells will give the maximum value. Generally, it could happen that all three neighbour cells would lead to the maximum value. *Note that in the figure not all arrows are drawn.*

Example

Let us find the value for

$$H_{8,7} = \max[H_{8,6} - g, H_{7,7} - g, H_{7,6} + R_{q_8 d_7}] = \max[0 - 1, 2 - 1, 1 + 0] = 1.$$

We see that the maximum value (1) is found from both $H_{7,7}$ and $H_{7,6}$. Therefore, there are two arrows to $H_{8,7}$. △

Algorithm 1.1 shows the dynamic programming procedure for global alignment.

Algorithm 1.1. Dynamic programming for global alignment.
Aligning sequences q and d of length m and n, respectively, with linear gap penalty

const.
g penalty for one blank
R_{ab} the score of aligning a and b
var
H the dynamic programming matrix
begin
       for $i := 0$ **to** m **do** $H_{0,i} := -ig$ **end**   initialize
       for $j := 1$ **to** n **do** $H_{j,0} := -jg$ **end**
       for $i := 1$ **to** m **do**
          for $j := 1$ **to** n **do**
             $H_{i,j} := \max[H_{i-1,j} - g, H_{i,j-1} - g, H_{i-1,j-1} + R_{q_i d_j}]$
          end
       end
end

It follows directly that the time complexity of the algorithm is $O(mn)$, the number of cells. The space complexity is the same, but by means of a more complex storage administration one can achieve linear space (see Bibliographic notes).

One can give a formal inductive proof that the algorithm above does find the maximum alignment score for a pair of sequences. The proof can be based on the fact that the scoring of a column is independent on how the other parts of the sequences are aligned.

1.4.3   Finding the alignments that give the highest score

The arrows constitutes paths in the matrix, and for finding the highest-scoring alignments, we can follow the paths from $H_{m,n}$ backwards to $H_{0,0}$. The arrows to follow for the example are shown in Figure 1.3(b).

From the arrows we can find the corresponding columns of the alignment. We remember that if the arrow comes from either the same row or the same column, a blank is introduced when extending the alignment to include (q_i, d_j), which means that the corresponding columns should contain a blank. Using those rules (Section 1.4.1), we find the column corresponding to cell $H_{i,j}$ as follows:

- if the arrow comes from $H_{i-1,j}$, the column is $(q_i, -)$;

- if the arrow comes from $H_{i,j-1}$, the column is $(-, d_j)$;

- if the arrow comes from $H_{i-1,j-1}$, the column is (q_i, d_j).

q\d			P	R	E	T	E	R	I	T
	i\j	0	1	2	3	4	5	6	7	8
	0	0	−1	−2	−3	−4	−5	−6	−7	−8
V	1	−1								
E	2	−2			$H_{i-1,j-1}$	$H_{i-1,j}$				
I	3	−3			$H_{i,j-1}$	$H_{i,j}$				
T	4	−4								
G	5	−5								
E	6	−6								
I	7	−7								
S	8	−8								
T	9	−9								

Figure 1.2 The dynamic programming matrix for the example sequences, and how the values of the cells are calculated. Row and column 0 are initialized for the score of a blank equal to −1. To calculate the value of $H_{i,j}$ one needs the values of $H_{i,j-1}$, $H_{i-1,j}$ and $H_{i-1,j-1}$.

Example

Let us find the column for $i = 8$, $j = 7$. Two arrows are coming in, from $i = 7$, $j = 6$ and from $i = 7$, $j = 7$. Backtracking to $H_{7,6}$ means that we go one position back in both sequences, hence the column becomes

    S
    I

Backtracking to $H_{7,7}$ means that we go one position back in q and none in d, hence the column becomes

    S
    −

△

Several alignments can give the highest score; in our example it is two (note that the alignments are found inverted of how it is presented):

```
q': V  E  I  T  G  E  I  S  T
d': P  R  E  T  -  E  R  I  T
    0  0  0  1 -1  1  0  0  1          Score 2


q': V  E  I  T  G  E  I  S  T
d': P  R  E  T  E  R  I  -  T
    0  0  0  1  0  0  1 -1  1          Score 2
```

In programs, the arrows can be represented by variables. An alternative is to not store this direction information in the forward process, but calculate the direction in the backward process. This is done in Algorithm 1.2.

Algorithm 1.2. Backtracking for the best global alignments.

The best alignments are stored in B, one at the time

proc *backtrack(i, j, B, k)* recursive procedure

                             called the first time as *backtrack(m, n, B, 1)*

const

R_{ab} the scoring matrix

g penalty of one blank, linear gap penalty

var

B the alignment is filled in table B in reversed order,

                               q' in row 1, d' in row 2.

k column in B

i, j indices for q and d

begin

    if $i = 0$ **then**   gaps at the beginning of q'

        while $j > 0$ **do** $B_{1,k} =' -'; B_{2,k} = d_j; k := k + 1; j := j - 1$ **end**

        write(B)   one best alignment found

    elseif $j = 0$ **then**   gaps at the beginning of d'

        while $i > 0$ **do** $B_{1,k} = q_i; B_{2,k} =' -'; k := k + 1; i := i - 1$ **end**

        write(B)

    else

        if $H_{i,j} = H_{i-1,j} - g$ **then**

            $B_{1,k} = q_i; B_{2,k} =' -'; backtrack(i - 1, j, B, k + 1)$ **end**

        if $H_{i,j} = H_{i,j-1} - g$ **then**

            $B_{1,k} =' -'; B_{2,k} = d_j; backtrack(i, j - 1, B, k + 1)$ **end**

        if $H_{i,j} = H_{i-1,j-1} + R_{q_i d_j}$ **then**

            $B_{1,k} = q_i; B_{2,k} = d_j; backtrack(i - 1, j - 1, B, k + 1)$ **end**

    end

end

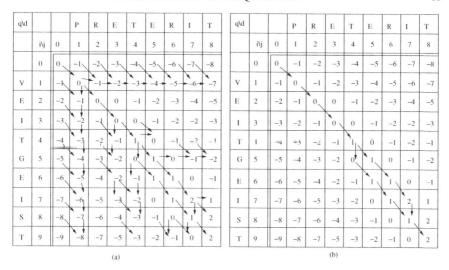

Figure 1.3 (a) The filled-in matrix. The arrows show which cells are used for finding the maximum score. Note that not all arrows are drawn. (b) The arrows showing the paths giving alignments with highest score.

1.4.4 Gaps

In the example there are only single blanks. Generally, more blanks might follow each other for getting the best alignment. One or more following blanks is called a *gap*. Also, there might be several gaps in an alignment.

Example

An example of an alignment with more gaps is

```
AC--GRTV
ACMTG-TV
```

△

1.5 Scoring Matrices

The scoring used in Section 1.4 is too simple to be used when aligning real protein or DNA sequences. The main issue of a database search is to find sequences homologous to a query sequence, and this is done by scoring a similarity between the query and the database sequences. Hence, a scoring scheme should be based on the similarity of the residues occurring in the sequences. For two residues (q_i, d_j), we need a measure of the probability (or likelihood) that they have a common ancestor, or that one is a result of one or several mutations of the other. The position of the residues is ignored, and a general measure for the similarity of the occurring amino acids is used. This measure can then be given as a $l \cdot l$ *scoring matrix*, where l is the number of amino acids. If we claim that the scoring for $a \to b$ should be equal to the scoring for $b \to a$ (a reasonable claim), the scoring matrices must be symmetrical, hence a triangular $l \cdot l$ matrix is sufficient.

The most common scoring matrices are the PAM and BLOSUM series. Those are developed based on observed mutations in the nature, and are explained in Chapter 5. Figure 1.1 shows one of the PAM matrices. Note the great variation in the scoring values, and that both positive and negative values occur. Note also that the score of aligning equal amino acids vary, aligning A with A scores 2, but aligning W with W scores 17.

Example

The score of the columns for one of the alignments found in Section 1.4 is by use of the 250 PAM matrix:

$$
\begin{array}{ccccccccc}
q': & \text{V} & \text{E} & \text{I} & \text{T} & \text{G} & \text{E} & \text{I} & \text{S} & \text{T} \\
d': & \text{P} & \text{R} & \text{E} & \text{T} & - & \text{E} & \text{R} & \text{I} & \text{T} \\
    & -1 & -1 & -2 & 3 & & 4 & -2 & -1 & 3
\end{array}
$$

The score for gap is not specified here, but is discussed in the following section. $\triangle$

1.6   Scoring Gaps: Gap Penalties

Deciding how to score gaps is perhaps the most difficult task in performing sequence alignments. Usually, a local form of gap penalty is used, which means that the penalty of a gap is found independently of other gaps in the alignment. Here we only treat local gap penalties.

Scoring gaps should mirror the model we use for constructing alignments. A gap might be the result of one or several mutations (insertions or deletions). Below we will assume that a gap has occurred by a single mutation.

Example

Let an alignment be

```
ASDEDFGH
AS----GH
```
,

We assume that the deletion (or insertion, if the evolution has gone the other way) of the four amino acids has happened in one mutation. Another way of modelling, for example, could allow two mutations, first deletion of DE and then of DF. $\triangle$

Following our model, the penalty for a gap of length four should be less than or equal to the penalty of, for example, two gaps each of length two. This can be formulated generally as a constraint on the penalty g_l for a gap of length l:

$$\forall r : 0 < r < l : g_l \leqslant g_{l-r} + g_r. \tag{1.1}$$

Table 1.1 Scoring matrix for the evolutionary distance of 250 PAM, rounded to one digit The amino acids occur in alphabetic order of their full names.

	A	R	N	D	C	Q	E	G	H	I	L	K	M	F	P	S	T	W	Y	V
A	2																			
R	-2	6																		
N	0	0	2																	
D	0	-1	2	4																
C	-2	-4	-4	-5	12															
Q	0	1	1	2	-5	4														
E	0	-1	1	3	-5	3	4													
G	1	-3	0	1	-3	-1	0	5												
H	-1	2	2	1	-3	3	1	-2	7											
I	-1	-2	-2	-2	-2	-2	-2	-3	-2	5										
L	-2	-3	-3	-4	-6	-2	-3	-4	-2	2	6									
K	-1	3	1	0	-5	1	0	-2	0	-2	-3	5								
M	-1	0	-2	-3	-5	-1	-2	-3	-2	2	4	0	6							
F	-4	-5	-4	-6	-4	-5	-5	-5	-2	1	2	-5	0	9						
P	1	0	-1	-1	-3	0	-1	-1	0	-2	-3	-1	-2	-5	6					
S	1	0	1	0	0	-1	0	1	-1	-1	-3	0	-2	-3	1	2				
T	1	-1	0	0	-2	-1	0	0	-1	0	-2	0	-1	-3	0	1	3			
W	-6	2	-4	-7	-8	-5	-7	-7	-3	-5	-2	-3	-4	0	-6	-3	-5	17		
Y	-4	-4	-2	-4	0	-4	-4	-5	0	-1	-1	-4	-2	7	-5	-3	-3	0	10	
V	0	-3	-2	-2	-2	-2	-2	-1	-2	4	2	-2	2	-1	-1	-1	0	-6	-3	4

This constraint also tends to prefer one longer gap over several neighbouring short ones.

Formulae for gap penalties satisfying Equation (1.1) are said to be *concave gap penalties*.[1]

The linear gap penalty function ($g_l = gl$), which we have used previously, is concave. Biologically (and following our model), extending a gap should be penalized less than opening one, hence a better formula for the gap penalty should be an *affine* gap penalty function (which is also concave). The function for the affine gap penalty is either $g_l = g_{open} + lg_{extend}$ or $g_l = g_{open} + (l-1)g_{extend}$, meaning that in some programs the penalty for opening a gap is $g_{open} + g_{extend}$, in other programs it is g_{open}. Some also argue that the penalty for extending should decrease with the length; $g_l = g_{open} + \log l$ is an example of such a function.

For completeness we also mention the *constant* gap penalty function ($g_l = g$), where the penalty is independent of the gap length.

Example

The alignment of the insulin proteins in Section 1.2 was found by using the PAM 250 matrix, and a linear gap penalty of $g = 5$. It has nine gaps. If we change to an affine gap penalty, $5 + (l-1)0.5$ we get the alignment:

```
MAVWLQAGALLVLLVV-SSVSTNPGTPQHLCGSHLVDALYLVCGPTGFFYNPK--RDVE-PLL
MALWTRLVPLLALLALWAPAPAHAFVNQHLCGSHLVEALYLVCGERGFFYTPKARREVEGPQV

GFLPPKSAQETEVADFAFKDHAELIRKRGIVEQCCHKPCSIFELQNYCN
GALELAGGPGAG----GLEGPPQ---KRGIVEQCCAGVCSLYQLENYCN
```

which contains fewer (four) gaps.

Changing the gap penalty to $1 + (l-1)0.1$ results in the alignment:

```
MALWTRL-V-PLLALL---ALWA--P-APAHAFVNQHLCGSHLVEALYLVCGPTGFFYNPK--R
MAVW--LQAGALLVLLVVSSV-STNPGTP------QHLCGSHLVDALYLVCGERGFFYTPKARR

DVE-P----L-LAGGPGAGG-LEGPP---Q------FAFKDHAELIRKRGIVEQCCHKP--CSI
EVEGPQVGALEL-------GFL--PPKSAQETEVAD----------KRGIVEQCC--AGVCSL

FELQNYCN
YQLENYCN
```

                                                                          △

The example illustrates that the problem of determining the gap penalty is difficult. Affine gap penalties are the most used, and typically $g_{open} \approx 10 g_{extend}$. We will discuss this more in the context of local alignments (see Chapter 2.3).

Another aspect to discuss is if gaps at the end shall have the same penalties as gaps not at ends.

---

[1] Gap penalties satisfying Equation (1.1) satisfy the definition of the concave gap function as defined in Waterman (1995).

Example

Assume two sequences AGVARTLR and AGTLR, and make two alignments:

$$
\begin{array}{cc}
\text{AL1} & \text{AL2} \\
\text{AGVARTLR} & \text{AGVARTLR} \\
\text{AG---TLR} & \text{---AGTLR}
\end{array}
$$

AL1 will get the highest score if the end gaps have the same penalty as other gaps. However, often one of the sequences is a subsequence of the other, and AL2 would here be the correct one in that case (and would be found if end gaps were not penalized).
△

1.7 Dynamic Programming for General Gap Penalty

The recurrence formula presented for dynamic programming is only valid for linear gap penalties. The reason is that we have assumed that each blank in a gap has the same penalty, independent of how long the gap is.

For finding the value (score) in $H_{i,j}$ when general gap penalties are used, we must compare

- the score if the subalignment ends with the pair q_i, d_j,

- the score if the subalignment ends with a gap in q of length l, $1 \leqslant l \leqslant j$,

- the score if the subalignment ends with a gap in d of length l, $1 \leqslant l \leqslant i$.

Figure 1.4(a) shows which elements must be used to calculate $H_{i,j}$.
The recurrence formula for this is

$$H_{i,j} = \max\left[H_{i-1,j-1} + R_{q_i d_j}, \max_{1 \leqslant l \leqslant j}(H_{i,j-l} - g_l), \max_{1 \leqslant l \leqslant i}(H_{i-l,j} - g_l)\right]. \quad (1.2)$$

The time complexity of this recursion can be found by noting that the number of cells examined for finding $H_{i,j}$ is $1 + i + j$, hence the total number of cells examined is

$$\sum_{i=1}^{m}\sum_{j=1}^{n}(1 + i + j) = mn + \sum_{i=1}^{m}ni + \sum_{i=1}^{m}\sum_{j=1}^{n}j$$
$$= mn + O(nm^2 + mn^2) = O(nm^2 + mn^2).$$

Figure 1.4(b) shows some of the values in the dynamic programming (DP) table for an example of using affine gap penalty. The scoring scheme is defined in the figure's text.

We see that the work for finding the best alignment when a general gap penalty is used is an order larger than when using a linear gap penalty.

Note that the general recurrence formula supports gap penalties that are not concave (see Section 1.6).

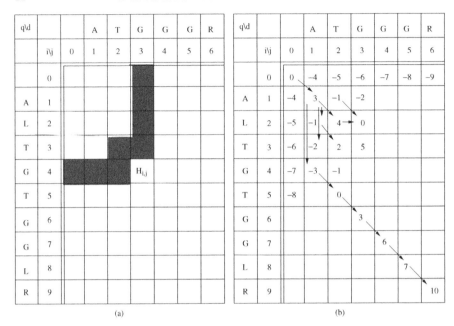

Figure 1.4   (a) Figure showing which cells to use for calculating $H_{i,j}$ when a general gap penalty function is used. (b) Dynamic programming using the gap penalty $g_l = 3 + 1l$, and scoring 3 for equal, 1 for unequal. The best value for $H_{4,1}$ is $H_{1,1} - g_3 = 3 - 6 = -3$. Following the arrows backwards, we get the same alignment as AL2 in the second example in Section 1.6.

Example

A gap of length 5 can be modelled as either one mutation of length 5, two mutations of lengths 2 and 3, three mutations of lengths 1, 1, 3 or of 1, 2, 2, four mutations of lengths 1, 1, 1, 2, or five mutations each of length 1. A gap penalty of $g_l = 3 + l^2$ will, for example, model three mutations (of length 1, 2, 2) for a penalty of $4 + 7 + 7 = 18$. The other five possibilities result in higher penalties.   △

Note, however, that there is a general agreement that the penalties should model the one-mutation philosophy, hence using a concave gap penalty.

There exist techniques for reducing the real running time (with the same worst-case time complexity); this will be treated more generally in Chapter 4 on multiple alignment methods. For an affine gap penalty, we can still achieve an algorithm of $O(mn)$.

1.8   Dynamic Programming for Affine Gap Penalty

Let the affine gap penalty be $g_l = g_{\text{open}} + l g_{\text{extend}}$. We can look at the algorithm for the linear gap penalty (Section 1.4.1), and see how it must be changed in order to

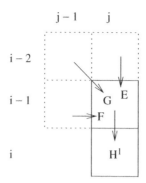

Figure 1.5 Illustration for the algorithm for affine gap penalty. See text for the explanation of E, F, G, H^1.

use affine gap penalties. When a blank is to be inserted, we must find if it is the start of a gap ($g_{open} + g_{extend}$), or an extension (g_{extend}). For determining $H_{i,j}$ we looked at the three neighbouring cells $H_{i-1,j-1}$, $H_{i,j-1}$ and $H_{i-1,j}$. The formula for using $H_{i-1,j-1}$ is

$$H_{i,j}^{(3)} = H_{i-1,j-1} + R_{q_i d_j},$$

and this can still be used, since it involves no gap (see Section 1.4.1 for $H_{i,j}^{(3)}$).

For calculating $H_{i,j}^{(1)}$ (the alternative via $H_{i-1,j}$), we must take into account how the alignment for $(q_{1...i-1}, d_{1...j})$ can end. Three cases have to be considered (see Figure 1.5).

(a) Let $E_{i-1,j}$ be the score at $i-1$, j when coming from $i-2$, j. Then

$$H_{i,j}^{(1),a} = E_{i-1,j} - g_{extend}.$$

(b) Let $F_{i-1,j}$ be the score at $i-1$, j when coming from $i-1$, $j-1$. This is unlikely, for it would produce an alignment ending in

$$\begin{matrix} \cdots & - & q_i \\ \cdots & d_j & - \end{matrix}$$

It is more likely that the two last columns would be one, without a blank. But it must be considered, hence

$$H_{i,j}^{(1),b} = F_{i-1,j} - g_{open} - g_{extend}.$$

(c) Let $G_{i-1,j}$ be the score at $i-1$, j when coming from $i-2$, $j-1$. Then

$$H_{i,j}^{(1),c} = G_{i-1,j} - g_{open} - g_{extend}.$$

So the maximum score when coming to cell (i, j) from $(i-1, j)$ is

$$H_{i,j}^{(1)} = \max[E_{i-1,j} - g_{extend}, F_{i-1,j} - g_{open} - g_{extend}, G_{i-1,j} - g_{open} - g_{extend}].$$

Therefore, three variables have to be saved at $H_{i-1,j}$ to be able to calculate the correct value of $H_{i,j}^{(1)}$. For finding the correct value of $H_{i,j}^{(2)}$ by use of symmetry we can conclude that three variables have to be saved at $H_{i,j-1}$. As a conclusion, for changing the procedure for a linear gap so that it can handle an affine gap, it is only necessary to introduce three variables in each cell, and change the assignment equations. Hence, the algorithm is still of order $O(mn)$.

1.9   Alignment Score and Sequence Distance

In the preceding subsections we have shown how to score the *similarity* of sequences. We can also measure the *distance* between two sequences. The *edit distance* is a common measure for strings: the edit distance between two strings is the minimum number of operations for transforming one of the strings to the other, where the operations are substitution, deletion and insertion of single symbols.

Example

Using our sequences $q =$ VEITGEIST, $d =$ PRETERIT, we can transform q to d by the following six operations:

$$\rightarrow P; V \rightarrow R; I \rightarrow; G \rightarrow; \rightarrow R; S \rightarrow$$

This is the minimum number of operations, hence the edit distance between them is six. $\triangle$

We can now define a scoring scheme that allows us to find the edit distance between two strings from an alignment with maximum score. Let the scoring scheme be as follows:

- $R_{ab} = 0$ for $a = b$; -1 for $a \neq b$, and $g = 1$;
- let T be the score of the best alignment.

Then there are $-T$ columns containing either a mismatch or a blank, and this is the minimum number of operations for transformation; hence the edit distance is $-T$.

Example

The best alignment of q, d, using the scoring scheme defined above, becomes

```
q': -VEITGE-IST
d': PRE-T-ERI-T
```

with score $T = -6$, so the edit distance is six, as found in the example above. $\triangle$

Often, the distances between objects constitute a *metric space*. A set X of elements is said to be a metric space if for any two elements x and y there is a real number d_{xy} called the distance from x to y, such that

1. $d_{xy} = 0$ for $x = y$,

2. $d_{xy} > 0$ for $x \neq y$,

3. $d_{xy} = d_{yx}$,

4. $d_{xy} \leqslant d_{xz} + d_{zy}$ for any $z \in X$ (the triangle inequality).

The edit distance constitutes a metric space. Note, however, that the minimum transformation (the transformation with the minimum number of operations) between a pair of strings is not necessarily unique.

For comparison of biological sequences, the edit distance can be used under the assumption that each observed difference in the sequences represents one mutation, which can be reasonable for very similar sequences. However, when the number of observed changes is large, there might be several mutations to each observed change.

Example

Assume an evolutionary history:

$$\text{AKLDC} : \text{K} \to ; \text{L} \to \text{V}; \to \text{R} ; \text{V} \to \text{M} : \text{AMRDC}$$

The edit distance between the two sequences is two, corresponding to the first alignment below. The correct alignment (corresponding to the history) is, however, the second alignment below, showing three mutations, but the correct number of mutations is four.

```
AKLDC              AKL-DC
AMRDC              A-MRDC
```

$\triangle$

The example shows that there might be several substitutions in one site: here $\text{L} \to \text{V} ; \text{V} \to \text{M}$.

For comparing distances between different pair of sequences, it is common to divide the observed distance by the length of the longest sequence, resulting in (relative) distances not greater than 1. Also, it is not unusual to only count the columns in the alignment which do not contain a blank, and divide by the number of those columns.

Several models for correcting for multiple mutations are presented. Of course, the growth of the function for the corrected distance must increase with the observed distance, and most models result in a formula with a logarithmic function. Let D be the observed (relative) distance; then a common model for finding the corrected (relative) number of mutations is

$$K = -a \ln(1 - f(D)),$$

where a is a constant, and $f(D)$ is a positive function less than 1. One simple formula used for proteins (when columns with blank are ignored) is $f(D) = D + \frac{1}{5}D^2$ (Kimura 1983). (Note, however, that this cannot be used for large D (D greater than

0.85, since then $f(D)$ becomes greater than 1.) Using these values for a and $f(D)$ gives us an expression for K, the number of estimated substitutions per column as

$$K = -\ln(1 - D - \tfrac{1}{5}D^2). \tag{1.3}$$

This can be greater than 1. For example, if the observed value is 0.8 (eight of ten columns have different amino acids), then the number of estimated substitutions becomes 2.6 substitutions per column during the evolutionary time since the two sequences diverged.

An analogue to a metric space for similarity would inverse the triangle inequality: $R_{ab} \geqslant R_{ac} + R_{cb}$. This is generally not satisfied when using scoring matrices such as the PAM series. For example, in the PAM 250 matrix, $R_{GR} = -3$, $R_{GA} = 1$, $R_{A,R} = -2$, hence $R_{GR} < R_{GA} + R_{AR}$.

1.10 Exercises

1. Let two sequences be q = CDAA and d = AEECA, and a scoring matrix:

	A	C	D	E
A	2	-2	-2	-1
C		1	0	0
D			2	-2
E				2

(a) Find the highest score by aligning q and d when the gap penalty is $g_l = 2l$. Then find the best alignments.

(b) Now use gap penalty $g_l = 1.8 + 0.4l$. The dynamic programming table will be partly filled as below, using the general DP procedure:

q\d			A	E	E	C	A
		0	1	2	3	4	5
	0	0.0	-2.2	-2.6	-3.0	-3.4	-3.8
C	1	-2.2	-2.0	-2.2	-2.6	-2.0	-4.2
D	2	-2.6	-4.2	-4.0	-4.2	-2.6	-4.0
A	3	-3.0	-0.6	-2.8	-3.2		
A	4	-3.4	-1.0	-1.6			

Note especially how the value -3.2 in $H_{3,3}$ is found, the value is $H_{3,1} - g_2 = -0.6 - (1.8 + 0.8)$. Fill in the rest of the table, and find the best alignment(s).

(c) Compare the alignments found under (a) and (b), and find for each of them the minimum number of mutations which might have occurred, when we suppose that only one residue is included in a substitution.

2. In some cases one wants to score gaps at the ends of an alignment as 0.

(a) In what cases is this reasonable (what is the relation between the two sequences)?

(b) The general procedure for dynamic programming can be changed in the following way to take care of this:

- initialize all cells in row and column 0 to 0;
- blanks in last row and column shall score 0.

Explain why these changes will produce the best alignment.

(c) Change Algorithm 1.1 to take into account end gaps with zero score.

(d) We have the sequences q = ART and d = AARRTRT. Use score 1 for equal symbols, -1 for unequal, and a (linear) gap penalty of 1. Find the best alignments when a score of 0 is used for the end gaps.

3. How would you find the alignments if a constant gap penalty is used?

4. Change Algorithm 1.1 so that it can be used for an affine gap penalty. Use the method explained in Section 1.8.

5. Suppose q = LARKTLVAKVLSV, d = KLVASTVLRKRSA. By using a score of -1 for mismatches and blanks, and 0 for matches, one best alignment is

$$q': \quad \text{-L-ARKT-LVAKVLSV}$$
$$d': \quad \text{KLVAS-TVL-RKR-SA}$$

(a) What is the edit distance between the sequences?

(b) Estimate the relative evolutionary distance between them using Equation (1.3). What does the distance found mean?

(c) Could other best alignments result in other relative evolutionary distances?

1.11 Bibliographic notes

The first to use dynamic programming for comparing biological sequences was Needleman and Wunsch (1970). A linear space algorithm is presented in Hirschberg (1975).

A discussion of gap penalties can be found in Pascarella and Argos (1992) and Benner et al. (1993).

Formulae for correcting the sequence distance for multiple mutations (for DNA or proteins) are presented in Kimura (1980, 1983), Li (1993, 1997), Li and Gu (1996) and Swofford et al. (1996).

2

Pairwise Local Alignment and Database Search

When a new protein sequence is determined, the most immediate task is to see if there are any other sequences that are homologous to it. As mentioned in Chapter 1, this is because homologous sequences have similar structures and often related functions. The search for homologous sequences can be made by comparing the new sequence to all known sequences using the alignment method described in Chapter 1. However, for distantly related proteins, the accumulation of mutations might have occurred unevenly across the sequences leading to a similar segment remaining in otherwise dissimilar sequences. The same result might also have arisen more directly through the rearrangement of gene segments in the genome giving rise to related subsequences in two otherwise unrelated genes. In biological terms, such a local similarity could correspond to a shared motif or domain associated with a structure or function that is robust to mutations. Using the dynamic programming method to find a global alignment in this situation might fail to identify a significant match as the island of similarity might be lost against the background of random residue matches. Instead, we require a method to find similar segments (subsequences) in the two sequences.

The known sequences are collected in biological databases, and there are a lot of publicly available biological databases around the world. This includes the databases of DNA sequences (EMBL (Europe), DDBJ (Japan), GenBank (USA), which all contain the same sequences), databases of protein sequences (e.g. PIR and SwissProt) and databases of protein structures (PDB). Other databases, for instance, TransFac (Transcription Factor Database), are more specialized and are of most interest to a smaller group of researchers.

An entry in one database may contain information related to an entry in another database. For example, an EMBL (DNA) entry can contain the region coding for one particular protein described by one particular SwissProt entry. The databases should allow users to perform queries to retrieve data satisfying specified criteria. For example, it should be possible to retrieve from a sequence database all the sequences from the organism *Rattus Norvegicus* or all entries that have been updated after 14 August 1995 and cite a paper with author 'Aasland'. The querying systems should also allow

Protein bioinformatics: an algorithmic approach to sequence and structure analysis
I. Eidhammer, I. Jonassen and W. R. Taylor   © 2004 John Wiley & Sons, Ltd   ISBN: 0-470-84839-1

users to utilize data from several databases, for example, to find all three-dimensional structures existing for proteins encoded by sequences from *Rattus Norvegicus* in the DNA database. This is possible utilizing the links between different databases.

For the database search task described in this chapter, the goal is to find those sequences {d} in a database D which are homologous to a query sequence q, in a reasonable time. Three aspects are important: the program's ability to present all evolutionary homologous sequences, its ability to not present nonhomologous sequences, and the time it takes. By this we mean that the algorithm should find all the database sequences with sufficiently local similarity to q, such that the similarity is unlikely to have arisen by chance.

We first describe finding local alignments for two sequences, and then an efficient program for database searching. The statistical significance of this process is treated in the next chapter.

2.1   The Basic Operation: Comparing Two Sequences

The task can be formulated as follows. Given a *query sequence q* and a database D of sequences {d}, find those d which share the highest similarities with q, and present the similarities and their statistical significances. Statistical significance is important. This is used to calculate whether a similarity is likely to be caused by chance or is a result of homology.

The basic operation is a comparison of $q_{1...m}$ with a database sequence $d = d_{1...n}$. The task is to find the segments (subsequences) of these two sequences with highest degree of similarity, and to calculate the statistical significance of the similarity. The result of the task is called a *pairwise local alignment*. Note the meaning of the concept of subsequence here: it has the same meaning as substring in the context of strings (where subsequence has another meaning). The reason for this, perhaps confusing, terminology is that we talk about sequences and not strings. We will use subsequence and substring interchangeably, and specify if we use subsequence with another meaning.

We now summarize some definitions.

Definition

- A *segment* is a subsequence (substring) of q or d. A segment does not contain gaps.

- A *segment pair* is a pair with one segment from each of q and d (they need not be of equal length).

- A *local alignment* is an alignment of a segment pair.

$\triangle$

Example

A global alignment of the insulin from sheep with the 'probable insulin-like peptide 2' from the fruit fly is (spaces at the ends indicate no penalties for end gaps)

```
MALWTRLVPLLALLALWAPAPAHAFVNQ-HLCGSHLVEALYLVCGERGFF
    MSKPLSFISMVAVILLASSTVKLAQGTLCSEKLNEVLSMVCEE----

YTP--KARREVEG--PQVGALE--------------------LAGGPG
YNPVIPHKRAMPGADSDLDALNPLQFVQEFEEEDNSISEPLRSALFPGSY

AGGL--------EGPPQKRGIVEQCCAGVCSLYQLENYCN
LGGVLNSLAEVRRRTRQRQGIVERCCKKSCDMKALREYCSVVRN
```

A local aligning, using the same scoring matrix and gap penalties as for the global case, results in the local alignment:

```
 85 GIVEQCCAGVCSLYQLENYCN    105
113 GIVERCCKKSCDMKALREYCS    133
```

(The numbers are the first and last residues in the segments.) The local alignment corresponds mainly to the A chains of the two proteins. Note that in this case, the local alignment corresponds exactly to a part of the global alignment. This will naturally not always be the case. △

2.2 Dot Matrices

Historically, the first technique used for discovering local similarities was that of dot matrices (also called dot plots). An $m \cdot n$ matrix is constructed with the amino acids of q along one side (vertical), and those of d along the other side (horizontal). The values of the matrix are either a dot or a space. A dot in the cell (i, j) means that $q(i) = d(j)$. Figure 2.1(a) shows a dot matrix, where lines are drawn showing common substrings, and reversed substrings (dotted lines).

Some of the properties of a dot matrix are

- it is easy to understand, giving a visual picture;

- it is easy to find common substrings, appearing as contiguous dots along a diagonal, a, b, c, d in Figure 2.1;

- it is easy to find reversed substrings, appearing as contiguous dots along antidiagonals, e, f, g in Figure 2.1(a);

- it is easy to discover internal exchanges of subsequences—in Figure 2.1 the strings a and b are, for example, exchanged;

- it is easy to discover displacements—in Figure 2.1 a is at the beginning of one of the strings, and at the end in the other;

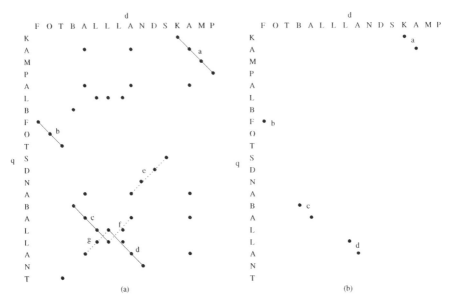

Figure 2.1 (a) A dot matrix, showing common substrings of length at least 3, and reversed substrings (dotted lines). (b) A filtered dot matrix, with window length 5, and threshold 60%.

- it can contain much noise with most of the dots not being in common substrings;

- the matrices can be large.

It is also possible to include mismatches in the interpretation; for example, g in Figure 2.1(a) could be extended to five letters, allowing a mismatch between B and N.

It is also sometimes easy to combine contiguous dots from two different diagonals, constructing a local alignment with gaps. c and d could, for example, be combined into an alignment with gap in q:

$$q: \quad \text{BALL-AN}$$
$$d: \quad \text{BALLLAN}$$

Figure 2.2 is the dot matrix for the last part of the insulin sequences used in the introduction to this example. It is easy to see the common A chains at the bottom right corner (indicated by the slashes).

2.2.1   Filtering

Real dot matrices for biological sequences will contain a lot of dots, many of which can be considered as noise, and it is not always easy as a result of all these dots to discover similarities. To reduce the noise one can compare the sequences using

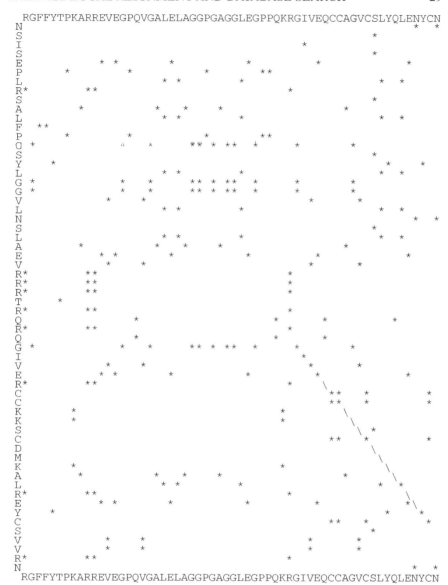

Figure 2.2 A simple dot matrix for the end parts of the insulin sequences in the introduction to this chapter. The slashes indicate the identified local alignment.

overlapping windows of fixed length k, $k > 1$, and claim that the comparison within the windows must match in a fixed number of positions in order to write a dot. First, $q_{1...k}$ is compared with $d_{1...k}$, then $q_{1...k}$ to $d_{2...k+1}$, etc. Then continue with the next position of q (row two in the dot matrix): $q_{2...k+1}$ to $d_{1...k}$, etc.

The result of a comparison is here also a dot or blank, showing either a match or no match. Usually, an exact match is not required: a match is registered if $C\%$ of the letters are equal.

Example

Let the window length be five, and the requirement for a match be three, then

```
    ALLLA              TBALL
    ALLAN              ABKLL
```

will result in matches, but not

```
    ALKLA              TBALL
    ALLAN              ABKKL
```

$\triangle$

Figure 2.1(b) shows the dot matrix when a window of size five is used, and at least three equal letters are required. We see that all noise disappears, but the sensitivity (the ability to discover weak similarities) is weakened.

Instead of using an equal number of letters, one can use a scoring matrix for giving scores to the comparison inside a window. A dot is drawn if the segment score is higher than a given threshold. The filtering procedure is shown in Algorithm 2.1.

Algorithm 2.1. Deciding the dots when using window.

const
k   window length
C   the requirement for match is $C\%$
var
D   the dot matrix
begin
    for $i := 1$ **to** $m - k + 1$ **do**
        for $j := 1$ **to** $n - k + 1$ **do**
            if sum$(q_{i...i+k-1}, d_{j...j+k-1} \geqslant C\%$ **then** $D_{i,j} =' \cdot'$ **end**
        end
    end
end

2.2.2   Repeating segments

Dot matrices can be used to discover repeating segments in a sequence. The sequence is compared with itself, and repeating segments appear as subdiagonals. If $d_{i...j} = d_{i+k...j+k}$, there will be dots on the subdiagonal from (i, j) to $(i + k, j + k)$ (see Figure 2.3). It is easy to see the repetitions since they are exact, and there is no noise. Looking for approximate repeating segments is more difficult, especially if the matrix includes much noise.

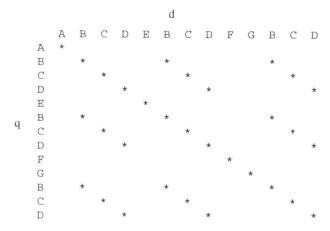

Figure 2.3 Repeating segments are illustrated on subdiagonals.

2.3 Dynamic Programming

Dynamic programming is used to find the best global alignment. Dynamic programming can also be used to find the best local alignment, but some changes have to be made in the algorithms for global alignment. It was first used for local alignment by Smith and Waterman (1981).

Example

Let a scoring scheme be 0.5 for equal symbols, -0.3 for unequal, and let the gap score be $-0.5l$. Furthermore, let the two sequences be

$$q:\quad \text{ACEDECADE}$$
$$d:\quad \text{REDCEDKL}$$

We do not know at what residues the best (highest scoring) local alignments end; therefore, every pair of residues must in principle be investigated. Let us first look at q_8, d_6. The best alignment, from the beginning, ending with these residues (the best alignment of $q_{1...8}, d_{1...6}$) is found in Figure 2.4(a), with a score of 0.4:

$q'_{1...8}$	A	C	E	D	E	C	A	D
$d'_{1...6}$	R	–	E	D	–	C	E	D
column score	-0.5	-0.3	0.5	0.5	-0.5	0.5	-0.3	0.5
align. score	-0.5	-0.8	-0.3	0.2	-0.3	0.2	-0.1	0.4

We see that the first two pairs score negative, hence by removing those we will achieve a better local alignment ending with q_8, d_6, namely

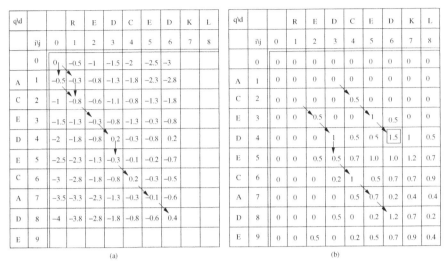

Figure 2.4   (a) Alignment of $q_{1...8}, d_{1...6}$. (b) Local alignment of q, d.

$q'_{3...8}$	E	D	E	C	A	D
$d'_{2...6}$	E	D	–	C	E	D
column score	0.5	0.5	-0.5	0.5	-0.3	0.5
align. score	0.5	1.0	0.5	1.0	0.7	1.2

$\triangle$

The example shows that prefixes (coming first in the alignment) which score negative should be removed. That means, to find the best alignment ending with q_i, d_j, we must examine where in the preceding residues to start. It then follows that the score of the best alignment ending with q_i, d_j is

$$\max_{k,l}\{S(q_{k...i}, d_{l...j}) : 1 \leqslant k \leqslant i, 1 \leqslant l \leqslant j\},$$

where S is the alignment score.

Fortunately, there exists a dynamic programming procedure which can be used to find this maximum value for the residue pairs. If we look at the procedure for finding global alignments, we can remove the contribution from negative prefixes by not allowing negative values in the DP matrix. Negative values are replaced by 0. Let now $H_{i,j}$ be the highest score of any local alignment where q_i and d_j are the last symbols, or 0 if this score is < 0:

$$H_{i,j} = \max\left\{0, \max_{k,l}\{S(q_{k...i}, d_{l...j}) : 1 \leqslant k \leqslant i, 1 \leqslant l \leqslant j\}\right\}. \tag{2.1}$$

$H_{i,j}$ can now be calculated for all i, j, and for a linear gap penalty ($g_l = lg$) the recurrence becomes

$$H_{i,j} = \max[0, H_{i-1,j} - g, H_{i,j-1} - g, H_{i-1,j-1} + R_{q_i d_j}]. \tag{2.2}$$

The score of the best local alignments can now be found as the maximum value in H. Note that these alignments will never end with gaps, since gaps give negative contributions. (The best local alignments ending with other q_i, d_j can, however, end with gap, but these will not correspond to the highest value in H.)

2.3.1 Initialization

In global alignments, column and row zero were initialized to $-gl$. As the cells for local alignment cannot be negative, they are initialized to 0, hence prefixes with gaps are ignored:

$$H_{0,j} = H_{i,0} = 0, \quad 0 \leqslant i \leqslant m, \ 0 \leqslant j \leqslant n. \tag{2.3}$$

The DP matrix for the local alignment of q, d is shown in Figure 2.4(b). The score of the best alignment is 1.5, and ends with q_4, d_6. The arrows for the best alignment ending with q_8, d_6 are also shown for illustration.

2.3.2 Finding the best local alignments

The best alignments are now found by following the arrows from the cells with maximum value, until a cell with value 0 is reached. For the example one easily find the local alignment:

$$
\begin{array}{ll}
q'_{2...4} & \text{CED} \\
d'_{4...6} & \text{CED}
\end{array}
$$

Remember that

- several cells might contain the maximum value,
- there might be several paths to such a cell.

2.3.3 Algorithms

It is now straightforward to change the algorithms for global alignment to algorithms for local alignment. In the forward procedure it is only necessary to change the initialization, and include 0 in the test for the maximum. In the backtracking procedure, one must find all cells with maximum score (or one if one alignment is enough), backtrack from each of those, and stop each backtracking when a cell with value 0 is reached.

The outlined algorithm returns the highest-scoring local alignment (or all the alignments if several share the highest score, but that is rare in real sequences). However, two sequences may share more than one region, and one may want, for example, to find the k best, which are not trivial variants of each other, e.g. having no columns in common. This might be done by recalculating H for each new alignment, giving special small values to the cells being in the already identified alignments.

2.3.4   Scoring matrices and gap penalties

The scoring system should be such that the expected scoring of aligning two arbitrary residues should be negative, otherwise alignments close to the global alignment will often be found as the best local alignment.

The value of the gap penalty, relative to the scoring matrix, determines the number of gaps that will occur in the best local alignment. Too low a gap penalty will result in too many gaps. On the other hand, with very high gap penalties, the best local alignment tends to be ungapped. When the penalties are reduced, a boundary is crossed into the region of gapped alignments. The 'correct' gap penalties seem to lie close to this boundary.

Example

Let the score between equal symbols (amino acids) be x, and the score between unequal symbols be $-x$. Assume two sequences CDECE and HCDCA, and a linear gap penalty $g_k = 10xk$. Then the optimal local alignment will be (CD CD). Reducing the penalty to $g_k = xk$ results in two optimal alignments (CD CD) and (CDEC CD-C). Reducing the penalty further (while keeping $g_k \geqslant 0$) will result in (CDEC CD-C) being the only optimal alignment. We see that in this simple example the boundary is crossed when the score for a gap of length one is equal to $-x$. $\triangle$

The example indicates that the score for a gap of length 1 should be equal to the average score between unequal amino acids, which is proposed in several analyses (also showing again that the gap penalty should depend on which scoring matrix is used). Programs for doing alignments usually allow the user to specify which scoring matrix to use, and then automatically propose a gap penalty in accordance to that matrix.

2.4   Database Search: BLAST

There exist many database search programs based on dynamic programming, and also specially made computers for efficiently performing dynamic programming search. As explained above, using dynamic programming is guaranteed to find the best alignments for a given scoring matrix and gap penalty. However, dynamic programming is time expensive, especially for searching in databases with many thousands of sequences. In order to speed up the search for general purpose computers, search programs using heuristics have been developed. The most popular of these is the group of programs called FASTx (Lipman and Pearson 1985; Pearson 1990), and BLASTx (Basic Local Alignment Search Tool) (Altschul et al. 1990, 1997). The main heuristic approach is the same for both sets of programs, but it is differently implemented. Here we will describe the program BLASTP, and an extension of it will be described in Chapter 6. In some of the technical details BLASTP works slightly differently than presented; this is for reasons of clearness.

The main principle for BLAST is to first use a fast search method to find approximately equal small segments in the two sequences, and then extend the local alignments formed by such segment pairs. A few definitions for use in the description later are given first.

Definition

- A *maximal segment pair* (MSP_{qd}) is a pair of identical length segments chosen from the sequences q and d, which when aligned have the highest possible score obtained for local ungapped alignment of q and d.

- A *high-scoring segment pair* (HSP) is a segment pair which does not increase its score while either extending or shortening its length. Also called a local maximal segment pair (LMSP).

- A *word* is a segment of fixed length w.

- A *word pair* is a pair of segments of fixed length w.

$\triangle$

The problem can now be described as finding those database sequences d for which MSP_{qd} scores over a threshold V (we assume a normalized score, see Chapter 3.3.2). V is determined such that there is reason to believe homology between the q and d if $MSP_{qd} > V$.

The main idea is then as follows. *A segment pair with a score of not less than V must, with high probability, contain at least one word pair of length w with score of not less than T.* The first step of the method is to find those word pairs (of length w) scoring at least as high as T, and then extending them to high-scoring segment pairs, seeing if they score over V. T is chosen as a compromise between the desires of fast search and high sensitivity:

- the running time of the search program will decrease with increasing T, as fewer word pairs are found and extended;

- the sensitivity of the program will also decrease with increasing T, since word pairs might be overlooked (causing homologous sequences to be discarded).

Definition

Word pairs scoring at least as high as T are called *hits*. $\triangle$

Example

Let $V = 20$, $w = 4$, $T = 3$, which means that *only word pairs of length four with a score of not less than* 3 *are extended to make HSP.*   △

In the last version of BLAST, the procedure above has been changed to require two hits before extending. Requiring two hits means that it is assumed that an HSP with statistical significance includes at least two high-scoring word pairs not far from each other. The distances and the order between the hits on the two sequences must be equal, since no gaps are allowed in this step.

Example

Let $w = 2$ and exact equality be required for hits. Let $q = $ ..ADRTSEVL.. and $d = $ ..VTDRSEEVKH.... There are three hits (DR, SE, EV). The two hits (DR, EV) can be combined (DRTSEV, DRSEEV), possibly extending to an HSP. The hit of SE cannot be combined with any of the other two without introducing gaps.   △

When the two-hits constraint is used instead of one hit, the threshold parameter T must be lowered to retain comparable sensitivity. Many more hits are then found, but this is more than compensated for by fewer extensions.

2.4.1 The procedure

Let the alphabet be $\mathcal{C}$ with cardinality r ($r = 20$ for protein sequences, four for DNA sequences), and word length w. Then there are r^w different words. For protein sequences w is typically equal to 3, and for DNA typically 12.

The query q has to be compared with a large number of database sequences. In order to make the comparison efficient, q is first *preprocessed*, where the important information of q is saved in an easy-to-use data structure. During the preprocessing, the words o^T (of all possible words of length w, i.e. r^w words) which satisfy $S(o, o_q) \geqslant T$ for at least one word $o_q \in q$ (S is the score) are found, and are represented in the data structure.

The main procedure then consists of searching in the database sequences for *occurrences* of the words o^T (those found in the preprocessing), hence finding hits. The comparison of q to each sequence d in the database is then done in three steps.

1. Search in d for occurrences (o_d) of the words (o^T) that are in the data structure (find the hits).

2. Extend (heuristically) to a high-scoring segment pair, those pairs of (non-overlapping) succeeding hits which satisfy the distance constraint (make HSPs).

3. Perform dynamic programming alignment around those HSPs scoring more than a given threshold, allowing for introduction of gaps.

2.4.2 Preprocess the query: make the word list

The parameters here are a scoring scheme, the query sequence q, the word length w and the threshold T. We explain with an example.

Example

1. Alphabet $\mathcal{C} = \{A, C, D\}$. With word length $w = 2$ there are $3^2 = 9$ possible words.

2. Let the score be

$$
\begin{array}{cccc}
 & A & D & C \\
A & 1 & 0.5 & -0.5 \\
D & & 1 & 0 \\
C & & & 1
\end{array}
$$

3. Let $q = \text{ADACCC}$, $w = 2$, $T = 1.5$.

The first word in q (AD) clearly matches AD (score 2) but also AA with a score of 1.5. ((A, A) scores 1 and (A, D) scores 0.5). We then see

- the word starting at residue 1 in q matches AD, AA, DD with a score of not less than T;

- the word starting at residue 2 matches DA, AA, DD;

- the word starting at residue 3 matches AC, DC;

- the word starting at residue 4 matches CC;

- the word starting at residue 5 matches CC.

$\triangle$

The matching words must then be organized in a data structure, such that scanning the sequences in the database can be done in an efficient way. The idea is that, for each word o_d in d, it should be easy to find the matching words in q. Two alternative techniques were implemented.

Table

A table of r^w entries are used, one entry for each possible word. An entry contains the indices in q for the matching words, and the scores. The table for the example above is shown in Table 2.1.

We see that not all of the possible words have matching words in the query. Typically, for protein sequences ($r = 20$) and $w = 4$ only a few thousand of the 20^4 possible words have matching words in q. Therefore, some clever use of indexing can be used to reduce the size of the table, but still maintaining the possibility of fast look-up.

Table 2.1   A table for saving the matching words of q in the example.
It shows, for example, that AA has matching words at index 1 and 2 in q.

Word	Matching words (index in q : score)	
AA	1 : 1.5	2 : 1.5
AC	3 : 2.0	
AD	1 : 2.0	
CA		
CC	4 : 2.0	5 : 2.0
CD		
DA	2 : 2.0	
DC	3 : 1.5	
DD	1 : 1.5	2 : 1.5

Finite-state machine

One way of using a finite-state machine is to have r^{w-1} states, and from each state have r edges to r different states. Each state represents a current reading (sub)word of length $w - 1$. Each edge represents a word, the word resulting from concatenation of the (sub)word of the state and the letter on the edge. The edges are further labelled with the indices of the matching words in q. Note that this implements a Mealy machine, with acceptance on transition (edges) (as opposed to a Moore machine with acceptance on states). The finite-state machine implementation of the example is shown in Figure 2.5. In addition to the indices in q, the labels on the edges could also contain the matching scores.

2.4.3   Scanning the database sequences

It is now easy to see how the scanning of a database sequence can be performed, using either the table or the state machine. Altschul et al. have found the finite-state machine implementation to be fastest.

Example

Let a database sequence d be CDDACACD, and $w = 2$. First use the table, and the word CD is processed (finding no hits), and then move one position to right, processing the word DD (finding two hits).

Suppose now the state machine is used. Since C is the first symbol, we start in state C. D is then read, and we follow the edge labelled D to state D. There are no indices on the edge, hence no matches. Then D is read, and we follow the edge labelled D back to state D. There are two indices (1 and 2) on the edge, showing that there are two matches in q to DD. A is then read, and so on.   △

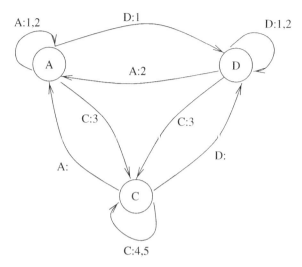

Figure 2.5 A finite-state machine for the example. For clarity, the states are labelled. We see that input of A in state D will output the word in q matching DA (the word at position two).

```
                            d
                L U K A L W Y A R . . .
        i\j     1 2 3 4 5 6 7 8 9
        1   E
        2   A         *
        3   L           *
     q  4   C
        5   K     *
        6   A       *h=-2        *
        7   R                 *h=2
        8   V
        9   A                   *
       10   R                 *h=-1

            .
```

Figure 2.6   Figure illustrating the two-hit method.
h is diagonal number; see text for more explanation.

To explain the search it will help to make an (imaginary) table (dot matrix) as shown in Figure 2.6. In the example illustrated by Figure 2.6, $w = 2$ and an exact match is required for a hit.

Let i be the index of q, and j of d. Then the diagonals in the matrix can be numbered $h = j - i$. For each diagonal h, the start position of the current hit (the last found) is stored in $i'(h)$ (an array). As d is scanned, the hits on the diagonals are found (by use of the state-diagram or the table). The first hit is KA on diagonal -2 ($j - i = 3 - 5$), and 5 (the index of q) is saved in the array i', such that $i'(-2) = 5$. The next hit is AL, saved as $i'(2) = 2$. When AR is found as hits on diagonal 2 and -1, it is recognized that a hit already exists on diagonal 2. The difference between the start of these two hits ($i''(2) - i'(2) = 6 - 2 = 4$) is compared with the threshold, to see if they should be combined for extension to an HSP. The match on diagonal -1 is saved as $i'(-1) = 9$.

We see that the diagonals can be processed in parallel while d is scanned.

2.4.4   Extending to HSP

Extending to HSP is straightforward; however, in order to be sure of finding the HSP one has to extend to the end of the sequences, which can be several hundred residues. Therefore (for reasons of efficiency), the extension is stopped (in any direction) if the score falls a certain distance below the best score found for shorter extensions. In this way, finding the correct HSP is not guaranteed, but experience has shown that distances can be chosen which give a small probability for losing the best (e.g. probability of 0.001), and still saving considerable time.

Example

Let the query q be CCAACCDACCACD, the database sequence d be ADAADACACA, with the scoring scheme as in the example in Section 2.4.2. Suppose we treat the second word, DA, which will first have a match at index three in the query with score 1.5 (AA DA). We will extend this hit (using only one hit in this example), and let the cut-off distance be 1. Extending to right gives the following:

```
From q:          ... A    A    C    C    D    A    C   C   A   C   D
From d:          ... D    A    A    D    A    C    A   C   A
 Pairwise score  0.5  1.0 -0.5  0.0  0.5 -0.5 -0.5
 Sum score            1.5  1.0  1.0  1.5  1.0  0.5
```

The extension stops at the second (C, A) match, since the score has dropped below the threshold (1). Two segment pairs with score 1.5 are found (AA, DA) and (AACCD, DAADA). Note, however, that these are not (really) local maximals, since further extension (with CA, CA) would result in a higher score (2.5).   △

2.4.5   Introducing gaps

BLAST might find two HSPs, near in sequence, but not on the same diagonal, and where each HSP does not by itself score high enough for being presented as significant. However, taken together they might achieve a high enough score.

Example

Let the two sequences be:

q': ..ACYGWVAKKYLAGQAKLVRST...
d': ..KRYKDVAAKAGRAELAVQR....

Assume two HSPs are found:

    YGWVAKK   AGQAKL
    YKDVAAK   AGRAEL

and that they are not statistically significant separately. Looking at the sequences, the two might be grouped into one by introducing gaps as

    YGWVAKKYLAGQAKL
    YKDVAAK--AGRAEL

which might be significant.   △

Since BLAST can report several HSPs from each sequence, those HSPs can be analysed (in a postprocessing step) to check if they can be combined to a pair (local alignment) with gaps. However, this postprocessing suffers from two weaknesses:

- it is time-consuming;

- all of the actual HSPs must be reported for postprocessing.

Therefore, the BLAST program is extended by a dynamic programming procedure which is applied around HSPs. However, to save time, the following is implemented.

- The HSP must have a high score (they have chosen a score such that the extension is performed for only one per 50 database sequences, which corresponds to a score of approximately 22 bits for a typical-length query sequence (see Section 3.3)).

- The most common method for introducing gaps to an ungapped (local) alignment is to perform dynamic programming in a band around the ungapped alignment (in the DP table). Let the diagonal of the alignment be h, then a dynamic programming is performed in the band restricted by the diagonals $[h - k, h + k]$.

In BLAST, the extension around an HSP is done in another way. The extension is restricted such that only alignments that do not drop in score more than a threshold below the best score (of the current segment pair) are considered.

The extension starts by choosing a *seed point* from the HSP (one residue from each of the segments in (q, d)). This pair will be a pair in the final alignment, hence the selection of the seed point is critical. Dynamic programming then proceeds both forwards and backwards from this point, and those two processes can be done independently of each other. Cells from any diagonal might be expanded, but the number of cells expanded on each row tends to remain small or become zero. Consider a step in the forwarding dynamic programming, and let the best gapped alignment found (until now) have score S'. Let T be the

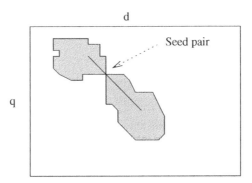

Figure 2.7   The figure illustrates the expanded cells in a dynamic programming procedure for finding the best gapped local alignment around the HSP. The line represents the HSP. See the text for explanation.

threshold, and H be the 'DP matrix'. If all of $H_{i,j-1}$, $H_{i-1,j-1}$, $H_{i-1,j}$ have scores less than $S' - T$, then $H_{i,j}$ will not be calculated. Figure 2.7 illustrates the expanding cells and seed point.

2.4.6   Algorithm

A coarse algorithm for the method used in BLAST is shown in Algorithm 2.2. Note that, to decide if two hits should be extended to an HSP, in this algorithm a score is given for the substrings constituting of the two hits and the symbols between them.

Algorithm 2.2. BLAST approach.
A coarse algorithm for the method used in BLAST for comparing a query to a database sequence, after preprocessing of the query.

const

w	the word length
W	a current word
thr1	threshold for deciding near hits
thr2	threshold for deciding dynamic programming around an HSP

var

q	the query sequence, it has been preprocessed
i	index in q
d	a database sequence
j	index in d
curr_hit(h)	a table with an entry for each diagonal h
	shows where the last found hit on diagonal h is
HSP	High-Scoring Segment Pair

proc

score(q',d')	scoring substrings for deciding if extending to HSP

begin
    for *each diagonal h* **do** curr_hit(h)=0 **end** initialize, no hits yet
    for $j := 1$ **to** $n - w + 1$ **do**   for each symbol in d
      $W :=$ *the word starting in d_j*
      for *each word in q matching W* **do**   a hit is found
        $i :=$ *the position of the matching word in q*
        $h := j - i$ diagonal for matched word
        if curr_hit$(h) \neq 0$ **then**   two hits on same diagonal?
            $l := curr\_hit(h)$ index in d
            if score$(q_{i-(j-l)...i+w-1}, d_{l...j+w-1}) \geqslant$ thr1 **then**
                              two 'near' hits found
                extend the segment pair $(q_{i-(j-l)...i+w-1}, d_{l...j+w-1}$ *to an HSP*
                save the HSP for possible dynamic programming
            end
        end
        curr_hit$(h) = j$ new current hit
      end
    end
    for *each found HSP* **do**
      if score(HSP) $>$ thr2 **then**
        perform dynamic programming around HSP
      end
    end
end

2.5 Exercises

1. (a) Explain how dot matrices can be used to find substrings of a sequence q which also exists reversed in another sequence d.

   (b) Try this on the sequences $d =$ DABADCE and $q =$ EDABCBDA when the length of the substrings must be at least three.

2. (a) Explain how dot matrices can be used to find common subsequences of two sequences, when by *subsequence* we here mean the more formal understanding of it: it need not consist of consecutive symbols, but the symbols must come in the same order as in the sequence. ECAEDA is, for example, a subsequence of AECEAECEDAA.

   (b) Find a sequence of length seven which is subsequence of both $q =$ AECEAECEDAA and $d =$ DECECAEDA.

   (c) Let (i_1, i_2) be the first and last position in q of the subsequence found in (b), and (j_1, j_2) be the same for d. Show an alignment of $q_{i_1...i_2}$ and $d_{j_1...j_2}$ where all seven symbols are aligned. Discuss if this is the 'best' alignment of $q_{i_1...i_2}$ and $d_{j_1...j_2}$.

3. Let a scoring scheme be 1 for equal amino acids, -1 for unequal amino acids, and 1 for the gap penalty (linear). Two sequences are given: AARRTRT and ART.

   (a) Find the highest-scoring local alignments (you should find four).

   (b) You have probably not found the alignment

$$
\begin{array}{l}
\texttt{ARRT} \\
\texttt{A-RT}
\end{array}
$$

   which also scores maximum. Explain why the procedures used in this book do not find this.

4. Write out the algorithms for finding the best local alignments using a linear gap penalty.

5. Generalize Equation (1.2) to be valid for finding a local alignment with general increasing gap penalty function. Show that if $H_{i,\,j-l} = 0$, then it is not necessary to examine $H_{i,\,j-t}$ for $t > l$.

6. Consider the search method BLAST with search sequence $q = $ SILSALIISL and database sequence $d = $ ASSLAIAIA (notice that only the amino acids A, I, L, S occur). Use $w = 2$, $T = 5$ and the scoring matrix BLOSUM62 (Table 5.4).

   (a) Make a table M with all possible words ($= 4^2 = 16$ words).

   (b) Examine q, and for each word o in q, insert the position of o into M where there are words matching o.

   (c) Examine d, and for each word, find matching words in the table M. Record for each word the position and scoring of the matching words.

   (d) You will now find that there is a word pair with score eight, and several with score six. Expand (heuristically to the right), to HSP, the word pair that contains SL from d and has score six. Stop the expansion if you obtain a score that is at least two below the highest found until now. Record the HSP you have found, with the score. Which HSP (with score) would you have found if you did not stop by using the cut-off value (two)? Be aware that we only demand one hit for expansion in this example.

7. (a) Now detail some of the statements in the high-level algorithm for BLAST. Assume that the result from the preprocessing of the search sequence q is in the table M. M can be indexed with words as index. The algorithm is to return in the table SEG, the segments that contain at least two hits, where the distance between the start of the succeeding two hits has to be less than or equal to a constant k. Assume a word length of w.

(b) If there are three hits at the same diagonal, and near in the sequences, they will be treated by the algorithm as two occurrences of two hits. Change the algorithm such that three or more such hits will result in one segment pair (to be extended to HSP).

2.6   Bibliographic notes

Waterman (1995) is a comprehensive book treating local alignments. The first to use dot matrices for biosequences was Gibbs and McIntyre (1970). A discussion of dot matrices is also found in Argos (1987) and Vingron and Argos (1991). Several programs exist on the Web, for example, a program by Sonnhammer:

http://www.cgr.ki.se/cgr/groups/sonnhammer/Dotter.html.

The alignment problem has close connections to the 'longest common subsequence problem', and several algorithms for solution are proposed (see Stephen 1994).

A lot of articles have been published that describe variants of the dynamic programming algorithms (for global or local), being less time expensive in practice, and reducing the space complexity (see, for example, Chao et al. 1994; Gotoh 1982; Myers and Miller 1988). Variants for finding near optimal alignments are described in Waterman and Eggert (1987), Zuker (1991), Barton (1993) and Chao (1994).

The empirical determination of gap penalties is described in Reese and Pearson (2002). FASTx is described in Pearson (1990), and Gapped BLAST in Altschul et al. (1997).

3

Statistical Analysis

Comparing a query sequence (q) with each sequence (d) in a database gives rise to many local alignments (q, d), each with a score. It must then be decided whether the score is high enough to suggest homology, and to do this we need to know what is expected by chance. Similarities between local alignments with scores having low probabilities to occur by chance are said to have high statistical significance, and, in this situation, it is reasonable to think that the sequences are evolutionarily related (homologous). However, it must be remembered that low significance does not necessarily mean no evolutionary relation. Homologous sequences having few sequence similarities are said to be in the 'twilight' zone (or beyond!) and the similarities are difficult to discover using sequence information alone.

In this chapter we first describe how to test for significance of an alignment of two sequences, and then expand this for database searching. Then we describe how to assess and compare programs for database searching.

3.1   Hypothesis Testing for Sequence Homology

When a best local alignment is found, the next task is to assess its biological relevance. This is most often done based on hypothesis testing. Hypothesis testing can briefly be described as follows.

1. A *null hypothesis* H_0, the validity of which we will test, is given. An alternative hypothesis, H_1, is also given.

2. Perform a relevant experiment for testing H_0, and record the result.

3. Find the probability, p for the result, given that H_0 is valid. If p is less than a given threshold (e.g. 0.02 (2%)), there is reason for rejecting H_0 (at the 2% level) and accepting H_1.

With sequence comparison, one wants to decide if the highest-scoring segment pair found in (q, d) is significant, i.e. if the two sequences are homologous. The procedure is then as follows.

Protein bioinformatics: an algorithmic approach to sequence and structure analysis
I. Eidhammer, I. Jonassen and W. R. Taylor   © 2004 John Wiley & Sons, Ltd   ISBN: 0-470-84839-1

1. Formulate the null hypothesis H_0: the two sequences (q, d) are *not* homologous. The alternative hypothesis is that they are homologous.

2. Decide what experiments to perform: find the segment pairs from (q, d) with the highest score (highest significance).

3. Determine the probability of the result, given H_0. This is often done by finding the probability distribution for the highest-scoring segment pairs in randomly generated sequences (see Section 3.1.1). A large number of such sequences are generated, and each compared with q, and the scores of these comparisons is the basis for the probability distribution of the scores.

4. Determine the rejection level for H_0, for example, 0.5×10^{-5}.

5. Perform the experiment chosen in (2): find the segment pair from (q, d) with the highest score and record the result.

6. Determine the probability of achieving the result or higher, given H_0 (use the probability distribution found above), and compare with the rejection level for H_0.

3.1.1   Random generation of sequences

For the random generation of sequences, a frequency distribution of the occurrences of the amino acids has to be used. The amino acid in each position (of the random sequence) is drawn using this distribution, often independent of the position and which amino acids are in the other positions.

Example

Assume an alphabet of four symbols $\{A, C, D, E\}$ and the sequences

$$q = \text{ACADAEA} \quad \text{and} \quad d = \text{ECAEDACECE},$$

and we find the best local alignment to be

```
CA-DA
CAEDA
```

with score $S = 6$.

We then make random sequences, with the same amino acid distribution and length as d. The frequency distribution to use when generating random sequences is then $\{f_A = 0.2, f_C = 0.3, f_D = 0.1, f_E = 0.4\}$. E will be drawn with a probability twice the probability of drawing A.

Assume we make a lot (e.g. 10 000) of random sequences, and for each we make a local alignment to q and note the score of the highest-scoring local alignment. We then find the distribution of these scores; suppose it is the distribution function shown in Figure 3.1. We see that this curve has its maximum for a score of 1. The whole

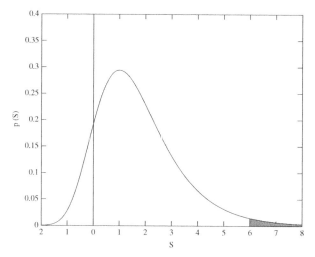

Figure 3.1 An example probability distribution function. The area under the curve is 1, and the shadow area is 0.0181. This means that the probability for getting a value $S \geqslant 6$ is 0.0181. The distribution is an extreme value distribution with parameters $u = 1, \lambda = 0.8$.

area under the curve is 1, and the area under the curve from 6 to ∞ is the probability of finding a local alignment with score at least 6, when q is aligned with a random sequence with the same amino acid distribution and length as d. This value is 0.0181, and it is hardly low enough to indicate homology between q and d. $\triangle$

The amino acid frequency distribution can be found in one of several ways.

1. *Universal.* A universal distribution is used, e.g. the distribution of the amino acids over all known sequences.

2. *Global.* A distribution of the amino acids occurring in a group of sequences of which the actual sequences are members, e.g. a superfamily.

3. *Local.* The distribution of the amino acids in the actual sequences (q, d) is used. Instead of really making a distribution and performing a drawing, a shuffling of one (or both) of the sequences is done. This will generate random sequences with the same distribution as q and d. A large number of shufflings of, for example, d, is done, and each compared with q. The scores are used in estimating parameters for the probability distribution.

A *restricted* version of shuffling can also be carried out. With restricted shuffling, the sequence is divided into regions, and each region is shuffled independently. This means that the amino acid distribution in a region is the same before and after the shuffling and this can be useful if the distribution of the amino acids varies in regions of the sequences. With ordinary shuffling this distribution will be spread, and statistically significant segment pairs might be overlooked, or vice versa. Let, for example, positions 1–15 in q have many A and C, but no D, and positions 31–45 only two A and two C. Let a segment, ACDAC, in d exactly match a segment in positions 31–45 in q. It has obvious statistical significance, but with an ordinary shuffling, all A and C might be spread, so that the segment might occur several places in the shuffled sequence. Restricted shuffling will counteract this, as the amino acids in position 1–15 will still be in these positions (but in shuffled order).

What frequency distribution to use depends on the context. The distribution of the amino acids in D should be used to assess the score obtained by comparing q with D (either a single sequence or a set of sequences).

3.1.2   Use of Z values for estimating the statistical significance

Z values are sometimes used for estimating the significance of found correlations. A procedure for this is as follows.

1. Compare the two sequences, and record S', the score of the highest-scoring segment pair.

2. Make a distribution of the scores for random sequences, as explained above, or use a known distribution. Let μ be the mean of the random scores, and σ the standard deviation. The standard deviation is a measure of variance, and for discrete values defined as

$$\sigma = \sqrt{\frac{1}{n-1}\sum_{i=1}^{n}(y_i - \mu)^2},$$

   where y_i are the scores obtained for randomized sequences.

3. Find the Z value of the highest-scoring segment pair, with scoring S'. The Z value is the number of standard deviations that the score S' is above the mean value, calculated as

$$Z(S') = \frac{S' - \mu}{\sigma}.$$

   A threshold, Z_t, is determined, such that a score higher than $Z_t\sigma + \mu$ should indicate statistical significance, hence that the sequences are biologically related (homologous). Experiments have shown that $Z_t = 7$ is appropriate for protein sequence comparison and protein database search.

Example

Assume we have found a score $S' = 17.2$. We must then calculate the Z value for S'. Let the probability distribution for the scores from random sequences have mean and standard deviation as $\mu = 4.2, \sigma = 3.4$. By use of the formula above, we find the Z value for S' to be 3.8. This is much less than 7, hence not convincing enough to indicate homology between the two sequences.   △

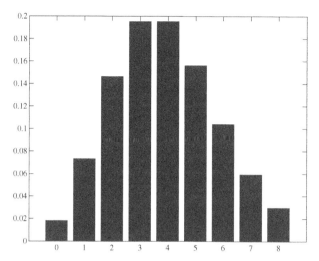

Figure 3.2 Poisson distribution for $v = 4$.

Use of the Z value is less informative when the random score distribution deviates from a symmetrical distribution. In the following sections it is shown that the distribution of scores of searching against random protein sequences is not symmetrical.

3.2   Statistical Distributions

It is shown that the statistical analysis for assessing the results of sequence alignments is based on Poisson and extreme value distributions. We will therefore briefly describe them.

3.2.1   Poisson probability distribution

The possible values for the Poisson distribution are the nonnegative integers. The distribution has a parameter v, which is also the *expected value*, and the *variance*. The distribution is given by (the probability that the stochastic variable X will have the value x)

$$P[X = x] = \frac{v^x}{x!}e^{-v}. \tag{3.1}$$

From this we see that

$$P[X \geqslant x] = 1 - \sum_{i=0}^{x-1} \frac{v^i}{i!}e^{-v}. \tag{3.2}$$

Figure 3.2 shows an example of a Poisson distribution.

3.2.2 Extreme value distributions

- Let $\{X_1, \ldots, X_n\}$ be independent, identically distributed random variables.

- Let Y_n be the maximum of the random variables. The cumulative distribution function for Y_n is then (since the X are independent of each other)

$$F_{Y_n}(y) \overset{\text{def}}{=} P[Y_n \leqslant y] = (P[X_i \leqslant y])^n = [F_{X_i}(y)]^n.$$

We sketch very informally that the distributions of the scores of comparing a sequence q with a random sequence may approximate an extreme value distribution. We can look upon a sequence as consisting of a set of (overlapping) segments. Let the segments of q be numbered $1, 2, 3, \ldots$. Each segment of q is compared with a (random) segment of d, and let the score of this comparison for segment i be X_i. Now let Y_n be the maximum of these scores. A distribution for Y_n can then be found by comparing the segments of q to many random segments of d. Note, however, that the X_i are not independent in this case, since the segments are overlapping.

Several types of extreme value distribution exist, dependent on the distribution of X_i, e.g. the Normal distribution. The type we are going to treat here has an exponential distribution in the upper end for $F_{X_i}(y)$ and can be written as $F_{X_i}(y) = 1 - e^{-g(y)}$, where $g(y)$ is increasing. A Normal distribution of X_i results in such an extreme value distribution.

When $n \to \infty$ the following can be shown ($Y_n \to Y$):

$$F_Y(y) = P[Y \leqslant y] = \lim_{n \to \infty} F_{Y_n}(y) = \lim_{n \to \infty} (1 - e^{-g(y)})^n = \exp[-e^{-\lambda(y-u)}] \quad (3.3)$$

for fixed λ, u.

The probability (density) distribution of Y is

$$f_Y(y) = \frac{dF_Y(y)}{dy} = \lambda \exp[-\lambda(y - u) - e^{-\lambda(y-u)}].$$

The form of $f_Y(y)$ depends on λ and u.

- u is the modal value of the distribution. It is the value where the density function has its maximum value; u is also called the *characteristic value*.

- λ is a measure of the variance, and is also called the *decay constant*.

The extreme value distribution is not symmetrical. An example is shown in Figure 3.1. From Equation (3.3) we get

$$P[Y > y] = 1 - F_Y(y) = 1 - \exp[-e^{-\lambda(y-u)}]. \quad (3.4)$$

Often, $F_Y(y)$ is unknown, hence there is a problem: *calculate λ and u*. One way of solving the problem with unknown u and λ is to try to fit the distribution to observed data. The procedure can be as follows.

1. Perform n experiments, and find the maximum value for each.

2. Calculate the middle value m_Y and the variance σ_Y^2 in the distribution of the maximum values found in 1.

There is a relation between m_Y, σ_Y^2 and u, λ in the corresponding extreme value distribution (0.577 is Euler's constant):

$$m_Y = u + \frac{\gamma}{\lambda} \approx u + \frac{0.577}{\lambda},$$

$$\sigma_Y^2 = \frac{\pi^2}{6\lambda_2} \approx \frac{1.645}{\lambda^2} \quad \Rightarrow \quad \lambda \approx \frac{1.282}{\sigma_Y},$$

$$\sigma_Y \approx \frac{1.282}{\lambda},$$

$$u \approx m_Y - \frac{0.577}{\lambda}.$$

Practical application

Assume that the score of the highest local alignment between q and d is $S' = 6$, and assume that the distribution of the scores for q against random sequences is the distribution in Figure 3.1. Then the probability for finding a local alignment with a score of not less than S' corresponds to the shaded area, which is 0.0181. △

3.3 Theoretical Analysis of Statistical Significance

Karlin and Altschul (1990) have done theoretical analysis for local alignment without gaps.

The details in the theory by Karlin and Altschul is beyond the scope of this book; however, it is useful to have a basic understanding of it, since it is the basis for the significance calculated for sequence database searching. To sketch the theory we make some definitions and present the assumptions for the theory. The theory is general; we shall explain it by using amino acids.

- $\mathcal{M} = \{a, b, \dots\}$ is the alphabet of the amino acids.

- $s_{1\dots m}$, $\bar{s}_{1\dots n}$ are two sequences of amino acids, *randomly generated by using the frequencies* $\{p_a\}$, $\{\bar{p}_a\}$ (often $p = \bar{p}$).

- R_{ab} is the scoring matrix.

- At least one of the scores R_{ab} is positive.

- The expected score for aligning two randomly chosen amino acids,

$$E = \sum_{a,b \in \mathcal{M}} p_a \bar{p}_b R_{ab},$$

must be negative.

- μ is the unique positive value for y that satisfies the equation:

$$\sum_{a,b \in \mathcal{M}} p_a \bar{p}_b e^{R_{ab} y} = 1.$$

- K is a constant which can be calculated from $\{R_{ab}\}$ and μ (how this is done is beyond the scope of this book).

- $Z_{S'}$ is the number of segment pairs from $s_{1...m}$ and $\bar{s}_{1...n}$ with a score of at least S'.

- S_M is the maximum score for a segment pair from $s_{1...m}, \bar{s}_{1...n}$.

If now

- m and n are sufficiently large,

- $\{p_j\}$ and $\{\bar{p}_j\}$ are sufficiently similar,

- S' is sufficiently large,

then they find the following.

- $Z_{S'}$ has an approximate Poisson distribution with parameter $v = Kmne^{-\mu S'}$. Since the parameter of the Poisson distribution is the expectation, it follows that the E-value of the score S', defined as *the expected number of segment pairs scoring at least S'*, is

$$E(S') = Kmne^{-\mu S'}. \tag{3.5}$$

- By setting $X = Z_{S'}$ in Equation 3.2 we get

$$P(Z_{S'} \geqslant x) \approx 1 - e^{-v} \sum_{i=0}^{x-1} \frac{v^i}{i!}, \tag{3.6}$$

where $v = Kmne^{-\mu S'}$. This is the probability of finding x or more distinct segment pairs, all with a score of at least S' ('$\approx$' is used because of the approximate Poisson distribution).

- By setting $x = 1$ in Equation (3.6) we get the P-value of the score S', defined as *the probability of finding at least one segment pair with a score greater than or equal to S'*:

$$P(S') = P(S_M \geqslant S') = P(Z_{S'} \geqslant 1) \approx 1 - e^{-v} = 1 - \exp(-Kmne^{-\mu S'})$$
$$= 1 - \exp(-E(S')) \tag{3.7}$$

Note that $\exp(x)$ is equivalent to e^x.

- By expanding Equation (3.7) into a power series, we get

$$P(S') = 1 - \exp(-E(S')) = 1 - \left(1 - \frac{E(S')}{1!} + \frac{E(S')^2}{2!} + \cdots \right) \approx E(S') \quad (3.8)$$

for small E values (large S').

We see that for small values, the E value and the P value are similar.

Practical application

We have two sequences q and d of lengths m and n, an equal amino acid distribution $\{p_i\}$ and a scoring matrix R.

1. We find the best local (ungapped) alignment which has score S'.

2. We find the statistical significance in the following way.

    - Calculate μ and K from the scoring matrix and the amino acid frequency distribution.

    - Find the probability of S' occurring by chance: the P value $\approx E$ value $\approx K m n e^{-\mu S'}$.

$\triangle$

3.3.1   The P value has an extreme value distribution

Let the highest segment pair score found by comparison of two sequences be S'. From Equation (3.7), we get the probability for achieving this score or higher by chance as

$$P(S') = P(S_M \geqslant S') \approx 1 - e^{-v} = 1 - \exp(-K m n e^{-\mu S'})$$

and from this we get

$$P(S_M \geqslant S') \approx 1 - \exp(-e^{\ln(Kmn)} e^{-\mu S'}),$$
$$P(S_M \geqslant S') \approx 1 - \exp(-e^{-\mu(S' - (\ln(Kmn))/\mu)}).$$

By setting $\lambda = \mu$ and $u = (\ln(Kmn))/\mu$ we get

$$P(S_M \geqslant S') \approx 1 - \exp(-e^{-\lambda(S' - u)}),$$

which is similar to Equation (3.4); hence we have the two parameters (λ and u) characterizing an extreme value distribution.

Note that the development is valid for S' sufficiently large.

3.3.2   Theoretical analysis for database search

We here describe how the theoretical analysis for ungapped local alignments for two sequences can be expanded to database searching. This is used for BLAST, as explained in Altschul et al. (1997). The following description is based on this article.

For a score S', the E value (the expected number of segment pairs in two unrelated sequences with scores of at least S') is given in Equation (3.5).

With a database search, the same theoretical analysis can be used, but now n must be replaced by N, the length of the whole database defined as the total number of residues, resulting in

$$E(S') = KmNe^{-\mu S'}. \tag{3.9}$$

Ideally, N should be the number of residues in a nonredundant database (a database where no homologous sequences exist). Note that the E value refers to the number of segments, but for very small E values this can be regarded as the number of sequences.

The equation for the E value can also be derived in another way. Let L be the number of sequences in the database. The P value for two sequences is

$$P(S') = Kmne^{-\mu S'}. \tag{3.10}$$

This P value gives the probability of obtaining a score of S' for two sequences. The expected number of segments obtaining this score when L sequences (independent and of same length n) are compared with the query is L multiplied by this P value, resulting in the same Equation (3.9) (since $N = Ln$).

We have shown that, for small values, the P- and E values are equal. Experience has shown that, for a database search, this equality seems to hold for P values less than 0.01. Then the E value begins to increase faster than the P value.

The expression shows that the P value depends on the parameters μ and K, which are calculated from the scoring matrix used, and the *a priori* probability for each amino acid (background probabilities). This makes it impossible to compare scores when different scoring matrices (and background probabilities) are used. Therefore, we cannot conclude anything about the statistical significance by looking at the scores alone. It is, however, possible to convert S' into a *normalized* score S'' such that the P value depends only on this scoring. Define

$$S'' = \mu S' - \ln(KmN). \tag{3.11}$$

Then

$$S' = \frac{S'' + \ln(KmN)}{\mu}. \tag{3.12}$$

Replacing S' by S'' in Equation (3.10) results in

$$P = KmNe^{-\mu S'} = KmNe^{-(S'' + \ln KmN)}$$

$$= KmNe^{-S''}e^{-\ln KmN} = \frac{KmNe^{-S''}}{KmN} = e^{-S''}. \tag{3.13}$$

(It can be shown that this normalized score corresponds to using an extreme value distribution with $u = 0$, $\lambda = 1$ (see Exercise 6).)

The normalized scores is denoted by *bits*. From Equation (3.13) we find $S'' = -\ln P$, which is the normalized score required to achieve a particular P value.

For each scoring matrix and typical amino acid distribution, μ and K are calculated. When a score S' is achieved for a database search, the normalized score S'' can therefore be calculated by Equation (3.11), and the P value found by Equation (3.13).

3.4 Probability Distributions for Gapped Alignments

The statistical theory above is developed for ungapped local alignments. No proof is known that the theory also holds for gapped local alignments. However, several computational experiments strongly suggest that it is also approximately valid for alignments with gaps.

Evaluating the statistical significance by the method described requires calculating the two parameters μ and K. For ungapped alignments these can be found analytically for any scoring matrix. This is not (yet) possible for gapped alignments. Two different methods for estimating the value of the parameters for gapped alignments are used.

Perform simulation. This means generating random sequences from a typical amino acid distribution. Then, for a scoring matrix, apply dynamic programming on a number of the random sequences. A distribution of the scores is made, and the mean and variance calculated. The parameters of the extreme value distribution can then be found as explained at the end of Subsection 3.2.2. Note that simulations must be made for each scoring matrix that will be used. This procedure is used for (gapped) BLAST. A drawback is that statistical significance cannot be given if there is no simulation for the scoring matrix used.

Estimate the values 'on the fly' for each search. The scores, achieved by comparing the query sequence q with each database sequence, are noted. Assuming that q is homologous to a maximum of, for example, 4000 sequences in the database, the 4000 highest scores are thrown away, the rest can be regarded as comparing the query with a set of random nonhomologous sequences. An extreme value distribution can be fitted to those scores, as explained above.

Pearson (1998) has investigated several techniques for estimating the statistical parameters 'on the fly' for use in the database search program FASTA. The simplest technique calculates the mean μ and variance σ^2 from all the similarity scores and then calculates a Z value for each score S' ($Z = (S' - \mu)/\sigma$). Similarity scores from related sequences are then removed by excluding sequences with Z values greater than 7.0 or less than -3.0, and the process of estimation and exclusion is repeated up to five times. The remaining scores can now be adapted to extreme value distributions, as explained in Subsection 3.2.2. Pearson, however, finds the P value of a score by converting the corresponding Z value to a probability using an extreme value distribution.

Table 3.1   Explanation of the terms TP, FP, FN and TN, related to a threshold T.

	Homologous sequences	Nonhomologous sequences
Score at least threshold T	TP	FP
Score under threshold T	FN	TN

One reason why this cannot be calculated 'on the fly' for BLAST is that alignments are only found between q and some of the database sequences (those assumed best).

3.5   Assessing and Comparing Programs for Database Search

Suppose we have a database search program using a query sequence q, and that the program classifies the database sequences as either homologous to q or not homologous. Let us call the sequences which the program finds as homologous to q *positive sequences* and the others *negative sequences*. However, the program might classify wrongly.

A sequence classified as positive is called a *true positive* sequence if it is known to be homologous to q; otherwise it is called a *false positive* sequence. A sequence classified as negative is called a *true negative* sequence if it is nonhomologous (to q); otherwise it is called a *false negative* sequence. As the similarities between q and each of the database sequences are given a score, a simple method for classification is to say that a score over or equal to a threshold T means a positive sequence, and those with score below T are negative.

We can then define the following functions of T.

- $TP(T)$: the number of true positive sequences.

- $FP(T)$: the number of false positive sequences.

- $TN(T)$: the number of true negative sequences.

- $FN(T)$: the number of false negative sequences.

See Table 3.1 for an illustration. When it is otherwise clear, T is often left out.

The functions depend on T such that, for example, increasing T will increase FN and decrease FP (as shown in Figure 3.3(a)).

Example

To illustrate some of the concepts explained in this chapter, let us imagine a database of sequences in which 20 of them are homologous to q. We further assume that two programs, P1 and P2, are used for searching the database with query q. Let the score

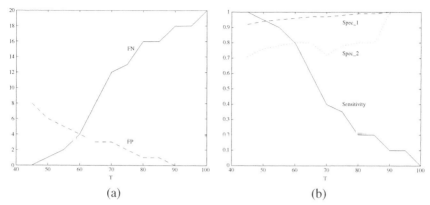

Figure 3.3 Illustration of results from program P2 in the example. (a) shows FP and FN as functions of T. Note that FP = FN when they both have the value 4. (b) The sensitivity and the two forms of specificity as function of the threshold T. TN + FP is assumed to be 100.

of the highest-scoring database sequences be as below, where underlining means the score of a sequence not homologous to q:

P1:

62 58 57 <u>55</u> 54 52 <u>52</u> 51 51 49 47 <u>46</u> 45 <u>44</u> 42 42 40 <u>39</u> 38 <u>38</u> <u>37</u> 36 35 <u>34</u> 33 33 32 <u>31</u> 29 <u>27</u> ...

P2:

97 96 <u>87</u> 86 85 <u>79</u> 78 78 75 73 <u>73</u> 69 69 66 65 <u>64</u> 63 63 61 60 <u>59</u> 58 56 <u>53</u> 52 <u>49</u> 47 <u>45</u>...

The result of the search can be illustrated as below, where H means homologous to q, and n nonhomologous:

    P1: HHHnHHnHHHHnHnHHHnHn|nHHnHHHnHnnn...
    P2: HHnHHnHHHHnHHHHnHHHH|nHHnHnHnnn...

The '|' marks where FP(T) = FN(T) (see Section 3.5.2).

Figure 3.3(a) shows how FP and FN depend on T for program P2.   △

3.5.1   Sensitivity and specificity

The concepts of sensitivity and specificity are central in discussing programs for database searching.

Sensitivity is a program's ability to discover homology between sequences which have long evolutionary distances between them (weak similarities). It is often measured by TP/(TP + FN), the proportion of the homologous sequences which are correctly classified. The sensitivity increases with decreasing threshold.

Specificity is a program's ability to not present nonhomologous sequences as homologous. Two different measures are in use:

- *Spec_1*: TN/(TN + FP), the proportion of the nonhomologous sequences which are correctly classified.

- *Spec_2*: TP/(TP + FP), the proportion of the positive sequences which are true positives. This formula is also known as the *positive predicted value*.

For database searching, both formulae show the same overall dependency of the threshold T. Since TN + FP is a constant, *Spec_1* is a nondecreasing function of increasing T. *Spec_2* shows the same overall trend, but local decreases might occur, as shown in Figure 3.3(b). Note that in database searching, there are generally many more nonhomologous sequences than homologous, and FP becomes small compared with TN. Hence the values for *Spec_1* are all close to 1, and *Spec_2* is the most widely used in such cases.

Correlation coefficient is

$$\frac{(TP \cdot TN - FP \cdot FN)}{\sqrt{(TP + FN) \cdot (TP + FP) \cdot (TN + FP) \cdot (TN + FN)}}.$$

This includes both sensitivity and specificity in one expression, and is in the range +1 (no false) to −1 (no true).

Ideally, we want FP = FN = 0, for which all the formulae give the value 1. Note that for a given query and database, the four formulae depend on the threshold. Figure 3.3(b) shows how three of the formulae depend on T for our search example, when we assume that the number of nonhomologous sequences in the database is 100.

3.5.2 Discrimination power

A search program's *discrimination* (or *classification*) *power* is how well it discriminates between homologous and nonhomologous sequences. A simple measure of the discrimination power (for given q and D) can be the value F where $F = FP(T) = FN(T)$ (i.e. equal numbers of sequences wrongly admitted and true sequences missed). A low value of F indicates that the program does a good job in discriminating between homologous and nonhomologous sequences. Where to find F for our search programs is indicated by the '|' in the search results. (Note that the actual value for T might be different for the two programs.) We see that $F = 6$ for P1 and $F = 4$ for P2. This means that for this example P2 has a better discrimination power than P1, when using this simple measure. The result is also illustrated in Figure 3.4(a).

The simple measure above corresponds to a single point on the curves in Figure 3.4(a), where FP = FN. Ideally, several points should be taken into account. This is done by using a method called receiver operating characteristic (ROC).

Receiver operating characteristic

An ROC curve is a sensitivity/specificity plot for an experiment, for example, treating healthy and sick patients. The numbers of healthy and sick patients in such experiments

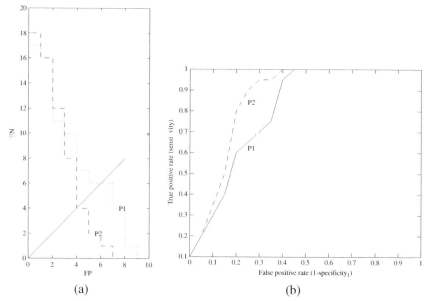

Figure 3.4 The plots illustrate discrimination power. (a) The result of two search programs (see the example) are plotted. The number of false negatives (FN) are plotted against the number of false positives (FP) as the threshold is decreased in each step. At each step either FN is decreased or FP is increased. The F values are found where the curves cross the line FP = FN (the solid line), six for P1 and four for P2. (b) ROC curve for the search results from the two programs in the example.

are usually not dissimilar. To use our search example for illustration, we first assume that the number of nonhomologous sequences (to q) is equal to 20. For the specificity, it is appropriate to use the *false positive rate*, which is

$$1 - Spec\_1 = 1 - \frac{TN}{TN + FP} = \frac{FP}{TN + FP},$$

the rate of the nonhomologous sequences which are classified as homologous). Note that the sensitivity is the *true positive rate* (the rate of the homologous sequences which are classified as homologous). The ROC curve for our example is shown in Figure 3.4(b).

An ideal curve would have a sensitivity of 1 for all the values of the false positive rate that correspond to all homologous sequences scoring higher than any nonhomologous sequence. From this we see that the closer the curve follows the left-hand and the top border, the better is the search. That means for our example that P2 is better than P1. The area under the curves can then be used as a measure of how well the different programs discriminate.

This form of the ROC curves is not, however, appropriate for database searching, since the number of nonhomologous sequences is very large compared with the number of homologous ones. We are primarily interested in how the programs classify the highest-scoring sequences, that is, we are most interested in the left-hand side of the ROC diagram (for a database of 50 000 sequences, this might, for example, be the area to the left of a false positive rate of 0.01). Therefore, we need a form which 'zooms in' to that area.

This can be done by counting, for each false positive, the number of true positives found, and summing up. Let P be the number of homologous sequences (20 in our example), and TP_i the number of homologous sequences found before the ith false positive. In the example above, TP_3 for P1 is 9, and 8 for P2. For a fixed number n of false positives, ROC_n is defined as

$$ROC_n = \frac{1}{nP} \sum_{i=1}^{n} TP_i. \tag{3.14}$$

Example

For the example above, ROC_5 for the two programs is

$$ROC_5(P1) = \frac{1}{5P}(3 + 5 + 9 + 10 + 13) = \frac{40}{100},$$
$$ROC_5(P2) = \frac{1}{5P}(2 + 4 + 8 + 12 + 16) = \frac{42}{100}.$$

$\triangle$

We see that the highest-scoring homologous sequences have the highest contributions to the formula in Equation (3.14). Note that ROC_n is a value in the interval [0, 1], with 0 as worst and 1 as best.

ROC_n can be illustrated by a sensitivity/specificity curve, where TP is drawn as a function of FP. Figure 3.5(a) shows this for our search example. ROC_n is the part of the area $(0 \ldots n, 0 \ldots P)$ under the curve, as illustrated for P1 in Figure 3.5(b) for $n = 5$.

Which n to use for comparing real programs for database search depends on the practical application, but a value of between 50 and 100 is often used, depending on the number of homologous (to the query) sequences.

3.5.3 Using more sequences as queries

In the presentation we have used only one sequence as query. It is, however, straightforward to generalize to an analysis using several sequences from a family as queries.

3.6 Exercises

1. Assume that in a search between the sequences q and d, the best segment pair is given a score of 20. Sequence d is then shuffled 10 times, and the best segment

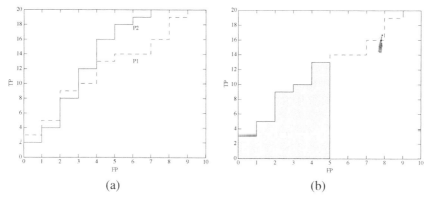

(a)   (b)

Figure 3.5 (a) TP as function of FP for our two example search programs. See the text for the correspondence to the ROC curve. (b) The filled area corresponds to ROC$_5$ for P1.

pair in the alignments between q and each of the shuffled sequences is found with the following scores: 8, 18, 2, 8, 4, 8, 4, 10, 12, 6.

(a) Estimate, using the Z value, whether q and d are homologous.

(b) Now assume the extreme value distribution. Estimate the values of the parameters u (the modal, or characteristic, value) and λ (the variance measure or decay constant), by the equations in Subsection 3.2.2. Then find the probability that a score equal to or above 20 does occur by searching in q and a random sequence, by using Equation (3.4).

2. Suppose we search in a database D with query q, and there are seven sequences in D homologous to q. We use the programs P1 and P2. Let H mean a sequence homologous to q, and n a sequence nonhomologous to q. Suppose that the search produced the following results, when the sequences are ordered after decreasing score:

```
P1: HHnnHHnnHHnHnn....
P2: HHHnHnHnnnnHnH....
```

(a) Let | mark where each of the programs sets its thresholds as

```
P1: HHnnHHn | nHHnHnn....
P2: HHHnHnHn | nnnHnH....
```

This means, for example, that P1 finds seven positive sequences. Find the sensitivity and specificity of the two programs. If this is the only information you have, which program would you recommend?

(b) Find the value $F = \text{FP}(T) = \text{FN}(T)$ for the two programs. Compare the result with what was found in (a).

(c) Find ROC$_n$ for the two programs for different values of n. Discuss the results.

3. When will the value of *Spec_2* decrease for increasing T, and what is the least local decrease?

4. Consider a family in the PROSITE database, for example, the family ZF_PHD_1 (PS01359) and a sequence in SwissProt belonging to that family, for example, YK09_YEAST (P36124). Search in SwissProt with the selected sequence by using the programs FASTA and BLASTP. You will find those programs, for example, at http://www2.ebi.ac.uk. Consider a sequence as homologous to the selected sequence if they belong to the selected PROSITE family. Calculate the sensitivity and specificity for both programs when the number of true positives are 2, 4, 6, 8, 10, 12, ... Plot the result in a sensitivity/specificity plot (sensitivity on the horizontal axis). Choose appropriate values for the scales on the axes. Discuss which program (with the chosen parameters) you regard to be the best.

5. Suppose that in a database with 10^6 residues, one searches with a query q of length 100. An affine gap penalty is used. To estimate the significance of the result, q is shuffled 10 000 times, and each shuffled sequence is used as a query to the database. From this the parameters in an extreme value distribution are estimated to be $\lambda = 0.252$ and $u = 0.035$. What score must be achieved in the search with q in order to obtain a P value no greater than 10^{-6}? Use Equation (3.4) in Subsection 3.2.2. (Remember that $\ln(e^x) = x$.)

6. From Subsection 3.3.1 we see that $P(S') = 1 - \exp(-KmNe^{-\mu S'})$. This corresponds to the extreme value distribution we considered with $\lambda = \mu$ and $u = (\ln(KmN))/\mu$. Consider the replacement in Equation (3.11). Show that $P(S'')$ will then have a normalized extreme value distribution ($u = 0$ and $\lambda = 1$).

3.7   Bibliographic notes

The statistical analysis is described in Karlin and Altschul (1990) and Karlin and Altschul (1993), and articles are referred to in Altschul et al. (1996, 1994). Analysis for gapped alignments and methods for estimating the parameters are to be found in Collins et al. (1988), Collins (1993), Altschul et al. (1996, 2001), Pearson (1996, 1998) and Bailey and Gribskov (2002). An approximate statistics for gapped alignments is in Mott and Tribe (1999). Webber and Barton (2001) estimate P values for global alignment.

Use of classification power can be found in Jonassen et al. (2000b) and Lindahl and Elofsson (2000). Use of ROC curves is shown in Gribskov and Robinson (1996) and Schæffer et al. (2001).

4

Multiple Global Alignment and Phylogenetic Trees

A multiple alignment is a natural extension of pairwise alignment, each column having m symbols, where m is the number of sequences. One tries to place the residues which have a common ancestor in the same column. Figure 4.1 shows a block from a multiple alignment of some insulin sequences.

There are several reasons for finding multiple alignments. Comparing several sequences can reveal what is common for a whole family (it has been said that 'two homologous sequences whisper, a multiple alignment shouts loudly'). In other words, properties shared by several sequences can become significant when all of the sequences are considered together, but need not appear significant when regarding only two (or a few) of them in the analysis. Furthermore, multiple alignments can show which residues are critical for the structure and functional of the proteins in the family (or superfamily). In this way, the analysis of a multiple sequence alignment can provide a basis for the prediction of protein structures. From multiple alignments one can find *motifs* and *profiles* describing signatures for families. Multiple alignments can also be used to estimate evolutionary histories. Due to the close connection between multiple alignments and evolutionary trees, phylogeny is also briefly treated in this chapter.

Several approaches for making multiple alignments have been proposed, and we will describe only a few of them. Others are cited in the bibliographic notes, where articles comparing different methods and reference to a benchmark database are also cited.

4.1 Dynamic Programming

The dynamic programming procedure used for pairwise alignment can be easily extended to multiple alignments, producing the best alignment in accordance with the given scoring scheme. Three types of column exist for the pairwise alignment:

Protein bioinformatics: an algorithmic approach to sequence and structure analysis
I. Eidhammer, I. Jonassen and W. R. Taylor   © 2004 John Wiley & Sons, Ltd   ISBN: 0-470-84839-1

```
            b>                                                           <a
XENLA1   YPKVKRDMEQALVSGPQD------NELDG--MQLQPQ--EYQKMKRGIVEQ
XENLA2   YPKIKRDIEQAQVNGPQD------NELDG--MQFQPQ--EYQKMKRGIVEQ
MOUSE1   TPKSRREVEDPQVEQLEL------GGSP---GDLQTLALEVARQKRGIVDQ
RAT1     TPKSRREVEDPQVPQLEL------GGGPEA-GDLQTLALEVARQKRGIVDQ
MOUSE2   TPMSRREVEDPQVAQLEL------GGGPGA-GDLQTLALEVAQQKRGIVDQ
RAT2     TPMSRREVEDPQVAQLEL------GGGPGA-GDLQTLALEVARQKRGIVDQ
CRILO    TPKSRRGVEDPQVAQLEL------GGGPGA-DDLQTLALEVAQQKRGIVDQ
RABIT    TPKSRREVEELQVGQAEL------GGGPGA-GGLQPSALELALQKRGIVEQ
BOVIN    TPKARREVEGPQVGALEL------AGGPG------AGGLEGPPQKRGIVEQ
SHEEP    TPKARREVEGPQVGALEL------AGGPG------AGGLEGPPQKRGIVEQ
PIG      TPKARREAENPQAGAVEL------GGGLGG---LQALALEGPPQKRGIVEQ
CANFA    TPKARREVEDLQVRDVEL------AGAPGE-GGLQPLALEGALQKRGIVEQ
HUMAN    TPKTRREAEDLQVGQVEL------GGGPGA-GSLQPLALEGSLQKRGIVEQ
PANTR    TPKTRREAEDLQVGQVEL------GGGPGA-GSLQPLALEGSLQKRGIVEQ
CERAE    TPKTRREAEDPQVGQVEL------GGGPGA-GSLQPLALEGSLQKRGIVEQ
AOTTR    APKTRREAEDLQVGQVEL------GGGSIT-GSLPP--LEGPMQKRGVVDQ
CAVPO    IPKDRRELEDPQVEQTEL------GMGLGA-GGLQPLALEMALQKRGIVDQ
CHICK    SPKARRDVEQP-LVSSPL------RGEAGV-LPFQQE--EYEKVKRGIVEQ
ORENI    NPR--RDVDPLLGFLPPKAGGAVVQGGEN---EVTFKDQMEMMVKRGIVEE
VERMO    TPK--RDVDPLLGFLPAKSGGAAAGG-ENEVAEFAFKDQMEMMVKRGIVEQ
BRARE    NPK--RDVEPLLGFLPPK------SAQETEVADFAFKDHAELIRKRGIVEQ
ONCKE    TPK--RDVDPLIGFLSPK------SAKENE--EYPFKDQTEMMVKRGIVEQ
            *       *    :                              ***:*::
```

Figure 4.1 A block from a multiple sequence alignment of some insulin proteins generated by the program CLUSTAL. The block is from the part between the two chains and some residues from the chains. b> and <a show where the chains are.

Table 4.1 The seven different column types in aligning three sequences: s^1, s^2, s^3.

s_i^1	—	s_i^1	s_i^1	—	—	s_i^1
s_j^2	s_j^2	—	s_j^2	—	s_j^2	—
s_k^3	s_k^3	s_k^3	—	s_k^3	—	—

no blank, or blank in one of the sequences. For m sequences, the number of different column types is (remember that columns with only blanks are forbidden)

$$\sum_{i=0}^{m-1} \binom{m}{i} = 2^m - 1, \tag{4.1}$$

where i is the number of blanks, and i blanks can be in $\binom{n}{i}$ different sets of sequences. Table 4.1 shows the seven alternatives for the three sequences: s^1, s^2, s^3.

The procedure can use an m-dimensional matrix for the calculations, and for each cell the number of values to compare is given by Equation (4.1). The number of cells

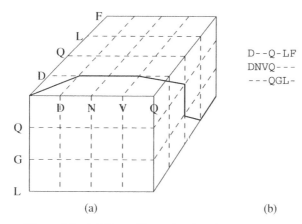

D--Q-LF
DNVQ---
---QGL-

Figure 4.2   (a) Figure illustrating the alignment in (b).

in the matrix is $n_1 n_2 \cdots n_m$, where n_i is the length of sequence i. An alignment can be illustrated in the m-dimensional space, as shown in Figure 4.2 for $m = 3$.

We see that the time and space complexity grows exponentially with the number of sequences ($O(n^m)$ for equal sequence lengths n), and for practical solutions of such computationally expensive problems, generally, two approaches are used.

- Try to reduce the running time by using pruning techniques (cut-offs) which still guarantee finding the highest-scoring alignment, but not reducing the worst-case complexity.

- Use heuristics. This means that some 'rules of thumb' are used in the solution, and the best (or correct) solution is not necessarily found. For some of the heuristic methods it is possible to find an upper bound on the deviation of the result from the correct one.

We will show examples of both of these approaches, but first we describe a scoring scheme that is suitable for use with the dynamic programming approach.

4.1.1   SP score of multiple alignments

For automatically finding the 'best' multiple alignment of several alternatives, we must be able to score them, and to also score subset alignments. By a subset alignment we mean an alignment of two or more of the m sequences we want to construct an alignment for. One intuitive scoring scheme is to build on the scores of the pairwise alignments defined by the multiple alignment, and simply sum them to obtain the score of the complete multiple alignment. This is called the sum-of-pairs score, SP score.

Let $\mathcal{A}$ be an alignment of the sequences $\{s^1, s^2, \ldots, s^m\}$ and $\bar{s}^i$ the *projection* of s^i, meaning s^i with the gaps as they are inserted in $\mathcal{A}$. Let $S(\bar{s}^i, \bar{s}^j)$ be the pairwise score of projections. Then the score of the alignment is

$$S(\mathcal{A}) = \sum_{i=1}^{m-1} \sum_{j=i+1}^{m} S(\bar{s}^i, \bar{s}^j).$$ (4.2)

The idea behind the SP score is that if $S(\bar{s}^i, \bar{s}^j)$ is the score of a reasonable pairwise score, the best multiple score should maximize the sum of all pairwise scores. The pairwise projection scores are complicated in that there might be columns with only blanks in the projection. These have no meaning for the pairwise alignment, and their score is usually set to 0.

If we use linear gap costs, the SP score can also be calculated as a sum of column scores:

$$S(\mathcal{A}) = \sum_{k=1}^{r} \sum_{i=1}^{m-1} \sum_{j=i+1}^{m} R_{\bar{s}_k^i \bar{s}_k^j},$$ (4.3)

where r is the number of columns in the alignment, $\bar{s}_k^i$ is the kth symbol of $\bar{s}^i$ and R is a scoring matrix or a gap score. The score of two blanks must be zero.

Example

Let an alignment be

```
A-VP-
A-V-T
PSVPT
```

and the scoring scheme be 1 for equal symbols, -1 for unequal, and -1 for blank (linear gap penalty), and remember 0 for two blanks. The score of the alignment using Equation (4.2) is (calculated row-wise) $0 + (-1) + (-1) = -2$, and by using Equation (4.3) it is (column-wise) $(-1) + (-2) + 3 + (-1) + (-1) = -2$.   △

Note that for a given scoring scheme the following is satisfied:

$$S(\bar{s}^i, \bar{s}^j) \leqslant S(s^i, s^j).$$ (4.4)

This follows from the fact that $S(s^i, s^j)$ is the highest score achievable by aligning those two sequences.

Example

The score of the projections of the two first sequences in the example above is 0, but the best alignment between the sequences is

```
AVP
AVT
```

which scores 1.   △

There is a drawback with the score defined in Equation (4.2): all sequences are weighted equally. There are biological reasons for letting the sequences have different weights. The sequences for which we will find a multiple alignment usually constitute a family or subfamily. Ideally, we want an alignment for this family, but the known sequences are usually only a subset of all sequences belonging to the family. The known sequences might give a biased 'picture' of the family, as there might be a lot of known sequences of one 'type' of the family and few of another 'type'. If we assume that the description should mirror a uniform distribution of the sequences, the biased representation could be corrected by giving the highest weights to the most isolated sequences and the lowest weights to those having many known similar sequences in the family. The SP score then becomes

$$S(\mathcal{A}) = \sum_{i=1}^{m-1} \sum_{j=i+1}^{m} S(\bar{s}^i, \bar{s}^j) w^i w^j, \tag{4.5}$$

where w^i is the weight of sequence i. Use of sequence weights is explained further in Subsection 4.4.3, where a method for calculating such weights is also presented.

4.1.2   A pruning algorithm for the DP solution

The main idea for reducing the running time is to avoid calculating cells which certainly cannot lie on the best alignment path. Let K be the score of a known alignment of the m sequences (if K is not known it is always possible to make a quick alignment and find its score). Consider a cell $v = (i_1, i_2, \ldots, i_m)$ of the DP matrix, and let the score of the best path (alignment) from the start cell to cell v be S_v (see Figure 4.3(a)). Furthermore, let the highest-scoring alignment from v to the end of the DP matrix be $\leqslant F_v$. Then we know that the score of the highest-scoring alignment going through v must be $\leqslant S_v + F_v$. If, therefore, $S_v + F_v < K$, we know that v cannot lie on the path of the best alignment. There is therefore no reason to use the value of cell v in later calculations, and this is used for pruning.

Example

We have three sequences: ARSTVK, ASVK and ARTR. The cell $v = (3, 2, 2)$ will then get the highest score for aligning ARS, AS, AR ($S_{(3,2,2)}$). We then have to determine an upper bound ($F_{(3,2,2)}$) for the alignment of TVK, VK and TR, and compare $S_{(3,2,2)} + F_{(3,2,2)}$ to K.   △

Different programs utilizing pruning techniques have been developed. Some use *static pruning*, in which cells that certainly do not lie on the best alignment path are found before the procedure is run. We will describe a method using *dynamic* pruning, in which the cells to avoid are found during the run. For using this efficiently, a dynamic programming procedure using *forward* recursion is used instead of *backward* recursion. When visiting a cell, instead of using values from the precedent neighbour

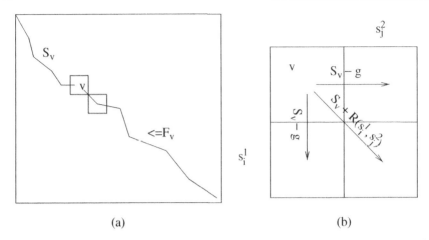

(a)   (b)

Figure 4.3 Illustrating the forward recursion procedure for two dimensions. (a) Test for pruning. (b) The sequences s^1, s^2 are aligned. From cell v values are sent to all its three forward neighbour cells. The scoring matrix is R and a linear gap penalty is assumed.

cells (as in backward recursion), a value is sent to its forward neighbour cells. Let v be a cell, and the score of the column from v to a forward neighbour cell w be $D(v, w)$. Then the value $S_v + D(v, w)$ is sent to w: the score of the best path to w via v. The value S_w is then the maximum of all values sent to w from its backward neighbour cells. This is shown for pairwise alignment in Figure 4.3. Note that the forward neighbour cells of v are found by increasing one or more of the indices of v by 1 (the linear gap penalty).

The cells must be visited in an order such that when a cell is visited, all backward neighbour cells have already sent their values. However, if a cell is found to certainly not lie on a best alignment path, there is no reason to send its value forward. Therefore, if a cell, when it is visited, has no values, we immediately know that it cannot be on a best path. Indeed, we can make the procedure such that cells which do not get any value will not be visited. This is done by use of a queue. When a cell is visited, it places its forward neighbours (to which it should send values) in the queue, in the order that they should be visited.

Example

Assume that we are comparing two sequences using dynamic programming and that the matrix is filled in row-wise. When visiting cell (i_1, i_2) its forward neighbour cells should be pushed in the order $(i_1, i_2 + 1), (i_1 + 1, i_2), (i_1 + 1, i_2 + 1)$.   △

Algorithm 4.1 shows the forward recursion with pruning.

Algorithm 4.1. Forward-recursion with pruning.

An algorithm for doing exact multiple alignment of m sequences
Forward recursion is used, with pruning of cells

const

h_0	the start cell of the DP matrix ($H_{0,0,...,0}$)
h_N	the end cell of the DP matrix ($H_{n_1,n_2,...,n_m}$)
K	a lower bound for the score of the whole alignment

var

u, v, w	denote cells
$S(u)$	the best score of an alignment (path) from h_0 to u
$P(u)$	the score of the best alignment from h_0 to u found so far
$D(u, v)$	the score for extending the alignment from cell u to cell v
Q	a stack of the cells u for which a value for $P(u)$ is found but u is not visited yet

proc

$F(v, h_N)$ a procedure which finds an upper bound of the score
           of the alignment from a cell v to the end-cell h_N

begin
    $v := h_0$; $P(v) := 0$; push(v, Q) push the start cell on the queue
    while $Q \neq \emptyset$ **do**
        pop(v, Q); $S(v) := P(v)$ v has got all values from its neighbours
        if $S(v) + F(v, h_N) \geqslant K$ **then**
            for *all forward neighbours w of v* **do** *in the right order*
                if $w \notin Q$ **then**
                    push(w, Q); $P(w) := S(v) + D(v, w)$
                else
                    $P(w) := \max(P(w), S(v) + D(v, w))$
                end
            end
        end
    end
end

Finding upper limits for scores

For any alignment $\mathcal{A}$ of sequences $\{s^1, s^2, \ldots, s^m\}$, the Equations (4.2) and (4.4) are valid when SP score is used. From this follows

$$S(\mathcal{A}) \leqslant \sum_{k=1}^{m-1} \sum_{l=k+1}^{m} S(s^k, s^l). \tag{4.6}$$

Let v be the cell $(i_1, i_2, \ldots, i_m)$. Then the procedure F should find an upper bound for the alignment of the subsequences $s^1_{i_1+1...n_1}, s^2_{i_2+1...n_2}, \ldots, s^m_{i_m+1...n_m}$. This can be done by using Equation (4.6) as

$$F = \sum_{k=1}^{m-1} \sum_{l=k+1}^{m} S(s^k_{i_k+1...n_k}, s^l_{i_l+1...n_l}). \qquad (4.7)$$

The score of each pairwise alignment is found in time $O(n^2)$, where n denotes an 'average' sequence length. Hence the complexity for finding F is $O(n^2 m^2)$.

Example

Again let the sequences be ARSTVK, ASVK and ARTR, $v = (3, 2, 2)$, and a scoring scheme of 1 for equal symbols, -1 for unequal and blank. An upper bound F for the alignment of TVK, VK and TR is then found from the alignments

TVK	TVK	VK
-VK	TR-	TR

giving the score $1 + (-1) + (-2) = -2$. The highest-scoring alignment is

```
TVK
-VK
T-R
```

scoring $1 + (-1) + (-3) = -3$.   △

It might be worthwhile calculating all pairwise alignment scores

$$S(s^k_{i_k+1...n_k}, s^l_{i_l+1...n_l}), \quad i_k = 0 \cdots n_k - 1, \ i_l = 0 \cdots n_l - 1$$

at the start of the procedure. This can be done then by performing pairwise alignment of each pair s^k, s^l, starting from the end symbols $s^k_{n_k}, s^l_{n_l}$, such that the complexity for finding all subset alignments for s^k, s^l is found in time $O(n_k n_l)$. The overall complexity of the method for finding all upper bounds is therefore $O(n^2 m^2)$.

The score of the alignment column corresponding to moving from v to w ($D(v, w)$) is calculated by the variant Equation (4.3) of the SP score. Note that the number of blanks in the column is m minus the number of different indices of v and w.

Although it is possible to reduce the practical running time of the DP procedure, experience has shown that when the number of sequences is over six the time tends to be prohibitive. Therefore, heuristic methods are used, and many of them use (rough estimates of) phylogenetic (or evolutionary) trees for help with the aligning.

4.2 Multiple Alignments and Phylogenetic Trees

One of the most important problems in biology is to find the evolutionary history of species existing today, and how they are related. This is usually done by constructing trees, with present-day species on the leaves (terminal nodes), and the interior

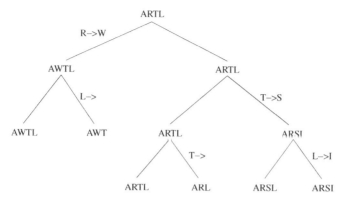

Figure 4.4 A phylogenetic tree constructed from the multiple alignment in the example. Sequences are constructed for the internal nodes, and the evolutionary changes are shown on the edges.

nodes representing hypothesized ancestors. Such a tree is called a phylogenetic (or evolutionary) tree, and usually all interior nodes in such trees have two 'children'. In the context of sequences, the nodes represent sequences, and the edges (branches) mutations, either explicitly or by a number indicating how many mutations have happened between the end nodes of an edge. In a tree constructed from protein (or DNA) sequences, an interior node can in principle represent the same sequence as a leaf node.

The close connection between multiple alignments and phylogenetic trees can be explained as follows. If a true phylogenetic tree for a set of sequences is known, it can be used as a guide in the construction of a multiple alignment, and knowing a true multiple alignment is of great help in constructing a phylogenetic tree.

Example

Consider a set of sequences {ARL, ARTL, ARSI, ARSL, AWTL, AWT}, and a (true) multiple alignment:

```
AR-L
ARTL
ARSI
ARSL
AWTL
AWT-
```

Our task is to construct a (true) phylogenetic tree. This is the same as finding the mutations which have occurred. One reasonable rule to use is to keep the number of mutations low. Looking at the second column, it seems that a mutation from W to R, or from R to W, has occurred in the past, dividing the set into two and giving rise to two subtrees. Looking at the third column in the largest subset, we see that there has probably been a mutation from S to T, or in the opposite direction. Noting that the

smallest subset has T in the third column, we choose the mutation to be from T to
S, dividing the set of sequences into even smaller subsets. Continuing this way, we
might end up with the tree in Figure 4.4. In the construction we have restricted the
number of mutations on each edge to be 0 or 1.

Now, going the other way, constructing a multiple alignment from the tree in Figure 4.4, we can perform alignment guided by the tree. This can be done by performing
a series of pairwise alignments between two sequences, one sequence and a (subset)alignment, or between two (subset)alignments. Following the tree in Figure 4.4,
first AWTL and AWT are aligned, then ARTL and ARl, then ARSL and ARSI, and
then the two subset alignments of ARTL and ARl and of ARSL and ARSI are aligned,
and so on. Doing this in a reasonable way will result in the alignments below:

```
AWTL     ARTL     ARSL          AWTL     ARTL          AWTL
AWT-     AR-L     ARSI    ->     AWT-     AR-L    ->    AWT-
                                          ARSL          ARTL
                                          ARSI          AR-L
                                                        ARSL
                                                        ARSI
```

                                                                                                    △

When neither a (true) multiple alignment nor a (true) phylogenetic tree is known,
we cannot use this approach directly. We can, however, approximate it. To construct a
multiple alignment, one can first construct a phylogenetic tree (PT) by other methods.
Similarly, to construct a phylogenetic tree, we can first construct a multiple alignment
(MA) by other methods. Combining these together in a cycle, we can iterate as

                    construct MA; construct PT; improve MA; improve PT

and hope to converge on the true alignment and tree together.

By reason of the close connection between phylogenetic trees and multiple alignments, we will briefly describe some methods for constructing phylogenetic trees.
For those wishing a more in-depth introduction, see the bibliographic notes.

4.3   Phylogeny

The purpose of phylogenetic studies of related objects are

  - to reconstruct the correct genealogical ties between them (the *topology*); and

  - to estimate the time of divergence between them since they last shared a com-
    mon ancestor (length of edges in the tree).

In phylogenetic studies, the objects are often referred to as *operational taxonomic
units* (OTUs). In our case the objects are protein or nucleic acid sequences. We will
denote the set of sequences we have at the start for the *original sequences*.

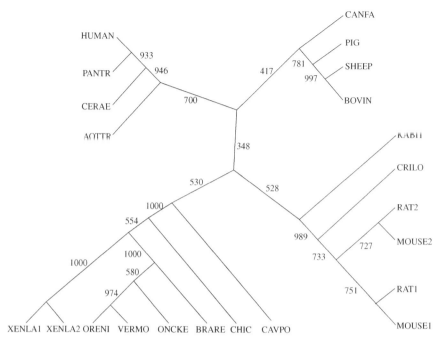

Figure 4.5 A phylogenetic tree (constructed by the neighbour-joining method using the program CLUSTAL) for a set of insulin proteins (the same as used for Figure 4.1). The numbers shows the result of using bootstrapping (see Section 4.3.7). Note that the tree is unrooted.

Hence the task can be formulated as follows. Given *m* original sequences, estimate the evolutionary relations between them as a phylogenetic tree. We consider only trees having the following properties.

- The tree has *m* terminal nodes (leaves), one for each original sequence.

- The tree has a set of internal nodes, which connect the terminal nodes and other internal nodes.

- The tree can be *rooted* or *unrooted*. If rooted, there is one node acting as the ancestor of all the other nodes. The root has two children, and the evolutionary direction is decided. (In unrooted trees, the direction is undecided.)

- The internal nodes of a rooted tree have two children; the internal nodes of an unrooted tree have three connected edges (branches).

- An internal node can (in some trees) have an explicit associated sequence.

- An edge can represent changes that have occurred between its nodes, either explicitly or by a number estimating the number of mutations.

Figure 4.5 shows an unrooted phylogenetic tree for a set of insulin proteins.

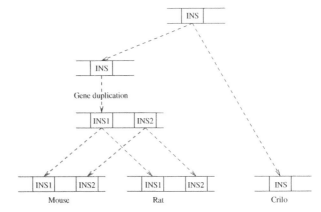

Figure 4.6   An evolutionary interpretation of part of the tree in Figure 4.5
illustrating paralog and ortholog. See the text for explanation.

Note that the mouse, rat and African clawed frog have two insulin genes. This tree
therefore gives us the opportunity to illustrate the concepts of *ortholog* and *paralog*.
Two homologous genes can be orthologs or paralogs. Two genes are orthologs if they
are derived from evolution of new species, but paralogs if they are derived from gene
duplication. If we interpret the tree in Figure 4.5 as the evolution shown in Figure 4.6,
we can say that

- INS1(Mouse) and INS2(Mouse) are paralogs;

- INS1(Mouse) and INS1(Rat) are orthologs;

- INS1(Mouse) and INS2(Rat) are only homologs;

- INS1(Mouse) and INS(Crilo) are orthologs.

We have assumed that the duplicated copies remain as active genes within the
species. If the new copy or copies are not active (transcribed) they are called *pseudo-
genes* (and slowly 'decay').

To illustrate the complexity of the problem of estimating phylogenetic trees, we
first look at the number of different tree topologies.

4.3.1   The number of different tree topologies

An unrooted tree (of the type we consider) has $m - 2$ internal nodes and a rooted has
$m - 1$. This can be used to find the number of different topologies. The number of
unrooted topologies for $m \geqslant 3$ original sequences is

$$T_{\text{unroot}}(m) = \frac{(2m - 5)!}{2^{m-3}(m - 3)!}. \quad (4.8)$$

For example, $T_{\text{unroot}}(10) = 2\,027\,025$. So, even for quite small m, it would be a large
number if all possible topologies had to be examined for estimating the 'best' tree.

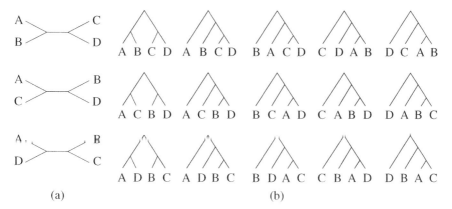

Figure 4.7 All topologies for four original sequences: (a) unrooted and (b) rooted.

The number of edges of an unrooted tree is $2m - 3$. An unrooted tree can be rooted by placing a root on one of its edges. Hence, the number of rooted topologies for $m \geqslant 2$ sequences can be found by multiplying $T_{\text{unroot}}(m)$ by $2m - 3$, resulting in

$$T_{\text{root}}(m) = \frac{(2m - 3)!}{2^{m-2}(m - 2)!}. \tag{4.9}$$

Hence $T_{\text{r0ot}}(10) = 34\,459\,425$.

Note that $T_{\text{unroot}}(m) = T_{\text{root}}(m - 1)$.

Figure 4.7 shows the different topologies for four sequences.

4.3.2 Molecular clock theory

The methods for constructing evolutionary trees are based on the observed mutations which have occurred. Although the sequences change at variable rates, it is simpler to assume a constant rate of change. Based on this assumption, molecular clock theory can be used to estimate the time since divergence of two sequences. Based on the difference between them, the number of mutations can be estimated (Section 1.9) to be t. Then the time since divergence corresponds to $t/2$ mutations. If we have an expression for the average number of mutations per position per time unit for such sequences, we can calculate the time in ordinary clock units. Some of the methods for constructing trees assume this theory, others do not. We will not pursue the details of this theory since absolute time is not relevant to multiple alignment methods.

4.3.3 Additive and ultrametric trees

Depending on what the tree is going to be used for, one can calculate and label the edges by values representing the evolutionary distance between the nodes. Usually, one assumes the edge lengths to be *additive*, meaning that the distance between any two

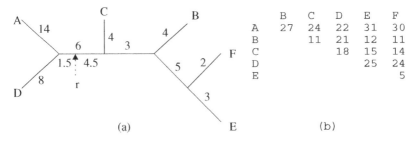

Figure 4.8   (a) An additive tree constructed for the sequences with the distances in (b). *r* shows where a root is placed.

nodes is the sum of the distances over the edges connecting the nodes. Figure 4.8(a) shows an unrooted additive tree constructed from the distances in Figure 4.8(b). The sums over the paths between any of the leaf nodes corresponds to the distances in the table.

Note that it is not always possible to construct an additive tree from the distances between the sequences, but there exists an elegant method for testing this.

Lemma

It is possible to construct an additive tree from the distances between the sequences (assuming metric space) if and only if for any four of them we can label them $i, j, k,$ l such that

$$D_{i,j} + D_{k,l} = D_{i,k} + D_{j,l} \geqslant D_{i,l} + D_{j,k}. \tag{4.10}$$

$\triangle$

We give an intuitive explanation of this statement using the unrooted trees for four sequences shown in Figure 4.7(a). We can label the nodes as in Figure 4.9(a) and can easily see that Equation (4.10) must be satisfied if the distances are additive. We also see that we can formulate a linear set of equations for finding the length of each edge. (Showing that Equation (4.10) implies additivity is beyond the scope of this book.)

Another property of evolutionary trees is if they are *ultrametric*. A tree is ultrametric if it is additive and the distances from two sequences to their common ancestor are equal. This means that molecular clock theory is satisfied. A simple test for ultrametricity is the following.

Lemma

It is possible to construct an ultrametric tree from the distances between the sequences (assuming metric space) if and only if, for every triple $i, j, k,$

$$D_{i,j} \leqslant \max(D_{i,k}, D_{k,j}). \tag{4.11}$$

$\triangle$

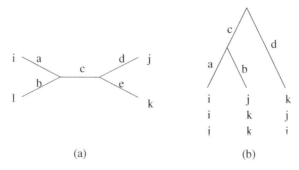

| (a) | (b) |

Figure 4.9   (a) Figure illustrating the additivity of the distances of four objects. (b) The three possible arrangements for a rooted tree of three objects. The ultrametric requirement is $a + c = b + c = d$.

Figure 4.9(b) illustrates this; Equation (4.11) is satisfied for all alternatives of the leaf nodes. Again, showing the other way—that Equation (4.11) implies that it is possible to construct an ultrametric tree—is more difficult. It is possible to show that Equation (4.11) implies Equation (4.10), hence ultrametricity implies additivity (as required in the definition).

The distances calculated from real data are very seldom additive; therefore, small, or even negative, lengths may be calculated for edge lengths.

4.3.4   Different approaches for reconstructing phylogenetic trees

Three main approaches are used for reconstructing phylogenetic trees:

- maximum parsimony methods;
- maximum likelihood methods; and
- distance based methods.

In addition, there exists a fourth, the methods of invariants, which is seldom used.

The first two are *character* or *site* based. In our context a site most often refers to a column of a multiple alignment, and we use sites and columns interchangeably. Selected (or all) of the columns are then used for the reconstruction. Also, those methods mainly use DNA sequences. In the following we will briefly explain the first two methods, and then the third in a little more depth. The reasons for this are that it is the distance method that is mainly used for the reconstruction of guide trees for driving multiple aligning, and this method is also used for protein sequences.

Maximum parsimony methods

The multiple alignment is used as input, and a search is performed to find a tree that minimizes the number of evolutionary changes necessary to explain the differences of the sequences. (The tree in Figure 4.4 was constructed using this idea for protein

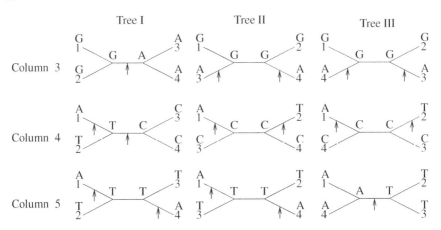

Figure 4.10   The three possible trees for columns 3–5 of the alignment in the text. The arrows show substitutions. We see that for column 3, tree I needs one substitution, the others two. That means column 3 favours tree I, and the column is informative. For column 4, all trees need two substitutions, column 4 is not informative. In the same way we see that column 5 is informative.

sequences.) The alignment is first inspected to find *informative columns*. A column is informative if it favours some tree topologies over the others. We illustrate this by an example. Suppose that we have four sequences, each of length seven:

```
                       Columns
                 1   2   3   4   5   6   7
    Sequence 1   C   T   G   A   A   T   A
             2   A   T   G   T   T   C   A
             3   A   T   A   C   T   G   T
             4   A   T   A   C   A   A   T
                         i       i       i
```

There are three different unrooted trees for four sequences. For each column we must find if it favours any of them. In Figure 4.10 we see that columns 3 and 5 are informative, column 4 not. In addition, column 7 is informative. We see that for a column to be informative it must have at least two different symbols, each occurring at least two times. We then use columns 3, 5 and 7 to decide the tree(s) with the minimum number of substitutions. Tree I has $(1 + 2 + 1)$ substitutions in those three columns, tree II has $(2 + 2 + 2)$, and tree III has $(2 + 1 + 2)$. Thus tree I is chosen and is said to be *supported* by two informative columns (3 and 7). Indeed, the trees which are supported by the largest number of informative columns are the maximum parsimony trees.

When the number of sequences grows, there are more possible trees, and the calculation is more complex. However, the main principle is the same.

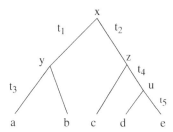

Figure 4.11 A tree for illustrating maximum likelihood methods.

Maximum likelihood methods

In principle, every possible tree is constructed, and the one that statistically has the highest probability for being correct (best suits the original sequences) is chosen. Therefore, this method is computationally expensive. Unrooted trees are the ones most often constructed. The methods require a probabilistic model for the substitutions. Suppose that a nucleotide is a at time zero. Then $P_{ab}(t)$ is the probability that the nucleotide is b at time t.

Let the tree in Figure 4.11 be a possible tree constructed from a site of five sequences, where the nucleotides of the internal nodes are known (x, y, z, u). The probability for the nucleotides a, b, c, d, e being at the leaves of this tree is

$$P_{xy}(t_1)P_{ya}(t_3)P_{yb}(t_3)P_{xz}(t_2)P_{zc}(t_4 + t_5)P_{zu}(t_4)P_{ud}(t_5)P_{ue}(t_5).$$

In practice, the nucleotides of the internal nodes are not known. They can be any of the four nucleotides, and the probabilities for each of them must be used and then summed up. Probabilities (or likelihoods) for each possible tree must then be calculated. These methods are the most useful when there are great variations among the sequences.

4.3.5 Distance-based construction

Distance-based methods require that the pairwise distances between the sequences are calculated. This can be done as discussed in Section 1.9, or by using more advanced methods. Depending on what the tree shall be used for, distances are either calculated for the edges or not calculated. If the tree is to be used as a guide tree for a multiple alignment, only the topology is required.

PGMA: PairGroup method using arithmetic mean

PGMA is a relatively simple distance-based method. First, a node is created for each sequence, and then the two nodes (u, v) with the most similar sequences are grouped, so that a new node is created with u and v as children, and the distances from the new node to all the other nodes (x) are calculated, as shown in Figure 4.12.

Note how *distance* is here used on the internal nodes. The distance between two nodes means the distance between the subtrees where the nodes are the roots.

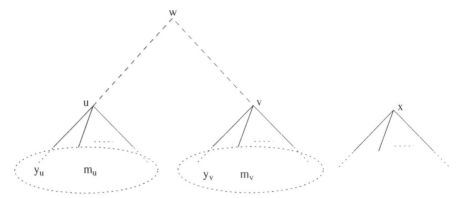

Figure 4.12 Figure illustrating the grouping of two trees, with roots u and v. y_u is a terminal node, and m_u the number of terminal nodes in the subtree with node u.

This pairwise grouping is then repeated, as shown in Algorithm 4.2. The method assume constant mutation rates along the edges, and hence ultrametric distances. If the distances are not ultrametric, some errors will occur (see the example below). An unrooted tree is constructed, but the place for a root can be calculated.

Algorithm 4.2. PGMA.

An algorithm for constructing an evolutionary tree based on arithmetic means of distances
const
m number of original sequences
var
U a set of current trees, initially, one tree for each original sequence.
D the distances between the trees in U
begin
    U:= *the set of one tree (each of one node) for each original sequence.*
    while $|U| > 1$ **do**
        $(u, v) :=$ *the roots of two trees in U with the least distance in D*
        make a new tree with root w with u and v as children
        calculate the length of the edges (v, w) and (u, w)
        for *each root x of the trees in $U - \{u, v\}$* **do**
            $D(x, w) :=$ *calculate the distance between x and the new node (w)*
        end
        $U := (U - \{u, v\}) \cup \{w\}$ update U
    end
end

Depending on how the distance between the new node w (with children u and v) to a root x in another (sub)tree is calculated, we get two variants of the procedure.

	B	C	D	E
A	3	7	8	10
B		6	8	7
C			4	5
D				6

(a)

	C	D	E
(A,B)	6.5	8	8.5
C		4	5
D			6

(b)

	(C,D)	E
(A,B)	7.25	8.5
(C,D)		5.5

(c)

	((C,D)E)
(A,B)	7.67

(d)

	B	C	D	E
A	3	7.66	7.66	7.66
B		7.66	7.66	7.66
C			4	5.5
D				5.5

(e)

Figure 4.13 (a) The distances between five sequences. (b) The distances when A and B are grouped, and a new node is created, using UPGMA. (c) In the next step C and D are grouped, and in the last (d) E is grouped with (C,D). (e) The distances in the constructed tree (see Figure 4.14).

UPGMA.

$$D_{w,x} = \frac{m_u D_{u,x} + m_v D_{v,x}}{m_u + m_v}.$$

Here m_u is the number of original sequences in the subtree with root u.

This means that the distances between the original sequences contribute equally. For this reason the method is called *unweighted* PGMA (UPGMA).

WPGMA.

$$D_{w,x} = \tfrac{1}{2}(D_{u,x} + D_{v,x}).$$

Here the number of leaves in u and v are ignored, hence the distances between the original sequences are differently weighted. This is the reason for calling this variant *weighted* PGMA (WPGMA).

Since the PairGroup method assumes ultrametric distances, the lengths on the paths from u to every leaf in the tree are equal, the same is true for every (sub)tree with root v. Then the length of the edges (v, w) and (u, w) can easily be calculated. Let y_v be a leaf of the tree with root v. The length of the edge (v, w) is then

$$L_{v,w} = \tfrac{1}{2}D_{u,v} - L_{v,y_v}.$$

Example

Assume five sequences, with distances given in Figure 4.13(a).

We use UPGMA to calculate the distances. First, A and B are grouped, and the distances from the new node (A,B) to each of the other nodes are calculated, as shown in Figure 4.13(b). Continuing in this way we get the tree shown in Figure 4.14, and the distances between the sequences in Figure 4.13(e).

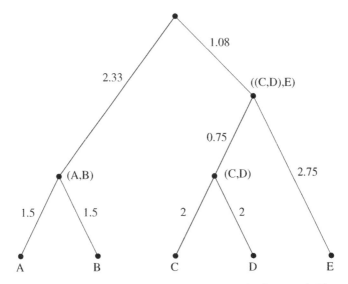

Figure 4.14   The tree constructed by using UPGMA on the distances in Figure 4.13.

Since the original distances (Figure 4.13(a)) are not ultrametric, the calculated distances in the constructed tree (Figure 4.13(e)) deviate from the original distances.

△

The neighbour-joining method

The neighbour-joining (NJ) method does not assume a constant evolutionary rate (the molecular clock theory need not be satisfied). In contrast to the PGMA method, which starts with m trees of single nodes and successively build larger and larger trees, the NJ method starts with a tree of all the sequences with the minimum number of edges (a star tree). Then the tree is successively changed by increasing by 1 the number of edges, until a final unrooted tree is constructed.

First, a star tree of degree m is constructed. Call the star node for X, as shown in Figure 4.15(a). New internal nodes are then successively created, and the degree of X is reduced by 1 in each cycle. The iteration stops when this degree becomes 3.

Let us call the (sub)trees connected to X for OTUs. A *neighbour pair* is a pair of OTUs. In Figure 4.15(b) ((A-B),C) is a neighbour pair, ((A-B),(E-F)) in (c), etc. In each cycle a neighbour pair is selected, and a new node created (Y), with edges to X and each of the OTU of the neighbour pair. The distances on the edges are then calculated.

In each cycle, we select the two OTUs which will result in the smallest total edge length. This means that it is *not* necessarily the pair with the least mutual distance which is selected.

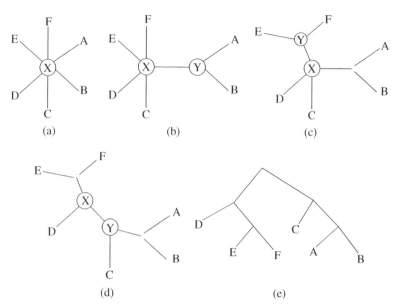

Figure 4.15 Illustration of the NJ method. (a) Six sequences are given, and a star tree of degree 6 is constructed. (b), (c) The degree of X is successively reduced. New internal nodes are marked by Y. (d) The iteration stops when X has degree 3. (e) A root is placed.

The sum of the distances over all edges in the initial star tree is

$$S_0 = \frac{1}{m-1} \sum_{i<j} D_{i,j}.$$

In the first cycle there are $m(m-1)/2$ possible choices for the neighbour pairs to select, and the sum over the edges for each possible tree must be calculated. Generally, in cycle i there are $(m-i+1)(m-i)/2$ choices. The sum for each tree is calculated by using the original distances, and the length of each new edge is also found. The formulae for these calculations can be found in Saitou and Nei (1987).

4.3.6 Rooting of trees

If a method produces unrooted trees, a simple method for constructing a rooted tree is to find the 'mid-point' and insert a root on an edge near this point. For the unrooted tree in Figure 4.8(a), a root can be inserted at point r, where the middle lengths from the root to the leaves are equal for the right and left subtrees. Figure 4.15(d) shows a possible resulting unrooted tree, and in Figure 4.15(e) a root is placed.

A root can also be determined by using an *outgroup*. An outgroup consists of one or several sequences which are known to have been separated from the others at an early evolutionary time. An unrooted tree (including the outgroup) will then probably

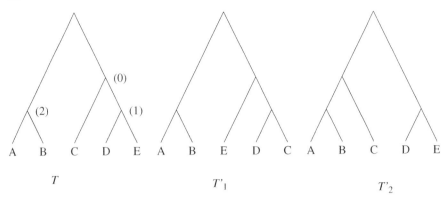

Figure 4.16 Two pseudo-trees are made for the original tree T. The nodes in T are labelled by numbers. Each node v is labelled by the number of pseudo-trees that has a node in the top of a subtree with the same leaves as the subtree below v. For example, the subtree consisting of the leaves A and B and the node combining them also exists in each of the pseudo trees, hence (2) is associated with that node.

contain an edge separating the outgroup from the rest of the sequences. The root is then placed on this edge.

4.3.7   Statistical test: bootstrapping

Bootstrapping is a general technique for estimating the correctness of some result, by seeing how robust (or stable) the result is to noise in the input. We will here sketch the method by showing how to test the reliability of a tree T constructed from an alignment $\mathcal{A}$, using some constructing method. From the alignment, a number (say 1000) of pseudo-alignments is constructed, this corresponds to introducing noise. Assume the alignment has n columns. A pseudo-alignment $\mathcal{A}'$ is constructed by randomly choosing n columns from $\mathcal{A}$. Each column in $\mathcal{A}$ can be chosen zero, one or several times. A pseudo-tree T' is constructed for each of the pseudo-alignments, and these trees are compared with the original constructed tree T. How this comparison is done depends on what sort of tree is constructed, rooted or unrooted. For a rooted tree, it is for each subtree (internal node) in the tree T, counted in how many pseudo-trees the same subtree occurs (containing the same set of leaves). This number is associated with each node as a measure of confidence, as the simple trees in Figure 4.16 show.

For unrooted trees, the confidence numbers are associated with the edges. For the tree in Figure 4.5, 1000 pseudo-trees were constructed. Let e be an edge, and L_1 and L_2 the set of leaves (here proteins) of the trees at each end of e, respectively. Then the number on the edge e tells in how many of the pseudo-trees there exists an edge splitting the leaves into L_1 and L_2.

```
     M1          M2                        A  L  G  H
                                           -  L  G  H

     FA-GH       ALGH           FFEF    ┌─────────────┐
     F-RGL       -LGH           A-LA    │             │
     ELRGH                      -RRR    │    -3/8      │
     FARGL                      GGGG    │             │
                                HLHL    └─────────────┘
```

(a)   (b)

Figure 4.17   Aligning (b) the two multiple alignments M1 and M2 in (a), the scoring scheme described in the example. The score for aligning the second column in M1 with the first in M2 is $-\frac{3}{8}$.

4.4   Progressive Alignment

Methods using progressive alignment are based upon the idea that an alignment can be looked upon as a generalized sequence, and one can align a sequence and an alignment, and also two alignments. The final alignment is constructed by successive pairwise alignments of simple and/or generalized sequences. The main drawback of the methods is that errors made in the early steps cannot be corrected later (as they can in stochastic methods). Wrongly included gaps are not removed ('once a gap, always a gap'). (Some methods try to overcome the strong dependency of the first alignments by doing some sort of iteration, sequences that may seem to be wrongly aligned are removed from the alignment, and added later.)

Techniques from clustering or phylogenetic analysis are often used in deciding which pairs to group in each cycle. Especially if a reasonable phylogenetic tree exists, it can be used to guide the aligning.

Example

Four sequences are given: s^1 = ALVK, s^2 = APFK, s^3 = ALFVK, s^4 = APFVK. Performing alignment (with a reasonable scoring scheme) in different orders will result in

$(s^1, s^2), (s^3, s^4)$ or $(s^1, s^2), (s^3, s^4)$ $(s^1, s^3), (s^2, s^4)$

```
     ALV-K              AL-VK           AL-VK
     APF-K              AP-FK           APF-K
     ALFVK              ALFVK           ALFVK
     APFVK              APFVK           APFVK
```

Which of the three different alignments is 'best' is questionable, and it is often up to the biologists to decide.   △

Thus the first alignments in the process are in a sense the most important since they affect all subsequent steps, and usually the most similar sequences (assumed to be the nearest in evolutionary time) are aligned first. This idea can be used throughout the process if there exists (or can be constructed) a (phylogenetic) tree which can guide the order in which the aligning of subset alignments are done.

Algorithm 4.3 is a general algorithm for progressive alignment. Different methods can then be implemented depending on how A_p, A_q are chosen, and how the alignment is performed. In deciding which method to use for choosing the alignments, we can use techniques from the field of hierarchical clustering.

Algorithm 4.3. General progressive alignment.

Progressive alignment of the sequences $\{s^1, s^2, \ldots, s^m\}$
var
C current set of alignments
begin
    $C := \emptyset$
    for $i := 1$ **to** m **do** $C := C \cup \{\{s^i\}\}$ **end** one alignment of each sequence
    for $i := 1$ **to** $m - 1$ **do**
        choose two alignments A_p, A_q *from* C; $C := C - \{A_p, A_q\}$
        $A_r := \text{align}(A_p, A_q)$; $C := C \cup \{A_r\}$
    end C now contains the (single) final alignment
end

Since the progressive methods repeatedly align pairs of subset alignments, we first outline how this can be done.

4.4.1 Aligning two subset alignments

For the explanation in the next subsections, regard two (subset)alignments A_p, A_q, with the sequences $\{s^{p_1}, \ldots, s^{p_n}\}$ and $\{s^{q_1}, \ldots, s^{q_m}\}$, where sequence s^i has weight w^i. (Recall that $\bar{s}^{p_i}$ is s^{p_i} with the blanks inserted in an alignment.)

Most methods use one of the following two methods for aligning pairs of subset alignments: *complete alignment* and *pair-guided alignment*. In complete alignment all sequences of the two subset alignments are used in a dynamic programming procedure. The score between the columns (r, t) is shown by Equation (4.12), where R is the score between two symbols. $\bar{s}^i_r$ is the symbol (either amino acid or blank) for sequence $\bar{s}^i$ in column r. In order to obtain the SP score, $R(--)$ must be zero:

$$S(r, t) = \frac{1}{nm} \sum_{j \in \{p_1 \ldots p_n\}} \sum_{k \in \{q_1 \ldots q_m\}} R_{\bar{s}^j_r \bar{s} z^k_t} w^j w^k. \qquad (4.12)$$

Example

Consider the multiple alignment M1 and M2 in Figure 4.17, and let a scoring scheme be 1 for equal, -1 for unequal and blank, and let the sequence weights all be 1. To the right of the figure is the DP matrix illustrated, with the result for aligning the second column in M1 with the first in M2, which is

$$\tfrac{1}{8}((1-1)+(-1+0)+(-1-1)+(1-1)) = -\tfrac{3}{8}.$$

$\triangle$

Gap penalties

As mentioned earlier, choosing appropriate gap penalties is one of the most difficult problems when aligning sequences, and for alignment of two multiple alignments it is even more difficult. Biologists often try several gap penalties before they get an alignment they are happy with. Often the user can give values for the gap opening and gap extension penalties, and the program might modify these to suitable values for the actual set of sequences. See Section 4.4.4 for an example.

Pair-guided alignment

Pair-guided alignment is a simpler alternative to complete alignment. Two sequences are chosen in each cycle, one from each (subset)alignment. These two are aligned, and the final aligning is done following this alignment.

Example

Let the two alignments be

```
ALEE      A-ERE
A-EE      ALER-
-LEE
```

Aligning the first sequence from the first alignment, and the last from the second gives

```
ALEE-
ALER-
```

and the final alignment guided by this pairwise alignment is shown as A1 below. A2 is the alignment obtained if the first sequence from both (subset)alignments are used as guide:

```
        ALEE-              ALE-E
        A-EE-              A-E-E
A1:     -LEE-      A2:     -LE-E
        A-ERE              A-ERE
        ALER-              ALER-
```

$\triangle$

Pair-guided alignment is faster than complete alignment, but not all information in the subset alignments is used. As shown by the example, the results strongly depend on the sequences chosen as guides.

4.4.2   Clustering

Ideally, the progressive alignment should be guided by a true phylogenetic tree, where the given sequences are on the leaf nodes. Such a tree is however, very rarely known. An estimated tree can be constructed, as explained in Section 4.3. We now discuss methods where no tree is constructed in advance.

At each cycle in the progressive alignment procedure, two (subset)alignments must be selected for aligning. In order to decide which, pairwise scoring between the sequences of the alignments can be used. Such a score can be static or dynamic:

- a *static pairwise score* means that the scores are precalculated as $S(s^i, s^j)$;

- a *dynamic pairwise score* means that the scores of the projections are used, $S(\bar{s}^i, \bar{s}^j)$. These must therefore be calculated when the subset alignments in which they occur are to be evaluated for aligning.

Since $\bar{s}$ changes during the alignment, the use of dynamic pairwise scores is the most time-consuming, but (probably) the more correct, since family information, as represented by the alignments done, is taken into account.

In the hierarchical clustering literature, three techniques using pairwise scores between elements are mainly described. For alignments, a fourth is also used.

- The *average linkage method* means that the two subset alignments, where the average score between all pairs of sequences (one from each alignment) is highest, are chosen for aligning. (This corresponds in a way to the PGMA method for constructing tree.) The sequences can be weighted such that a weighted average is calculated, as in Equation 4.5. (Instead of the arithmetic mean as described $((x_1 + x_2 + \cdots + x_n)/n)$, one can use the square mean $((x_1^2 + x_2^2 + \cdots + x_n^2)/n)$, the geometric mean ($\sqrt[n]{x_1 x_2 \ldots x_n}$) or the harmonic mean ($n/(1/x_1 + 1/x_2 + \cdots + 1/x_n)$).)

- The *maximum (single) linkage method* means that the two subset alignments where the maximum score between all pairs of sequences is highest are chosen for aligning. This means that it is enough that two sequences (one from each subset alignment) are similar (has a high score).

- The *minimum (complete) linkage method* means that the two subset alignments, where the minimum score between all pairs of sequences is *highest*, are chosen for aligning. This means that all pairwise scores are taken into account, hence the name complete. This tends to keep strong relations inside the clusters.

- The *special pair linkage method* means that special techniques are used for finding the two subset alignments with the 'best' pair of sequences; see the MULTAL algorithm below.

Note that in the classical clustering field, *distances* are measured instead of similarities. Therefore, maximum corresponds to complete in that case.

Example

We have three alignments: $\mathcal{A}_1 = \{s^1, s^2\}$, $\mathcal{A}_2 = \{s^3, s^4\}$, $\mathcal{A}_3 = \{s^5\}$, with pairwise scores:

	s^2	s^3	s^4	s^5
s^1	—	7	5	3
s^2		6	4	8
s^3			—	7
s^4				6

The scoring between the alignments for the different clustering methods are as follows:

Average linkage	$S(\mathcal{A}_1, \mathcal{A}_2) = (7 + 5 + 6 + 4)/4 =$	5.5	
	$S(\mathcal{A}_1, \mathcal{A}_3) =$	5.5	
	$S(\mathcal{A}_2, \mathcal{A}_3) =$	6.5	best
Maximum linkage	$S(\mathcal{A}_1, \mathcal{A}_2) = \max(7, 5, 6, 4) =$	7	
	$S(\mathcal{A}_1, \mathcal{A}_3) =$	8	best
	$S(\mathcal{A}_2, \mathcal{A}_3) =$	7	
Minimum linkage	$S(\mathcal{A}_1, \mathcal{A}_2) = \min(7, 5, 6, 4) =$	4	
	$S(\mathcal{A}_1, \mathcal{A}_3) =$	3	
	$S(\mathcal{A}_2, \mathcal{A}_3) =$	6	best

As shown, the different methods can find different clusters to group. △

The best method is probably to use average scores for deciding which alignments to cluster, and then performing complete alignment. This is, however, the most time-consuming, and the other linkage methods are also used. When pair-guided alignment is used, the guide pair to use can be the one with the highest score. For the special pair linkage method, the sequences used as guide are chosen in accordance with special given rules.

Example

We have 11 original sequences $A, B, \ldots, K$, and use static pair score calculation. Let (A, B) mean $S(A, B)$, and the scores in decreasing order are (A, D), (B, D), (A, B), (A, C), (B, C), (C, D), (J, K), (I, J), (I, K), (F, H), (D, E), (F, G), (D, F), (H, I), (E, G), (G, H), (A, F), ...

Using the maximum linkage method will result in the tree in Figure 4.18(a).

For even speeding up the procedure, a special technique is used in the MULTAL algorithm. The algorithm calculates the order of alignments using a simplified maximum linkage method. The sequences of the subset alignments are ordered, and only the first and last sequences are used for deciding which alignments to use in the next cycle. Pair-guided alignment is then performed, using two of those sequences. Cut-off values can also be used, so that the clustering stops when the static pair score is less than the cut-off. If the cut-off is set to a value slightly less than the score (E, G), two

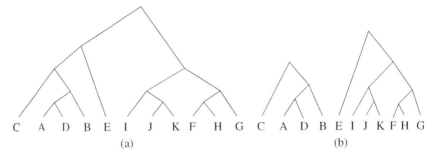

Figure 4.18   (a) The tree showing the clustering order of the sequences in the example using the maximum linkage method. (b) A special pair linkage method is used; see the example for explanation.

alignments are constructed, consisting of (A, B, C, D) and (E, F, G, H, I, J, K), as shown in Figure 4.18(b). The order of the clustering is

$$(C, ((A, D), B)) \quad \text{and} \quad (E, ((G, (F, H)), (I, (J, K))))$$

Note that (D, E) and (D, F) are not used in the clustering calculation, since D is not at the ends of the subset alignment in which it exists. $\triangle$

Linear clustering

Linear clustering is really a special case of hierarchical clustering, in that one of A_p, A_q must be a single sequence. It can be used for its simplicity. For the basic linear clustering, there is one current subset alignment, and sequences are added to this, one at a time, as shown in Algorithm 4.4. Pair-guided alignment is the most widely used method.

Algorithm 4.4. Basic linear clustering.

Basic linear clustering for aligning the sequences $\{s^1, s^2, \ldots, s^n\}$.

var
U   the set of sequences not aligned
$\mathcal{A}$ the current alignment
begin
    $U := \{s^1, s^2, \ldots, s^n\}$
    choose two sequences (the most similar) (s, t) from U
    $\mathcal{A} := \text{Align}(s, t); U := U - \{s, t\}$
    for $i := 1$ **to** $n - 2$ **do**
        choose a sequence $s \in U$; $U := U - \{s\}$
        $\mathcal{A} := \text{Align}(\mathcal{A}, s)$
    end
end $\mathcal{A}$ now contains the final alignment

Different methods for choosing the next sequence can be used, for example, the one most similar to the one already in the alignment, or the one most similar to an 'average' sequence of the alignment.

4.4.3 Sequence weights

The problem of determining sequence weights is loosely related to the problem of deriving a model giving a representative description of a class where only a limited set of examples (instances of the class) is available. The task is to make a description of the attribute values which are representative for the group. It should, for example, be used for later decisions if a new object belongs to the group.

Example

Assume that we want to estimate the mean height of school children in the first three year classes. Assume that the height of eight children are known, and we know that four of them are from the first year class, three from second-year class, and one from the third-year class. Shall all heights have equal weights? Assume that we do not know which year class each child belongs to, only the heights. How should we now weight the heights to get an estimate? △

In our case the objects are sequences, the class is a protein family, and the attributes are the amino acids. The description is the multiple alignment, and for each column we want the distribution of the amino acids for the whole family.

The multiple alignment should be representative for the sequences in the family, although not all of these sequences are known. As explained in Section 4.1.1, the known sequences might give a biased 'picture' of the family, as there might be a lot of known sequences of one 'type', and few of another 'type'. This biased representation could be corrected by giving the highest weights to the most isolated sequences, and the lowest weights to those having many known similar sequences in the family.

We here describe a method for calculating sequence weights using this approach, taken from CLUSTAL. A weighting scheme can be defined by using a guide tree constructed by using the known sequences. Sequences associated with the internal nodes need not be explicitly found. What has to be calculated, however, is a number representing the divergence of the sequences. These can be labels on the edges, and are estimates of the number of mutations which have occurred along each edge. Figure 4.19(b) shows such a tree.

The tree can be used to calculate the weights:

- the weights should increase with increasing difference (number of mutations) to the root (the distance to the 'middle sequence');

- the weights should decrease with increasing number of sequences in the neighbourhood (similar sequences).

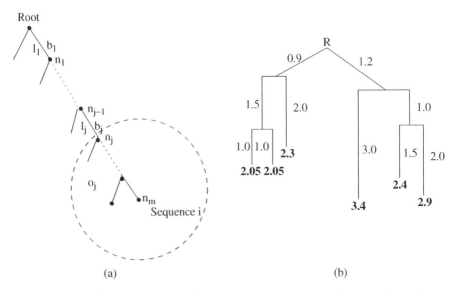

Figure 4.19 (a) Part of a guide tree illustrating how sequence weights are calculated. See text for the explanation of the variables. The dashed circle illustrates the subtree with root n_j. (b) The weights of each of the leaves (bold) of an example tree, calculated from the labels on the edges.

To describe how that can be done we introduce some variables, illustrated in Figure 4.19(a).

Let

- $b_1, b_2, \ldots, b_m$ be the edges on the path from the root to a sequence i;

- l_j be the value (number of mutations) on edge b_j;

- n_j be the node connected to the edge b_j farthest away from the root;

- o_j be the number of sequences in the subtree with root in the node n_j (o_m is therefore 1).

An expression for the weight of sequence i, taking the desired properties into account, can be

$$w_i = \sum_{j=1}^{m} \frac{l_j}{o_j}.$$

This means that the weight of a sequence is the sum of proportions of the weights (values) on the edges from the root to the leaf representing the sequence. The weight of each edge is equally shared between the sequences being leaves on the subtree below the edge.

In Figure 4.19(b) weights are calculated based on the tree. w_i can be normalized, such that the highest weight is, for example, 100.

4.4.4   CLUSTAL

CLUSTAL (Thompson et al. 1994b) is a popular program for progressive global alignment. The alignment is done in four steps (which can be repeated).

1. Calculate the (static) pairwise similarity scores for the sequences.

2. Construct a guide tree by use of the pairwise scores (NJ method).

3. Calculate sequence weights, using the guide tree.

4. Perform a progressive alignment, guided by the tree.

The multiple alignment in Figure 4.1 is generated by CLUSTAL.

Pairwise scoring

Static pairwise scores are used, and the user has the choice between a fast procedure for estimating the scores, and a procedure using full dynamic programming. The fast method has some similarities with the database search program, FASTA. Principally, a dot matrix is made with a window of length 1 or 2. Next, the t (e.g. five) diagonals with the highest number of dots are then used. For each of these diagonals a band is defined (e.g. two diagonals on each side), and dynamic programming is performed in each band. The score of the pairwise alignment is defined as the number of equal residues, minus the gap penalty. If there is a large difference in the lengths of the sequences, a better method is to transform the score to the percentage of the length of the smallest of the sequences (a score of 50 where the percentage length is 80 gives the score 62.5).

Progressive alignment

The constructed guide tree shows the order of the alignment, and the most similar sequences are aligned early. For the tree in Figure 4.15 the order can be (A, B), $((A, B), C)$, (E, F), $(D, (E, F))$, $(((A, B), C), (D, (E, F)))$. The quality of the alignment very often depends on the order, and the retention of errors introduced in early alignments. Alignment of isolated sequences is therefore often postponed. For example, all sequences with less than 40% identity against each of the other sequences can be aligned at the end of the process. Full dynamic programming is performed at each step, and sequence weights are calculated as explained above.

Scoring values

By default, the BLOSUM series of scoring matrices is used (see Section 5.3). (It is also possible to use other series of matrices, e.g. PAM or Gonnet.) For each alignment CLUSTAL assesses the similarity of the sequences being aligned, and chooses an appropriate matrix for the calculated distance. (The scoring matrices are transformed to positive values only.) Equation (4.12) is used for finding the scores between the columns in the two subset alignments.

Gap penalties

In the program CLUSTAL W the gap penalty is calculated by the use of several parameters. The user can first give two values:

- the GOP (gap opening penalty), which is a penalty for opening a gap of arbitrary length; and

- the GEP (gap extension penalty), which is the penalty for each extension (including the first).

The program then tries to modify these to suitable values for the actual set of sequences. Three types of modifications exist.

Sequence-independent modification.   The gap penalty increases with increasing average value of the scoring of unequal amino acids.

Position-independent modification   This modification depends on the number of columns in the subset alignments (since they use only positive scoring values), how similar the sequences are (higher penalty for strongly related sequences), and the difference in their lengths.

Position-dependent modification   The penalty is reduced for positions in which there are already gaps, and it is increased for positions near positions with gaps. The latter is to avoid small independent gaps near each other. Other biological knowledge is also used, by reducing the penalty in hydrophilic positions.

End gaps   A penalty is usually not given to gaps at the ends of the alignments, since it is often the case that some of the sequences are subsequences of the others, or are chopped off at one end. This can, however, have unwanted consequences. For example, an alignment with end gaps may be produced when the biologically most correct alignment should not have end gaps.

Improving the alignment

An alignment produced by CLUSTAL can often be improved. One way of doing this is to repeat the whole process, but now with the sequences containing the included gaps. This means that family information found is retained. Another way is to assess the alignment 'manually', and realign sequences which obviously are wrongly aligned.

Final phylogenetic tree

From the resulting multiple alignment, one can construct a new estimate of the phylogenetic tree by using the pairwise distances calculated from the alignment. The calculations can be done according to different rules.

1. One can use the projections, or one can neglect the positions with gap, regardless in which sequences the gaps occur. The latter means that the same columns are used in the calculation of all distances.

2. Adjust for multiple substitutions (mutations), since the number of mutations is probably higher than the number found (see Section 1.9). (The PAM matrices and the BLOSUM matrices take this into account.)

Bootstrapping is offered as an option.

4.5   Other Approaches

As explained before, the progressive approach has some drawbacks, especially that errors made in the first alignments do remain. Other approaches are developed for avoiding this dependency. One approach uses the idea that common local similarities (motifs, often descriptions of binding or active sites) should constitute a subalignment of the full alignment. The first step in these methods is therefore to discover such local similarities, align them, and then find the rest of the alignment under this constraint. How to describe and find local similarities is described in Chapter 7.

Other methods use search, especially stochastic search such as simulated annealing or genetic algorithms. Constructing good objective functions for those methods is extremely important, and some of them give the user the possibility of defining pairs or tuples of residues which should be aligned, absolutely or with high weight. References are given in the bibliographic notes.

4.6   Exercises

1. Generalize Equation (4.3) to allow for an affine gap penalty.

2. Explain why forward recursion is important for pruning the DP algorithm.

3. Suppose we have three sequences, s^1 = RAGT, s^2 = RATV, s^3 = RGTA, and that we will use three-dimensional dynamic programming to align them. Consider cell $v = (2, 3, 2)$ and find an upper limit for the path from v to the end cell, by using PAM 250 (Table 5.3). Choose an appropriate value for the gap penalty.

4. Draw all unrooted (binary) phylogenetic trees for five sequences.

5. Informative sites are defined as columns in an alignment that favours one tree topology over another. Find the informative sites in the following alignment:

        ATGTA
        TAGTA
        CGCTG
        GCCTG

Which tree will you find by the maximum parsimony method for this alignment?

6. Show that by using UPGMA for constructing evolutionary trees, each original distance contributes equally.

7. Try to explain (informally) that an ultrametric tree is additive.

8. Given a symmetrical distance matrix:

	A	B	C	D	E
A	0	2	1	2	1.5
B		0	2	0.5	2
C			0	2	1.5
D				0	2
E					0

(a) show that there exists an ultrametric tree for this matrix;

(b) construct such a tree.

9. Suppose we are given five sequences $s^1 = $ AECD, $s^2 = $ ACCDE, $s^3 = $ CEACD, $s^4 = $ CCDAE, $s^5 = $ ADCED. Let a scoring system have a linear gap score $g_l = 2l$ and a scoring matrix:

	A	C	D	E
A	0	2	1	3
C		0	2	3
D			0	3
E				0

In this way the scores can be regarded as distances between the sequences, and least score means least distance. The scores for the best pairwise alignments are as follows:

	s^1	s^2	s^3	s^4
s^2	5			
s^3	4	8		
s^4	8	4	9	
s^5	5	6	7	8

(a) For each pair show an alignment that gives these scores.

(b) Make a (static) guide tree based on these distances by using the UPGMA procedure. If there are several alternatives for joining (equal score), choose one of them arbitrarily.

(c) Make a multiple alignment based on the tree in (b). Use pairwise guided alignment using the sequences with least dynamic distance as guides. In the (dynamic) alignment, blanks can occur in different ways:

(i) two blanks existing in the sequences are aligned, use scoring 0;

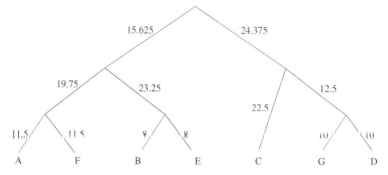

Figure 4.20   An evolutionary tree.

> (ii) one blank existing in one of the sequences is aligned against a new blank, use scoring 2;
>
> (iii) one existing blank is aligned to an amino acid, use scoring 2;
>
> (iv) a new blank is aligned to an amino acid, use scoring 2.
>
> (Try to find the best alignments without using dynamic programming.)
>
> Calculate the SP score of the multiple alignment.

> (d) Make an alignment using the minimum (complete) linkage method. Use pairwise guided joining; you can use the static calculated distances when finding the pairs. Calculate the SP score of the multiple alignment.

10. There is an evolutionary tree in Figure 4.20. This has been constructed using the pairwise distances between the sequences A, B, C, D, E, F, G. The values on the edges are the distances between the nodes. Find the weights of the sequences A, E, G when the method in Subsection 4.4.3 is used.

11. You will find the program CLUSTAL W at http://www.ebi.ac.uk/clustalw/. Pick some sequences (from SwissProt) that you will align.

> (a) Generate a guide tree. You will then get an unrooted tree. Make a drawing of it from the .dnd file.
>
> (b) Do a multiple alignment based on the guide tree.
>
> (c) Generate a new phylogenetic tree from the alignment. Compare it with the one in (a).
>
> (d) Generate a tree of the alignment where gaps are excluded. Compare it with the one in (c).
>
> (e) Do a new multiple alignment, after changing the GOP to 1. Compare it with the results from (b).

4.7   Bibliographic notes

A thorough discussion of multiple alignment is in Chan et al. (1992). Methods for speeding up the dynamic programming solution are discussed in Carrillo and Lipman (1988), Lipman et al. (1989), Gusfield (1993), Gupta et al. (1995), Reinert et al. (2000) and Zhang et al. (2000).

Determining sequence weights is explained in Altschul et al. (1989) and Thompson et al. (1994a). Vingron and Sibbald (1993) compare different methods for weighting aligned sequences.

The MULTAL program is described in Taylor (1990). A method based on simulating annealing is presented in Kim and Pramanik (1994), one using hidden Markov models in Eddy (1995), and one using genetic algorithms in Notredame and Higgins (1996) and Notredame et al. (2000, 1998). A method based on the Fourier transform is presented in Katoh et al. (2002). The use of iterative refinement is reported in Gotoh (1996). Althaus et al. (2002) describe a method using polyhedral combinatorics, and Lee et al. (2002) describe a method using partial order graphs.

An analysis of computational complexity is given in Just (2001), and a comparison of programs for multiple alignments is given in Thompson et al. (1999b). There is a benchmark database in Thompson et al. (1999a), and Karplus and Hu (2001) describe the use of it. Another program for the comparison of multiple alignments and the assessment of statistical significance is presented in Sadreyev and Grishin (2003).

A good book for evolutionary trees is Li (1997), which has many references.

5

Scoring Matrices

The aim of a database search is to find sequences homologous to a query sequence. This can be done by calculating the similarity between the query and each of the database sequences. Hence, a scoring scheme has to be defined. This is usually based on the similarity of the residues occurring in the sequences. For two residues (q_i, d_j), we need a measure of the probability that they have a common ancestor, or that one is a result of one or several mutations of the other. It is common to ignore the position of the residues, and use a general measure for the similarity of the occurring amino acids. This measure can then be given as an $n \times n$ *scoring matrix*, where n is the number of amino acids (20). It is generally claimed that the scoring matrices should be symmetrical, hence a triangular $n \times n$ matrix is sufficient. There also exists scoring matrices for pairs of amino acids (400×400).

The values in a scoring matrix do not necessarily define an analogue to a metric in the world of distance. For example, the score between (a, b) might be larger than the sum of the scores for (a, c) and (c, b) (or smaller than their difference).

The concept of a *substitution matrix* is also used for a scoring matrix. We will, however, make a distinction between them. By a substitution matrix we mean a matrix with values for the probability that an amino acid a is changed to another amino acid b in a certain (evolutionary) time. A substitution matrix is therefore not necessarily symmetric. The evolutionary time is often measured in numbers of mutations.

In this chapter we describe in some detail how two of the most widely used series of scoring matrices are developed, PAM and BLOSUM. References to other types of scoring matrices, and comparisons between them, are in the bibliographic notes.

A scoring matrix should reflect

- the degree of 'biological relationship' between the amino acids, and

- the probability that two amino acids occur in homologous positions in sequences that have a common ancestor, or that one is the ancestor of the other.

These two points are strongly related—see below.

Protein bioinformatics: an algorithmic approach to sequence and structure analysis
I. Eidhammer, I. Jonassen and W. R. Taylor © 2004 John Wiley & Sons, Ltd ISBN: 0-470-84839-1

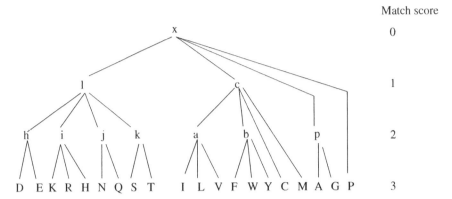

Figure 5.1 An amino acid class hierarchy used to construct AACH patterns during multi-alignment. Uppercase characters are the one-letter amino acid codes; lowercase characters are designated amino acid classes; x is the wildcard character representing one amino acid of any type. Reproduced from Smith and Smith (1990) by permission of the National Academy of Science.

5.1   Scoring Matrices Based on Physio-Chemical Properties

Several scoring schemes based on physio-chemical properties have been proposed.

Use of identity. A simple method is to score the alignment of unequal amino acids by 0 and the alignment of identical amino acids by 1. The score obtained when aligning two sequences can then be transformed to percentage identity by dividing the obtained score by either the length of the shortest sequence, that of the longest sequence or the average length.

Use of the genetic code. The score is based on how many mutations are needed in the nucleic acids for changing an amino acid into another (1, 2 or 3; 0 for equal). For example, two mutations are needed for transforming Phe (codes UUU, UUC) to Asn (codes AAU, AAC). Note that this defines a *distance*, but it is easy to define a scoring based on this. For small PAM values, this results in scores similar to the PAM matrices explained later in this chapter.

Use of classification of amino acids. Scoring matrices based more directly on the physio-chemical properties can be defined. Such properties can be hydrophobicity, polarity, charge, aromaticity, aliphacity, acidic/basic, size, H-bond donors, etc. Several methods of classification have been proposed. Smith and Smith (1990) proposed the classification AACH (amino acid class hierarchy), shown in Figure 5.1.

Taylor (1986) made a Venn diagram using these properties, which is shown in Figure 5.2.

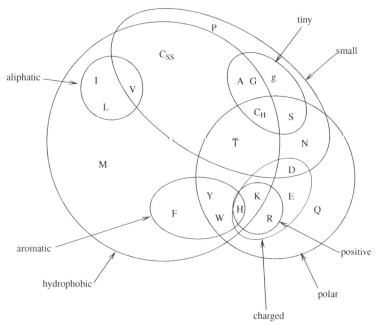

Figure 5.2 The Venn diagram shows the relationship of the 20 naturally occurring amino acids to a selection of physio-chemical properties, dominated by properties relating to size and hydrophobicity. The amino acids are divided into two major sets: one containing all the amino acids which contain a polar group and another containing amino acids which exhibit an hydrophobic effect. The set 'small' contains the nine smallest amino acids. Within this is an inner set of smaller residues, 'tiny', which have at most two side-chain atoms. The location of Cys is ambiguous as the reduced form (C_H) has similar properties to serine, while the oxidized form (C_{ss}) may be more equivalent to Val. Other sets include 'fully charged' (referred to as 'charged'), which contains the subset 'positive' ('negative' is defined by implication) and 'aromatic' and 'aliphatic'. Because of its unique backbone properties, proline was excluded from the main body of the diagram. An equivalent exclusive position is suggested for Gly by a small G. Reproduced from Taylor (1986) by permission of Elsevier.

5.2 PAM Scoring Matrices

It is strongly argued that the scoring matrices are best developed based on experimental data, thus reflecting the kind of relationships occurring in nature. The first scoring matrices developed from known data were the PAM matrices, developed by Dayhoff (1978). These matrices are still widely used.

Dayhoff estimated the substitution probabilities by using known mutational histories (by mutations we mean substitutions in this chapter). 34 superfamilies were used, divided into 71 groups of near homologous sequences (> 85% identity to reduce the number of superimposed mutations) and a phylogenetic tree was constructed for each group. Then the *accepted point mutations* on each edge were estimated. A mutation is *accepted* if it is accepted by the species. This usually means that the new amino acid must have the same effect (must function in a similar way) as the old one, which usually requires strong physio-chemical similarity, dependent on how critical the position of the amino acid is.

Let τ be a time interval of evolution, measured in numbers of mutations per residue. Then Dayhoff's procedure can be described by the following steps.

1. Divide the set of sequences into groups of similar sequences, and make a multiple alignment for each group.

2. Construct phylogenetic (evolutionary) trees for each group, and estimate the mutations on the edges.

3. Define an *evolutionary model* to explain the evolution.

4. Construct substitution matrices. The substitution matrix for an evolutionary time interval τ gives for each pair (a, b) an estimate for the probability of a to mutate to b in a time interval τ.

5. Construct scoring matrices from the substitution matrices.

Figure 5.3 shows part of an actual multiple alignment and an evolutionary tree. Figure 5.4 shows an example evolutionary tree, and the number of accepted point mutations.

5.2.1 The evolutionary model

The evolutionary model has the following assumptions: *the probability of a mutation in one position of a sequence is only dependent on which amino acid is in the position.* It is

- independent of position and neighbour residues (independence of neighbour residues means independence of which amino acids constitute the neighbouring residues (both in sequence and space)), and

- independent of previous mutations in the position.

Due to these assumptions, instead of writing 'an occurrence of an amino acid a in a position' we can just write a.

The *biological clock* is also assumed, which means that the rate of mutations is constant over time. Hence, the time of evolution can be measured by the number of mutations observed in a certain number of residues. This is measured in *point accepted mutations (PAMs)*, and 1 PAM means *one accepted mutation per 100 residues.*

5.2.2 Calculate substitution matrix

The substitution matrix is calculated by observing the number of accepted mutations in the constructed phylogenetic trees (1572 in the first experiment). If the tree contains

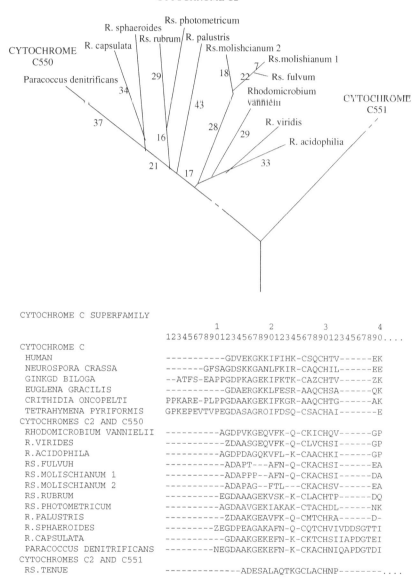

Figure 5.3 Part of a multiple alignment and the evolutionary tree for the Cytochrome C superfamily, used by Dayhoff. Redrawn from Dayhoff (1978) with permission of the National Biomedical Research Foundation.

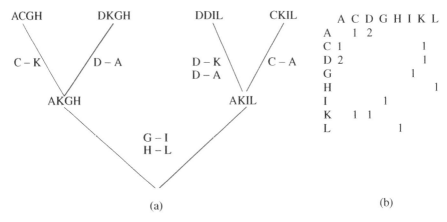

Figure 5.4 (a) A small phylogenetic tree of four observed sequences, and two derived parent sequences. (b) The mutations are on the edges. The numbers of different mutations are shown in the table.

ambiguity in the mutations, use the fraction representing the probability among the occurring amino acids. The task is then to calculate a value for the relation between the amino acids a and b in terms of mutations. This is done by first estimating the *probability* that a will be replaced by b in a certain evolutionary time τ, and denote this by M_{ab}^{τ} (hence M is also called the mutation probability matrix). τ is measured in PAMs, and first we look at $\tau = 1$, M_{ab}^{1}. When $\tau = 1$ the time specification is often omitted, and the probability denoted by M_{ab}. (M_{ab} is also used when the value of τ is inessential in the context.)

Note that M_{ab} need not be equal to M_{ba}. M_{ab} depends on

- The probability that a mutates (remember that in accordance with our model all occurrences of a have the same probability), and

- the probability that a mutates to b, given that a mutates.

The procedure can be described by the following steps.

1. Find all accepted mutations in the data. From this calculate

   - f_{ab}, the number of $a \rightarrow b$ or $b \rightarrow a$, where $\rightarrow$ means mutation (note that $f_{ab} = f_{ba}$);
   - $f_a = \sum_{b \neq a} f_{ab}$, the total number of mutations in which a takes part.
   - $f = \sum_a f_a$, twice the total number of mutations (since each mutation counts twice).

2. Calculate the frequency of p_a for all a. This is the relative occurrence of amino acid a in the data material, hence $\sum_a p_a = 1$.

3. Calculate the *relative mutability* m_a. This is a measure of the probability that a will mutate in the evolutionary time of interest. m_a therefore depends on

- f_a; m_a should increase with increasing f_a; a occurring in many mutations indicates high mutability;

- p_a; m_a should decrease with increasing p_a; many a indicate many a in mutations due to its frequent occurrences.

Hence m_a can be defined as $m_a = K f_a / p_a$, where K is a constant.

First look at $\tau = 1$ PAM; hence m_a means the probability that an arbitrary occurrence of a will mutate in 1 PAM.

The probability that an arbitrary mutation contains a is $f_a / (f/2)$. The probability that it is *from* a is (since $f_{ab} = f_{ba}$)

$$\frac{1}{2} \frac{f_a}{f/2} = \frac{f_a}{f}.$$

Among 100 residues there are $100 p_a$ occurrences of a, hence the probability for any one of these to mutate is

$$m_a = \frac{1}{100 p_a} \frac{f_a}{f}. \tag{5.1}$$

As a check we can find expected number of mutations per 100 residues:

$$\sum_a (100 p_a) m_a = \sum_a 100 p_a \frac{f_a}{f \, 100 p_a} = \frac{1}{f} \sum_a f_a = 1.$$

4. For determining M_{ab} we can now use the facts that

- the probability that a mutates (in time 1 PAM) is m_a, and

- the probability that a mutates to b, given that a mutates, is f_{ab}/f_a.

By use of the formula for conditional probability, there follows:

- $M_{ab} = m_a f_{ab}/f_a$, for $a \neq b$;

- $M_{aa} = 1 - m_a$ (the probability that a does not mutate).

Table 5.1 shows the matrix for 1 PAM calculated from data used by Dayhoff; for example, the probability for an A to be replaced by an L is 0.0003.

5.2.3 Matrices for general evolutionary time

In the derivation above we have used 1 mutation per 100 residues as the (evolutionary) time unit. If instead we use 1 mutation per 50 residues, the only change is to replace 100 by 50 in Equation (5.1). Note that this does not correspond to 2 PAM (2 mutations per 100 residues).

Table 5.1 Substitution (mutation probability) matrix for the evolutionary distance of 1 PAM. To simplify the appearance, the elements are shown multiplied by 10 000. The probabilities for not changing are replaced by *, the values vary between 9822 (N) and 9976 (W). An element of this matrix, M_{ab}, gives the probability that the amino acid in row a will be replaced by the amino acid in column b after a given evolutionary interval, in this case 1 accepted point mutation per 100 amino acids. Thus there is a 0.56% probability that D (Asp) will be replaced by E (Glu). The amino acids are alphabetically ordered on their names. Reproduced from Dayhoff (1978) with permission of the National Biomedical Research Foundation.

	A	R	N	D	C	Q	E	G	H	I	L	K	M	F	P	S	T	W	Y	V
A	*	1	4	6	1	3	10	21	1	2	3	2	1	1	13	28	22	0	1	13
R	2	*	1	0	1	9	0	1	8	2	1	37	1	1	5	11	2	2	0	2
N	9	1	*	42	0	4	7	12	18	3	3	25	0	1	2	34	13	0	3	1
D	10	0	36	*	0	5	56	11	3	1	0	6	0	0	1	7	4	0	0	1
C	3	1	0	0	*	0	0	1	1	2	0	0	0	0	1	11	1	0	3	2
Q	8	10	4	6	0	*	35	3	20	1	6	12	2	0	8	4	3	0	0	1
E	17	0	6	53	0	27	*	7	1	2	1	7	0	0	3	6	2	0	1	2
G	21	0	6	6	0	1	4	*	0	0	1	2	0	1	2	16	2	0	0	3
H	2	10	21	4	1	23	2	1	*	0	4	2	0	2	5	2	1	1	4	1
I	6	3	3	1	1	1	3	0	0	*	22	4	5	8	1	2	11	0	1	57
L	4	1	1	0	0	3	1	1	1	9	*	1	8	6	2	1	2	0	1	11
K	2	19	13	3	0	6	4	2	1	2	2	*	4	0	2	7	8	0	0	1
M	6	4	0	0	0	4	1	1	0	12	45	20	*	4	1	4	6	0	0	17
F	2	1	1	0	0	0	0	1	2	7	13	0	1	*	1	3	1	1	21	0
P	22	4	2	1	1	6	3	3	3	1	3	3	0	1	*	17	5	0	0	2
S	35	6	20	5	5	2	4	21	1	1	1	8	1	2	12	*	32	1	1	2
T	32	1	9	3	1	2	2	3	1	7	3	11	2	1	4	38	*	0	1	10
W	0	8	1	0	0	0	0	0	1	0	4	0	0	3	0	5	0	*	2	0
Y	2	0	4	0	3	0	1	0	4	1	2	1	0	28	0	2	2	2	*	1
V	18	1	1	1	2	1	2	5	1	33	15	1	4	0	2	2	9	0	1	*

M^z can, however, be found by matrix multiplications of M^1. This is due to the independent properties of the model (Markov model). We will show it for M^2.

Let the two mutations be denoted as mutations 1 and 2. An amino acid a can be replaced by an amino acid b after two mutations in the following ways.

1. a is mutated into b in the first mutation, and unchanged in the second. The probability for this is $M_{ab}M_{bb}$.

2. a is unchanged in the first mutation, but mutated into b in the second. The probability is $M_{aa}M_{ab}$.

3. a is mutated into an amino acid c in the first, which is mutated into b in the second. The probability is $M_{ac}M_{cb}$.

Point 3 can happen for all amino acids not a or b. The cases are independent, such that the final probability for a to be replaced by b in a time interval of 2 PAMs is

$$M_{ab}^2 = M_{ab}M_{bb} + M_{aa}M_{ab} + \sum_{c \notin \{a,b\}} M_{ac}M_{cb} = \sum_{c \in \mathcal{M}} M_{ac}M_{cb}. \qquad (5.2)$$

This is exactly the definition of the matrix multiplication $M^1 M^1$. M^{250} is shown in Figure 5.2. Note that after many multiplications, any errors in M^1 will be magnified.

5.2.4 Measuring sequence similarity by use of M^τ

M_{ab}^τ measures the equality of a and b as the probability that a will mutate to b in time τ (which may be different from M_{ba}^τ, the probability that b will mutate to a in time τ). The similarity of $q = q_1 q_2 \cdots$ and $d = d_1 d_2 \cdots$ (in time τ) is then $\prod_i M_{q_i d_i}^\tau$ (or $\prod_i M_{d_i q_i}^\tau$). Hence, this measure of equality is not symmetric, it does not take chance into account, and it has to use multiplication (as a consequence of the underlying probability model).

5.2.5 Odds matrices

We now use M as a *general substitution matrix*, without specifying τ.

M does not take chance into account: two amino acids can be opposite to each other in the alignment just by chance, and not as result of homology and mutations. Instead, we can give scores compared with what one can expect by chance. Consider an amino acid a in one of the sequences (q). The probability that it is aligned to an amino acid b in the other sequence (d) only by chance, depends on how many occurrences of b there are in d. One way of taking the effects of chance into account is to divide M_{ab} by the frequency of b in d. We then get an *odds matrix*:

$$O_{ab} = \frac{M_{ab}}{p_b}. \qquad (5.3)$$

Table 5.2 The mutation probability matrix for the evolutionary distance of 250 PAMs. To simplify the appearance, the elements are shown multiplied by 100. In comparing two sequences of average amino acid frequency at this evolutionary distance, there is a 13% probability that a position containing A (Ala) in the first sequence will contain A in the second. There is a 3% chance that it will contain R (Arg), and so forth. Reproduced from Dayhoff (1978) by permission of the National Biomedical Research Foundation.

	A	R	N	D	C	Q	E	G	H	I	L	K	M	F	P	S	T	W	Y	V
A	13	3	4	5	2	3	5	12	2	3	6	6	1	2	7	9	8	0	1	7
R	6	17	4	4	1	5	4	5	5	2	4	18	1	1	5	6	5	2	1	4
N	9	4	6	8	1	5	7	10	5	2	4	10	1	2	5	8	6	0	2	4
D	9	3	7	11	1	6	11	10	4	2	3	8	1	1	4	7	6	0	1	4
C	5	2	2	1	52	1	1	4	2	2	2	2	0	1	3	7	4	0	3	4
Q	8	5	5	7	1	10	9	7	7	2	6	10	1	1	5	6	5	0	1	4
E	9	3	6	10	1	7	12	9	4	2	4	8	1	1	4	7	5	0	1	4
G	12	2	4	5	2	3	5	27	2	2	3	5	1	1	5	9	6	0	1	5
H	6	6	6	6	2	7	6	5	15	2	5	8	1	3	5	6	4	1	3	4
I	8	3	3	3	2	2	3	5	2	10	15	5	2	5	3	5	6	0	2	15
L	6	2	2	2	1	3	2	4	2	6	34	4	3	6	3	4	4	1	2	10
K	7	9	5	5	1	5	5	6	3	2	4	24	2	1	4	7	6	0	1	4
M	7	4	3	3	1	3	3	5	2	6	20	9	6	4	3	5	5	0	2	10
F	4	1	2	1	1	1	1	3	2	5	13	2	2	32	2	3	3	1	15	5
P	11	4	4	4	2	4	4	8	3	2	5	6	1	1	20	9	6	0	1	5
S	11	4	5	5	3	3	5	11	3	3	4	8	1	2	6	10	8	1	2	5
T	11	3	4	5	2	3	5	9	2	4	6	8	1	2	5	9	11	0	2	7
W	2	7	2	1	1	1	1	2	2	1	6	4	1	4	1	4	2	55	3	2
Y	4	2	3	2	4	2	2	3	3	3	7	3	1	20	2	4	3	1	31	4
V	9	2	3	3	2	3	3	7	2	9	13	5	2	3	4	6	6	0	2	7

Remember that $\sum_{a \in \mathcal{M}} p_a = 1$ and $\sum_{b \in \mathcal{M}} M_{ab} = 1$; hence O is the ratio of two frequencies. The frequency in the numerator takes the biological relation between a and b into consideration, while the frequency in the denominator does not; hence it is independent of a.

From Equation (5.3) it follows that

- $O_{ab} > 1$: b replaces a more often in biologically related sequences than in randomly generated sequences where b occurs with the frequency p_b;

- $O_{ab} < 1$: b replaces a less often in biologically related sequences than in randomly generated sequences where b occurs with the frequency p_b.

Properties

The odds matrix O is symmetrical, which is shown by

$$O_{ab} = \frac{M_{ab}}{p_b} = \frac{f_{ab} m_a}{f_a p_b} = \frac{f_{ab} f_a}{f_a 100 f p_a p_b} = \frac{f_{ab}}{100 f p_a p_b} = \frac{M_{ba}}{p_a} = O_{ba}.$$

5.2.6 Scoring matrices (log-odds matrices)

The similarity between two sequences q and d now becomes

$$L(q,d) = \prod_i O_{q_i d_i}.$$

The measure of similarity between sequences does not need to be (or is not) absolute, but relative (we need to know, for example, that q is more similar to d_1 than to d_2. This relativity is kept if we use the logarithm of $L(q,d)$ instead of $L(q,d)$. As a result of this we can measure the similarity by adding the scores for the aligned residues, instead of multiplying:

$$L'(q,d) = \log L(q,d) = \sum_i \log O_{q_i d_i}. \tag{5.4}$$

Matrices of $\log O_{ab}$, *log-odds matrices*, are therefore developed as

$$R_{ab} = \log \frac{M_{ab}}{p_b}. \tag{5.5}$$

Table 5.3 shows the log-odds matrix for 250 PAM multiplied by 10.

5.2.7 Estimating the evolutionary distance

Suppose two sequences q and d have evolutionary distance τ (if q is an ancestor of d, this means that τ mutations per 100 residues have occurred in the transition from q to d). With

$$100\left(1 - \sum_{c \in \mathcal{M}} p_c M_{cc}^{\tau}\right) \tag{5.6}$$

Table 5.3 Log-odds matrix for 250 PAMs. Elements are shown multiplied by 10. The neutral score is zero. A score of -10 means that the pair would be expected to occur only one-tenth as frequently in related sequences as random chance would predict, and a score of $+2$ means that the pair would be expected to occur 1.6 times as frequently. The order of the amino acids has been arranged to illustrate the patterns in the mutation data (grouped according to the chemistries of the side groups). Reproduced from Dayhoff (1978) by permission of the National Biomedical Research Foundation.

	C	S	T	P	A	G	N	D	E	Q	H	R	K	M	I	L	V	F	Y	W
C	12																			
S	0	2																		
T	−2	1	3																	
P	−3	1	0	6																
A	−2	1	1	1	2															
G	−3	1	0	−1	1	5														
N	−4	1	0	−1	0	0	2													
D	−5	0	0	−1	0	1	2	4												
E	−5	0	0	−1	0	0	1	3	4											
Q	−5	−1	−1	0	0	−1	1	2	2	4										
H	−3	−1	−1	0	−1	−2	2	1	1	3	6									
R	−4	0	−1	0	−2	−3	0	−1	−1	1	2	6								
K	−5	0	0	−1	−1	−2	1	0	0	1	0	3	5							
M	−5	−2	−1	−2	−1	−3	−2	−3	−2	−1	−2	0	0	6						
I	−2	−1	0	−2	−1	−3	−2	−2	−2	−2	−2	−2	−2	2	5					
L	−6	−3	−2	−3	−2	−4	−3	−4	−3	−2	−2	−3	−3	4	2	6				
V	−2	−1	0	−1	0	−1	−2	−2	−2	−2	−2	−2	−2	2	4	2	4			
F	−4	−3	−3	−5	−4	−5	−4	−6	−5	−5	−2	−4	−5	0	1	2	−1	9		
Y	0	−3	−3	−5	−3	−5	−2	−4	−4	−4	0	−4	−4	−2	−1	−1	−2	7	10	
W	−8	−2	−5	−6	−6	−7	−4	−7	−7	−5	−3	2	−3	−4	−5	−2	−6	0	0	17

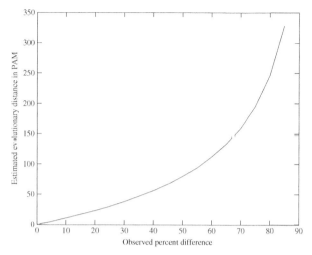

Figure 5.5   Correspondence between observed differences and
the evolutionary distances in PAM. Drawn using data from Dayhoff (1978).

we find how many residues in average are different per 100 residues. This is the
observed percentage difference in Figure 5.5. (Remember that a residue might mutate
several times, and can be mutated back into the original amino acid.)

For example, for an evolutionary distance between two sequences of $\tau = 112$,
it is found (by use of Equation (5.6)) that there should be a 60% difference in the
sequences. On the other hand, if a 70% difference is observed between two sequences,
it indicates an evolutionary distance of 159 PAMs.

5.3   BLOSUM Scoring Matrices

In the Dayhoff model the scoring values are derived from protein sequences with at
least 85% identity. Alignments are, however, most often performed on sequences of
less similarity, and the scoring matrices for use in these cases are calculated from
the 1 PAM matrix. Henikoff and Henikoff (1992) have therefore developed scoring
matrices based on known alignments of more diverse sequences.

They take a group of related proteins and produce a set of *blocks* representing this
group, where a block is defined as an ungapped region of aligned amino acids. An
example of two blocks is

```
KIFIMK        GDEVK
NLFKTR        GDSKK
KIFKTK        GDPKA
KLFESR        GDAER
KIFKGR        GDAAK
```

The Henikoffs used over 2000 blocks in order to derive their scoring matrices. For each column in each block they counted the number of occurrences of each pair of amino acids, when all pairs of segments were used. Then the frequency distribution of all $\frac{1}{2}(21 \cdot 20) = 210$ different pairs were found. A block of length w from an alignment of m sequences make $\frac{1}{2}wm(m-1)$ pairs of amino acids. For the first block above it is 60. We define

- h_{ab} as the number of occurrences of the amino acid pair (ab) (note that $h_{ab} = h_{ba}$);

- T as the total number of pairs:

$$T = \sum_{c} \sum_{e \geqslant c} h_{ce}, \quad c, e \in \mathcal{M},$$

where $\geqslant$ is interpreted as a total ordering over the amino acids (for example, $V \geqslant Y \geqslant W \geqslant \cdots$);

- $f_{ab} = h_{ab}/T$ (the frequency of observed pairs).

Example

For the last column in the first block we find $h_{KR} = 6$, $h_{KK} = 1$, $h_{RR} = 3$; all the other pairs are 0 for this column. In total for the two blocks $h_{KR} = 9$ is found. There are 110 pairs, hence $f_{KR} = \frac{9}{110}$.   △

5.3.1   Log-odds matrix

The observed frequencies must be adjusted by the effect of amino acids occurring by chance. For each (ab) the expected probability that they are aligned by chance, e_{ab}, must be calculated. Then

- $f_{ab} > e_{ab}$, the observed frequency is higher than expected by chance, which indicates a biological relation between the amino acids a and b;

- $f_{ab} < e_{ab}$, the observed frequency is less than expected by chance, which indicates a biological 'aversion' between the amino acids a and b;

- $e_{ab} = f_{ab}$, which indicates biological neutrality between a and b.

To calculate the expected number of occurrences of the amino acid pairs, assume that *the observed frequencies are equal to the frequencies in the actual population.* From this the expected probability that a specific amino acid a is in a pair can be calculated:

- the number of residues in the considered data is $2T$;

- amino acid a occurs $2h_{aa} + \sum_{e \neq a} h_{ae}$ times;

- amino acid a occurs with a frequency of

$$p_a = \frac{2h_{aa} + \sum_{e \neq a} h_{ae}}{2T} = f_{aa} + \sum_{e \neq a} \frac{f_{ae}}{2}.$$

Suppose now that all pairs are separated, and that new pairs are drawn according to the observed frequencies. The probability for drawing the pair (ab) is

- $e_{ab} = p_a p_a$, for $a = b$;
- $e_{ab} = p_a p_b + p_b p_a = 2 p_a p_b$, for $a \neq b$.

Example

Assume two blocks:

```
ACDD       ACAAD
AACD       AADAC
ACDD
```

$$T = 17, \quad f_{AA} = \tfrac{5}{17}, \quad f_{AC} = \tfrac{3}{17}, \quad f_{AD} = \tfrac{1}{17},$$
$$f_{CC} = \tfrac{1}{17}, \quad f_{CD} = \tfrac{3}{17}, \quad f_{DD} = \tfrac{4}{17},$$
$$p_A = \tfrac{7}{17}, \quad p_C = \tfrac{4}{17}, \quad p_D = \tfrac{6}{17},$$
$$e_{AC} = \tfrac{56}{289}, \quad e_{AD} = \tfrac{84}{289}, \quad e_{CC} = \tfrac{16}{289}, \quad e_{CD} = \tfrac{48}{289}, \quad e_{DD} = \tfrac{36}{289}.$$

$\triangle$

In order to obtain the log-odds matrix we need to calculate the ratio between the observed and the expected frequencies for each amino acid pair. Given the above this is simply f_{ab}/e_{ab}.

As explained in Section 5.2, it is more convenient to use the logarithm of the odds, such that the expression for the values in the scoring matrices becomes

$$R_{ab} = \log_2 \frac{f_{ab}}{e_{ab}}.$$

5.3.2 Developing scoring matrices for different evolutionary distances

When comparing two sequences q and d with an (evolutionary) distance X, one should use segment pairs corresponding to this distance for constructing an appropriate scoring matrix. As an example suppose that we have four segments in a block, S1, S2, S3, S4, and that we measure the distances in percentage identity. Let the distance between the sequences q, d correspond to a 30% identity, and the distances between the segments correspond to (in %):

```
        S2      S3      S4
S1      28      26      20
S2              90      28
S3                      25
```

Clearly, the occurrences of an amino acid pair from (S2,S3) should count much less than occurrences from the other pairs. This is done by 'collapsing' the segments (S2,S3) into one segment. This also takes care of another problem, that the segments can come from a biased distribution of sequences. So, for developing a matrix for an $X\%$ identity, similar blocks with X or higher percentage identity are grouped into one group, and treated as one segment. For example, if the percentage is 80, segments U and V are grouped if they have at least an 80% identity. Segment W is joined to the same group if it has at least an 80% identity to one of the segment in the group. In the extreme case, a block can be reduced to one segment, such that it does not give any contribution to the scoring matrix.

Depending on the identity percentage required, different BLOSUM scoring matrices are developed.

Example

Consider the block:

```
1  KIFIMK
2  NLFKTR
3  KIFKTK
4  KLFESR
5  KIFKGR
6  KRRESR
```

For BLOSUM 100, no segment will be grouped (none have a 100% identity), and all 15 segment pairs give equal contributions (all segments have equal weights). Let now $X = 60\%$. We see that segments one and three have a 60% identity, as also do segments three and five, and segments four and six. Therefore, segments $(1, 3, 5)$ are grouped to one segment, and also segments $(4, 6)$. The resulting block becomes

```
                  IMK
(1, 3, 5)   K  IF KTK
                  KGR

2           N  LF KTR

              LF
(4, 6)      K RR  ESR
```

This means that segments $(1, 3, 5)$ together has the same weight as segment two separately, and segments $(4, 6)$ together.

If we consider only column one in the block, there are 5 pairs with KN and 10 with KK; hence the pair frequencies for 100% are $f_{KN} = \frac{1}{3}$ and $f_{KK} = \frac{2}{3}$. For 60% there are only three symbols, and the pair frequencies become $f_{KN} = \frac{2}{3}$ and $f_{KK} = \frac{1}{3}$.

```
         PAM 120        PAM 160        PAM 250

      BLOSUM 80    BLOSUM 62    BLOSUM 45
```

Figure 5.6   Correspondence between the PAM and BLOSUM scoring matrices.

For the other columns, we must take into account that different amino acids are grouped   △

The procedure for developing a BLOSUM X matrix can then be described by the following steps.

1. Collect a set of multiple alignments.

2. Find the blocks (without gaps).

3. Group the segments with an $X\%$ identity.

4. Count the occurrences of all pairs of amino acids.

5. Develop the matrix, as explained.

The most common BLOSUM matrices are 45, 62 and 80. BLOSUM 62 is often used as the standard for ungapped alignments, and is shown in Figure 5.4. For gapped alignment, BLOSUM 50 is more often used.

5.4   Comparing BLOSUM and PAM Matrices

The basis for constructing the two sets of matrices is different. The PAM matrices are constructed from predictions of the first mutations when proteins diverge from a common ancestor, thus on an explicit evolutionary model. BLOSUM, on the other hand, is based on common regions in families, thus it is better designed to find conserved domains. Use of a low percentage for grouping segments when developing BLOSUM matrices results in higher contributions from the pairs of the more dissimilar sequences. Thus BLOSUM matrices with a low percentage correspond to PAM matrices for large evolutionary distances. By use of relative entropy (from information theory), it can be found that PAM 250 corresponds to BLOSUM 45, and PAM 120 to BLOSUM 80, as shown in Figure 5.6.

When comparing sequences it is always a question of which PAM or BLOSUM matrix to use, especially when the evolutionary distance between the sequences is unknown. Different studies have concluded that for the PAM matrices it is generally best to try PAM 40, PAM 120 and PAM 250. When used for local alignments, lower PAM matrices find short local alignments, but higher PAM matrices find longer but weaker local alignments. Often a quick alignment is done first (using, for example, the identity scoring matrix), and the evolutionary distance estimated, and the corresponding scoring matrix used. However, several different matrices should be used,

Table 5.4 The BLOSUM 62 scoring matrix (*lower*) and difference matrix (*upper*) obtained by subtracting the PAM 160 matrix position by position. These matrices have identical relative entropies (0.70); the expected value of BLOSUM 62 is −52; that for PAM 160 is −0.57. Reproduced from Henikoff and Henikoff (1992) by permission of the National Academy of Science.

	C	S	T	P	A	G	N	D	E	Q	H	R	K	M	I	L	V	F	Y	W
C	9	1	1	0	2	1	1	2	1	2	0	0	2	4	1	5	1	2	−2	5
S	−1	4	2	−1	0	−1	0	0	0	1	0	0	0	1	0	1	−1	1	1	−1
T	−1	1	5	2	2	−1	0	0	0	0	−1	0	0	0	−1	−1	0	0	1	3
P	−3	−1	−1	7	−2	−1	−2	0	0	0	−1	−1	0	−1	0	−1	0	−1	2	1
A	0	1	0	−1	4	2	0	0	0	0	0	0	0	0	0	0	0	0	−1	2
G	−3	0	−2	−2	0	6	3	2	−1	−1	0	0	0	0	0	0	−1	1	2	4
N	−3	1	0	−2	−2	0	6	2	−1	0	1	0	0	0	−1	0	−1	0	0	0
D	−3	0	−1	−1	−2	−1	1	6	1	0	−2	−1	−1	−1	−1	0	−1	2	1	0
E	−4	0	−1	−1	−1	−2	0	2	5	0	2	0	0	0	−1	0	0	2	2	3
Q	−3	0	−1	−1	−1	−2	0	0	2	5	2	−1	0	1	0	0	0	−1	2	2
H	−3	−1	−2	−2	−2	−2	1	−1	0	0	8	−1	0	0	−1	−1	0	−1	3	4
R	−3	−1	−1	−2	−1	−2	0	−2	0	1	0	5	1	0	−1	−1	0	2	3	3
K	−3	0	−1	−1	−1	−2	0	−1	1	1	−1	2	5	0	−1	−1	0	−1	2	3
M	−1	−1	−1	−2	−1	−3	−2	−3	−2	0	−2	−1	−1	5	0	0	1	2	−1	2
I	−1	−2	−1	−3	−1	−4	−3	−3	−3	−3	−3	−3	−3	1	4	0	1	−1	−1	−4
L	−1	−2	−1	−3	−1	−4	−3	−4	−3	−2	−3	−2	−2	2	2	4	0	−1	2	1
V	−1	−2	0	−2	0	−3	−3	−3	−2	−2	−3	−3	−2	1	3	1	4	−1	−1	4
F	−2	−2	−2	−4	−2	−3	−3	−3	−3	−3	−1	−3	−3	0	0	0	−1	6	−2	1
Y	−2	−2	−2	−3	−2	−3	−2	−3	−2	−1	2	−2	−2	−1	−1	−1	−1	3	7	2
W	−2	−3	−2	−4	−3	−2	−4	−4	−3	−2	−2	−3	−3	−1	−3	−2	−3	1	2	11

and the alignment that is judged to be evolutionarily the most accurate should be chosen.

5.5 Optimal Scoring Matrices

Sometimes one is searching for pairs of segments satisfying special properties, e.g. hydrophobic segments (segments containing an overrepresentation of hydrophobic amino acids). In the theory by Karlin and Altschul (1990) (see Section 3.3), an 'optimal' scoring matrix for recognizing such segment pairs (without gaps) is developed.

Let the two sequences have the background distributions $\{p_a\}$ and $\{\bar{p}_a\}$, respectively (p_a is the frequency for amino acid a). Karlin and Altschul showed that in a maximal segment pair from two random generated sequences (using $\{p_a\}, \{\bar{p}_b\}$) the amino acids a, b are aligned by a frequency approaching

$$h_{ab} = p_a \bar{p}_b e^{\lambda R_{ab}}. \tag{5.7}$$

λ is defined in Section 3.3. h_{ab} is called the *target* frequency.

Solving Equation (5.7) for the scoring matrix, we get

$$R_{ab} = \frac{1}{\lambda} \frac{\ln h_{ab}}{p_a \bar{p}_b}. \tag{5.8}$$

We also see

$$\sum_{a,b \in \mathcal{M}} h_{ab} = \sum_{a,b \in \mathcal{M}} p_a \bar{p}_b e^{\lambda R_{ab}} = 1,$$

following the constraints on λ (Section 3.3).

h_{ab} is an 'alignment distribution' for the types of segment we are searching for, and from these the 'optimal' scoring matrix is found using Equation (5.8).

Example

If we search for highly hydrophobic and similar segment pairs, aligning pairs from the set {A, F, I, L, M, P, V} should score best. We therefore define higher 'alignment contributions' for pairs of these amino acids than for others.   △

The PAM matrices are, in principle, generated this way and all scoring matrices created in different ways by similar processes are *log-odds* matrices. The target frequencies are, however, seldom explicit. For PAM matrices the target frequencies are defined by the set of alignments on which the calculation is based.

5.5.1 Analysis for one sequence

The theory can also be used for scoring in one sequence, when we search a sequence for segments containing overrepresentations of certain amino acids. Target frequencies and a scoring vector similar to the scoring matrix in Equation (5.8) are defined:

$$R_a = \frac{1}{\lambda} \frac{\ln h_a}{p_a}. \tag{5.9}$$

High h_i values are given to those amino acids i we want to be overrepresented.

5.6 Exercises

1. How large would a scoring matrix for pairs of amino acids be? Why might it be better than the matrix that only considered single amino acids?

2. Make a scoring matrix based on the genetic code.

3. Explain why replacing 100 by 50 in Equation (5.1) does not corresponds to calculating a substitution matrix for two PAMs.

4. Suppose that there are only five different amino acids, A, B, C, D, E. We will make a scoring matrix for these amino acids. The basis is a multiple alignment of four sequences:

       AEACA
       DEBCB
       EAECA
       DAAEA

   (a) Create a rooted phylogenetic tree for these sequences, and find the mutations that most likely have occurred. The four sequences should be leaves in the tree, and you should suggest sequences for the three internal nodes, including the root. Assume that all mutations are acceptable, and make a table of f_{ab} (the number of mutations between a and b) for all a, b. Then find f_a and f.

   (b) Assume that the relative frequencies of the amino acids are $p_A = 0.2$, $p_B = 0.3$, $p_C = 0.2$, $p_D = 0.1$, $p_E = 0.2$. The scoring matrix we wish to find, should be for 1 PAM. Find for each amino acid the relative mutability m_a.

   (c) Use this to calculate the values of the substitution matrix M_{ab}. It is sufficient that you calculate the row corresponding to A (substitution from A) and the column corresponding to E (substitution to E).

   (d) Transform the matrix M_{ab} to an odds matrix O_{ab}. If you have 0 in the M_{ab} matrix, then set the corresponding value in O_{ab} to be the smallest element in the whole O_{ab} (> 0).

   (e) Transform O_{ab} to a log-odds matrix R_{ab}, and multiply by 10.

(f) Consider the values you have in R_{ab}, and find where the highest value is ($a \neq b$). Explain by using the values of p_a and f_{ab} that it is reasonable that it is the found pair which has the highest value.

5. In the caption to Table 5.3 it is explained what the scores -10 and $+2$ mean. Show how this is deduced.

6. We have three sequences: q = RPAEKTNW, d_1 = RPAEKTNR and d_2 = RPPEKWNW.

    (a) On the web you will find the BLOSUM scoring matrices (for example, at http://helix.genes.nig.ac.jp/homology/swsearch-e_help.html). Calculate the scoring for (q, d_1) and (q, d_2) for BLOSUM 45 and BLOSUM 90, when no gap is inserted. Do the tables rank the alignments equally? Explain why the ranks are different.

    (b) Estimate the evolutionary distance in PAM between (q, d_1) and (q, d_2) when Equation (1.3) is used.

7. (a) Explain the shape of the curve in Figure 5.5. Estimate the observed difference in percentage at 100 PAM.

    (b) Consider 100 PAM, and suppose that we have a sequence of 100 residues, and that all the positions have the same probability of changing. Assume further that none of the positions which are changed are changed back to the initial value. Then find

      (i) the number of residues that have not changed;

      (ii) the maximal number of residues that might have changed twice;

      (iii) the minimum number of residues that have changed exactly once.

8. We have two blocks:

A	B
PEEKSAVTALWGK	EVGGEALGRLLVV
GEEKAAVLALWDK	EVGGEALGRLLVV
PADKTNVKAAWGK	EYGAEALERMFLS
AADKTNVKAAWSK	EYGAEALERMFLG
AAEKTKIRSAWAP	TSGVDILVKFFTS
EGEWQLVLHVWAK	GHGQDILIRLFKS
ESQAALVKSSWEE	KHTHRFFILVLEI

(a) Calculate the value in BLOSUM 100 for the amino acid pair (S, A).

(b) Show how the blocks become for BLOSUM 60 and BLOSUM 30, and find the values for (S, A) for these matrices.

5.7   Bibliographic notes

A finer alternative to the AACH classification is AACC (amino acid class covering) in Smith and Smith (1992). A method for defining distances between the amino acids is in Taylor and Jones (1993).

The description of PAM is from Dayhoff (1978) and BLOSUM is from Henikoff (1996) and Henikoff and Henikoff (1991, 1992).

A rapid method for generating mutation data matrices is in Jones et al. (1992a). Gonnet matrices are described and discussed in Gonnet et al. (1992) and Benner et al. (1993, 1994).

Theoretical analysis of scoring matrices is in Altschul (1991). Matrices for pairs of amino acids is in Gonnet (1994). An evaluation of scoring methodologies is in Johnson and Overington (1993) and some comparisons are to be found in Vogt et al. (1995).

6

Profiles

When a number of member sequences of a protein family has been found, one can search for additional family members in at least two different ways. The first is to search with each known member against a sequence database. The second is to gather information from all known members, form a *model* for describing the properties of the members, and match this against a database of sequences. The latter has been shown to be superior in detecting weak relationships, i.e. remote family members, when one manages to use the family-specific properties in the search. A number of different models has been applied. Each of these capture information about the family members and can be compared with sequences.

Motifs. These include *consensus sequences* and *regular expressions*, and are discussed in Chapter 7.

Position-specific scoring matrices (PSSMs) or weight matrices. These are constructed from multiple alignments and represent the variation found in the alignment columns. Gaps are not described as part of the PSSM and normally no gaps are permitted when aligning a PSSM and a sequence.

Profile. This is also made from a multiple alignment. A profile represents in matrix form the degree of conservation in a multiple alignment, and includes in addition to position-specific scores (position-specific) gap penalties to be used when comparing the profile to a sequence.

HMM profiles. These are conserved regions of multiple alignments represented as hidden Markov models.

A profile can be represented as a two-dimensional array, here denoted Prof, with one row per alignment column and one column per amino acid plus columns for specifying the position-specific gap penalties.

From a multiple alignment, the block of interest for making a profile is specified, and each row in the profile corresponds to a column in the block. The value in row r column a (Prof_{ra}) is the score of aligning amino acid a (from a sequence) to the position r of the profile, as shown in Figure 6.1.

Protein bioinformatics: an algorithmic approach to sequence and structure analysis
I. Eidhammer, I. Jonassen and W. R. Taylor © 2004 John Wiley & Sons, Ltd ISBN: 0-470-84839-1

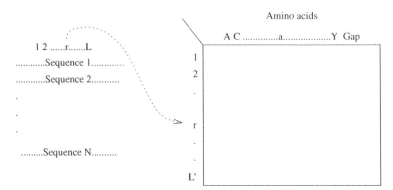

Figure 6.1 The figure shows the connection between a block of a multiple alignment and a derived profile. Positions $1 \ldots r \ldots L$ are positions in the profile. Gap specifies gap penalties for the gap in profile positions. (L' can be different from L if some of the columns are left out.)

Table 6.1 shows a profile generated from the first 33 columns of Figure 4.1 by the program ProfileWeight. Figure 6.2 shows some of the alignments resulting from searching SwissProt with that profile, using the program SearchWise. Note that the alignments are reported using a consensus sequence from the profile. This is only for illustration; the consensus is not used to calculate the alignment.

6.1   Constructing a Profile

The problem of constructing a profile can be treated as a special case of a more general problem.

> Given a family of which we know some members, but there also (probably) exist unknown members, how shall we use the known members to make a best possible description of the properties of the family, so that the description can be used to search for new family members?

We will first describe the creation of profiles with basis in ProfileWeight (Thompson et al. 1994a), which is an extension of a method by Gribskov et al. (1987). Then we describe an extension of BLAST (PSI-BLAST), and also give a brief introduction to HMM profiles.

When calculating the value Prof_{ra} from an alignment, several aspects should be taken into account.

1. The observed amino acids in position r in the alignment.

2. The number of independent 'observations' that has been used for constructing the alignment of position r.

3. The similarity of a to the amino acids observed in column r, to allow for not yet observed amino acids. Amino acid a is more likely to occur in unknown family

Table 6.1 The profile generated by ProfileWeight from the first 33 columns of the alignment in Figure 4.1. The scoring matrix used is PAM 250. Note that there are seven columns in the alignment for which there are no rows in the profile, due to many gaps. Also, three of the sequences are removed, each of these is identical to one of the remaining. 'Cons' is a consensus, one of the amino acids occurring most frequently: lowercase letters indicate gaps in the alignment, or that the precedent column is removed. 'Gap' is gap penalty and 'Len' for extending (Gep).

Cons	A	C	D	E	F	G	H	I	K	L	M	N	P	Q	R	S	T	V	W	Y	Gap	Len
T	1	-18	-4	-5	-12	-7	-5	-2	-6	-17	-11	0	-7	-11	-13	3	11	-3	-39	-8	100	100
P	10	-30	-10	-10	-50	0	0	-20	-10	-30	-20	0	60	0	0	10	0	-10	-60	-50	100	100
K	-10	-49	-1	-1	-47	-21	0	-18	46	-27	1	8	-9	9	31	0	-1	-18	-26	-39	100	100
a	7	-13	0	0	-20	2	-7	0	-5	-12	-6	0	1	-4	-9	5	7	2	-36	-20	50	80
r	-13	-31	-6	-6	-31	-20	12	-14	25	-22	0	1	-1	7	40	0	-6	-14	7	-29	50	80
R	-20	-40	-10	-10	-40	-30	20	-20	30	-30	0	0	0	10	60	0	-10	-20	20	-40	100	100
D	0	-49	34	33	-54	6	8	-20	0	-35	-25	14	-9	18	-10	-8	0	-19	-70	-40	100	100
V	0	-25	-18	-17	-8	-11	-18	29	-17	17	19	-17	-9	-15	-19	0	0	30	-54	-20	100	100
E	2	-50	31	38	-51	1	10	-20	0	-31	-21	11	-10	20	-10	0	0	-20	-70	-40	100	100
D	-1	-43	19	15	-52	3	11	-20	0	-31	-21	11	11	17	-2	1	-2	-17	-60	-41	100	100
L	-1	-41	-21	-17	-16	-16	-10	0	-18	11	7	-13	13	-8	-16	-7	-7	4	-42	-29	100	100
q	-5	-45	0	2	-23	-18	10	-2	-3	5	6	-3	-9	16	-3	-14	-10	-3	-37	-26	50	100
V	1	-26	-12	-14	-18	3	-19	17	-20	5	7	-14	-8	-16	-23	-5	-1	22	-58	-27	100	100
F	-3	-32	-11	-10	-5	-1	-9	-7	-17	-14	-11	-6	-11	-14	-18	-2	-6	-6	-39	-10	100	100
Q	0	-42	0	1	-28	-4	1	-9	-7	-4	-2	-1	-6	7	-10	-6	-6	-5	-44	-31	100	100
P	3	-26	-14	-12	-23	-6	-10	3	-13	-2	0	-7	14	-10	-11	0	0	7	-46	-30	100	100
E	3	-42	15	20	-49	0	8	-19	-2	-28	-18	6	13	15	-4	2	0	-16	-63	-42	100	100
L	-14	-56	-18	-13	-9	-27	-10	4	-5	22	19	-12	-22	-6	-11	-18	-12	4	-29	-21	100	100
g	6	-28	7	1	-41	21	-4	-22	-5	-32	-21	4	2	0	-9	9	0	-12	-48	-39	36	36
G	8	-32	8	4	-43	31	-14	-22	-14	-29	-19	0	-1	-4	-23	6	0	-7	-66	-43	100	100
g	2	-35	3	0	-35	17	-10	-17	-10	-19	-12	-1	-5	-2	-18	1	-2	-6	-53	-37	50	100
P	2	-39	4	7	-38	-2	0	-13	-8	-18	-12	-1	12	4	-9	1	-1	-8	-56	-37	100	100
g	6	-30	9	1	-39	29	-10	-20	-11	-31	-22	2	0	-5	-20	8	2	-8	-59	-38	50	81
a	6	-22	6	7	-25	4	-3	-5	-5	-14	-8	0	0	1	-11	2	4	-1	-45	-24	50	81
g	2	-28	-2	-4	-22	12	-12	-8	-11	-7	-1	-4	-4	-6	-16	0	-2	0	-43	-27	100	81
D	3	-34	17	15	-45	10	3	-18	-3	-29	-20	8	2	10	-7	3	0	-1	-54	-37	100	100

```
Profile   1 TPKARRDVEDLQVFQPELGGGPGAGD 26    E-value 2.851e-6
            |||:||:||::|| |:|||||||:|:
Rabit    51 TPKSRREVEELQVGQAELGGGPGAGG 76


Profile   1 TPKARRDVEDLQVFQPELGGGPGAGD 26    E-value 2.755e-5
            |||:||:||:||| | ||||:||: :
Mouse1   51 TPKSRREVEDPQVEQLELGGSPGDLQ 76


Profile   1 TPKARRDVEDLQVFQPELGGGPGAGD 26    E-value 3.03
            |||   |||:::::  |: :||:::|:
Vermo    52 TPK..RDVDPLLGFLPAKSGGAAAGG 75
```

Figure 6.2 Results of searching SwissProt for local alignments with the profile in Figure 6.1, using SearchWise provided at the Bioccelerator at EMBL. Three of the 26 highest-scoring alignments are shown. The profile is given by its consensus (only for illustration). The numbers show the positions in the profile and the sequences. The block (Figure 4.1) from which the profile is calculated starts with positions 51 or 52 of the sequences, and the segments of the sequences are found. Note, however, that the alignments of the profile to Mouse1 and to Vermo are not consistent with the underlying multiple alignment in Figure 4.1.

> members if there are many amino acids similar to a in the known sequences. Thus a 'background' scoring matrix should be used.

4. The background (*a priori*) distribution of the amino acids.

5. The diversity and similarity of the sequences, resulting in the importance (or weight) of each sequence. The known sequences are normally not uniformly distributed in the 'family space', and should have different weights in the calculation.

6. The number of gaps over column r and the neighbouring columns.

These points are not independent. How these aspects are treated varies with the different methods for profile construction.

6.1.1 Notation

The following notation will be used.

Prof_{ra}	the profile element to be calculated, r referring to a position in the alignment, and a to an amino acid
$\mathcal{M}$	the alphabet of all amino acids
$\bar{s}_r^i$	amino acid in position r in sequence s^i
T_{rb}	the number of occurrences of amino acid b in position r in the multiple alignment
R_{ba}	the scoring between amino acids b and a (a scoring matrix)
V_{rb}	the weight of amino acid b in position r (position weight)
w_i	the weight of sequence s^i (sequence weight)

m the number of sequences

m_r the number of residues (not gaps) in position r of the alignment

6.1.2   Removing rows and columns

Sequences which are identical to another sequence should be removed. Columns with many gaps should be less important than those with few or no gaps. An easy way of treating this is to remove columns where the number of gaps is over a threshold. In Table 6.1, seven of the columns containing gaps are removed. Note that when aligning a profile to a sequence, gaps can be included in both the profile and the sequence, as shown in Figure 6.2 for a gap in a sequence.

6.1.3   Position weights

In this section we ignore the sequence weights (setting $w_i = 1$ for all i). Prof_{ra} should then be a function of V_{rb} and R_{ba}, for all b. We further assume a model in which the columns are independent of each other. A reasonable function is a linear combination of the similarity of a to all occurring amino acids:

$$\text{Prof}_{ra} = \sum_{b \in \mathcal{M}} R_{ba} V_{rb}, \tag{6.1}$$

where V_{rb} depends on T_{rb} (the number of occurrences of b). Two constraints might be

- $V_{rb} = 0$ for $T_{rb} = 0$, and

- $V_{rb} = 1$ for $T_{rb} = m_r$.

However, it is sometimes appropriate to use pseudo-counts (see later in this subsection).

It must then be decided how V_{rb} should increase with increasing T_{rb}. Let b_1 and b_2 be two amino acids, and $T_{rb_1} < T_{rb_2}$. We can then define three ways in which there is dependency.

1. V_{rb} increases proportionally with T_{rb}, meaning

$$V_{rb_2} = \frac{T_{rb_2}}{T_{rb_1}} V_{rb_1} \quad \text{and} \quad V_{rb} = \frac{T_{rb}}{m_r},$$

hence

$$\text{Prof}_{ra} = \frac{1}{m_r} \sum_{b \in \mathcal{M}} R_{ba} T_{rb}.$$

This is illustrated by curve 1 in Figure 6.3.

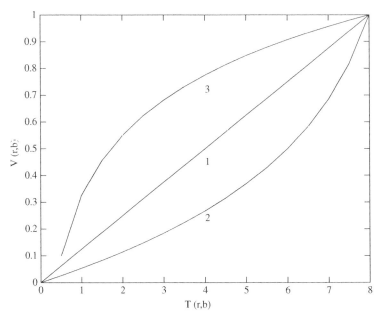

Figure 6.3   This figure shows different methods for calculating
V_{rb} from T_{rb} when $m_r = 8$. The numbers refer to the numbering in the text.

2. V_{rb} increases by more than T_{rb}, meaning

$$V_{rb_2} > \frac{T_{rb_2}}{T_{rb_1}} V_{rb_1}.$$

Amino acids occurring many times are 'favoured'. One equation satisfying this
is

$$V_{rb} = \frac{\ln[1 - T_{rb}/(1 + m_r)]}{\ln[1/(1 + m_r)]},$$

which is illustrated by curve 2 in Figure 6.3.

3. V_{rb} increases less than T_{rb} meaning

$$V_{rb_2} < \frac{T_{rb_2}}{T_{rb_1}} V_{rb_1}.$$

Amino acids occurring many times are 'punished'. One equation satisfying this
is

$$V_{rb} = \frac{1 + \ln T_{rb}}{1 + \ln m_r},$$

which is illustrated by curve 3 in Figure 6.3.

Statistical pseudo-count method

If the scoring matrices are of the 'log-odds' form (see Chapter 5), a simple equation for defining a profile for this form is due to Henikoff and Henikoff (1996):

$$\text{Prof}_{ra} = \log \frac{h_{ra}}{p_a},$$

where h_{ra} is the probability for a in position r, and p_a is the background probability for a. When the number of independent sequences is sufficiently high h_{ra} can be estimated as T_{ra}/m_r, but this does not work very well for small m_r, as the logarithm of 0 is indefinite. To avoid this, a *pseudo-count* is often used. An estimate for h_{ra} is then

$$h_{ra} \sim \frac{T_{ra} + Bp_a}{m_r + B},$$

where B is a pseudo-count for which a constant value should be chosen. It has been argued from experiments that $B = \sqrt{m_r}$ is a reasonable value, but this has been questioned by other investigations. Alternatively, more advanced solutions have been explored, some using Dirichlet mixtures.

Note that for Equation (6.1), the problem of nonoccurring amino acids is taken care of by using a scoring matrix.

6.1.4   Sequence weights

The sequences can be given weights, $\{w_i\}$, as explained in Subsection 4.4.3, and an expression for the position weight V_{rb} can be defined by use of the sequence weights. For a 'middle score method', this can be done by

$$V_{rb} = \frac{\sum_{i=1}^{m_r} w_i \delta_b}{\sum_{i=1}^{m_r} w_i}, \tag{6.2}$$

where

$$\delta_b = \begin{cases} 1 & \text{if } \bar{s}_r^i = b, \\ 0 & \text{if } \bar{s}_r^i \neq b. \end{cases}$$

We see that if $w_i = 1$, then, for all i, $V_{rb} = T_{rb}/m_r$, which is curve 1 in Figure 6.3.

6.1.5   Treating gaps

If a column in the alignment contains sufficiently many gaps, it is normally not included in the profile, and the penalty for opening a gap when aligning a sequence to the profile is reduced. This is illustrated in Table 6.1, where column 32 in Figure 4.1 is removed, and the penalty for opening a gap is lowered (50). Since the gap length is 1, the gap extension penalty is not lowered.

Generally, the penalty for introducing a gap during an alignment between a sequence and a profile should depend on whether there are gaps in the respective column

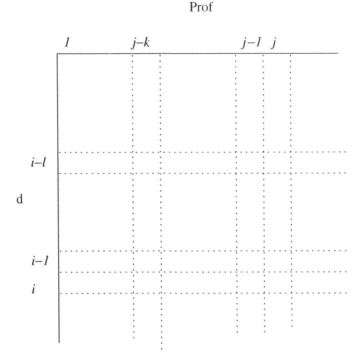

Figure 6.4 Schematic of the dynamic programming matrix when used for profile search.

of the alignment. This means that one needs position-dependent penalties for opening and extending gaps. A position-dependent gap penalty should be calculated as a function of the lengths and variations of the gap lengths in the multiple alignment, additionally the sequence weights should be used in the calculation, as shown in Table 6.1.

6.2   Searching Databases with Profiles

The task is to search the database for sequences that locally match the profile. This can be done by a Smith–Waterman dynamic programming algorithm. Conceptually, the profile can be placed at the horizontal edge of the 'dynamic programming matrix', H, and the database sequence d at the vertical edge, as shown in Figure 6.4. $H_{i,j}$ is then the best score which can be found in a local alignment between a part of the profile ending in position j (Prof_j) and a subsequence (substring) of d ending in position i (d_i). We here present a variant of the algorithm for finding the best local alignment. The entry $H_{i,j}$ gets the highest score possible for local alignments where the last column is (d_i, Prof_j) (not a gap).

The recursion equation can be written as

$$H_{i,j} = \max(0, \text{Prof}_{jd_i} + \max \begin{cases} H_{i-1,j-1}, \\ \max_{2 \leqslant k \leqslant j-1} (H_{i-1,j-k} - g_k^d) & \text{(gap in } d), \\ \max_{2 \leqslant l \leqslant i-1} (H_{i-l,j-1} - g_l^P) & \text{(gap in Prof).} \end{cases}$$

g_k^d and g_l^P are gap penalties:

- g_k^d is the penalty for a gap in d of length $k - 1$ with start at Prof_{j-k+1};
- g_l^P is the penalty for a gap in Prof of length $l - 1$ with start at d_{i-l+1}.

Generally, the gap penalties can be a combination of position-dependent and position-independent gap penalties, and the profile can contain GOP and GEP for both gaps in the profile and in the sequence for each profile position.

6.3   Iterated BLAST: PSI-BLAST

The advantage of using profiles has led to an extension of BLAST: position-specific iterated BLAST (PSI-BLAST) (Altschul et al. 1997). The main idea is to first use BLAST to search with the query q, and then make a profile of the results (the found sequences), then search with the profile, make new profile according to the results, etc. This is an automatic version of the procedure used by many bioinformatics practitioners in order to find remote homologies.

PSI-BLAST is a widely used method and it illustrates several important aspects of database-similarity search algorithms. PSI-BLAST uses the points in the list in Section 6.1 in a highly heuristic way. It is described here since it is an interesting extension to BLAST, and since it illustrates how several important aspects can be implemented and combined.

Algorithm 6.1. PSI-BLAST.
Search a sequence database

var
q the query
t a threshold for significance
begin
    $Q := \text{BLAST}(q, t)$ Q is the set of sequences found in the search
    loop
        $Q_1 := \text{Reduce}(Q)$ Remove equal sequences
        $M := \text{MultipleAlignment}(Q_1)$
        $P := \text{Profile}(M)$
        $Q := \text{ProfileSearch}(P, t)$ Search with the profile
    until *convergence* $(\text{Reduce}(Q) = Q_1)$ **or** *maximum number of cycles*
end

A high-level description of the procedure is given in Algorithm 6.1. ProfileSearch is a revised version of gapped BLAST, and it searches for local alignments using a profile. The profile P has a row for each residue in q, hence it is of the same length as q. Only the local alignments obtained in the search step are included in the multiple alignments. Position-specific gap scores are not derived; instead, the gap scores used in the first BLAST run are kept.

Note that Q in the first statement of the algorithm is the set of sequences that have local alignments with q with P value below a threshold.

6.3.1 Making the multiple alignment

Segments in Q which are (exactly) equal to the aligned segment in q are removed, and if two segments are at least 98% identical, one of them is removed.

The local alignments are then used as inputs to the multiple alignment procedure, with q as a template. Pairwise alignment columns that involve a gap in q are ignored. Consequently, M has exactly the same length as q. Because local alignments are used, the columns of M may involve a varying number of sequences.

Example

Let the query sequence be q = ACRAGTLRSH, and three database sequences are found, with the local alignments to q:

q:	C-RAGT-L	q:	R-AGT-LR	q:	CRAG-TLR
d_2:	DLRA-SRN	d_2:	RVA-TVNR	d_2:	C-AGTTLR

Then the multiple alignment becomes

```
                        r    s
         q:     ACRAGTLRSH
         d1:    .DRA-SN...
         d2:    ..RA-TNR..
         d3:    .C-AGTLR..
```

(r and s are for later reference to the example.) '.' is used to indicate positions outside the local alignment. Note that several of the columns only include the query.   △

6.3.2 Constructing the profile

The profile is constructed taking the points in the list of Section 6.1 into account. Some of the points are implemented in such a way that the profile values of one column also depend on other columns of the multiple alignment. This is realized through the construction of reduced alignments.

Reduced alignments

For each column r one reduced alignment M_r is constructed. Only sequences with a residue in column r are retained in M_r. The columns are the columns for which all the remaining sequences are represented in the original local multiple alignment.

Example

For the example above the reduced alignments M_r and M_s become

M_r	r		M_s	s
q:	RAGTL		q:	RAGTLR
d_1:	RA-SN		d_2:	RA-TNR
d_2:	RA-TN		d_3:	-AGTLR

$\triangle$

These alignments are used in later calculations.

Weighted frequencies

Weighted frequencies, $\{f_a\}$ (for each column), are calculated by use of the observed frequencies, the reduced alignments and the sequence weights. Sequence weights are found using a variant of the method explained in Subsection 4.4.3.

Number of independent observations

In addition to the frequencies, the number of independent observations is also important: a column consisting of a single V and a single I carries different information to one consisting of five independently occurring instances of each. A number m_r of independent observations constituted by the alignment M_r is therefore needed. A simple estimate is used: the number of different amino acids (including gap characters) in M_r.

The final profile values (scores)

As explained in Chapter 5, scoring matrices with the best theoretical foundations are of the form $\log (h_a/p_a)$, where p_a is the background probability for amino acid a, and h_a is the estimated probability for a to be in the considered position (the position r is implicit). Naturally, we could use the weighted frequency f_a to define h_a. However, the number of observations (sequences in the alignment) may be small, and prior knowledge of the relations among the amino acids should be used. Use of data-dependent pseudo-count frequencies is therefore implemented (see Subsection 6.1.3). A pseudo-count B_a is calculated for each amino acid, and is averaged with f_a to estimate h_a. For a given column, B_a is defined as

$$B_a = \sum_{b \in \mathcal{M}} \frac{f_b}{p_b} u_{ab}, \qquad (6.3)$$

where u_{ab} is the frequency implicit in the scoring matrix R_{ab} defined by

$$u_{ab} = p_a p_b e^{\lambda R_{ab}} \tag{6.4}$$

(see Equation (5.7) in Section 5.5). Next h_a is estimated as

$$h_a = \frac{\alpha f_a + \beta B_a}{\alpha + \beta}. \tag{6.5}$$

In order to obtain a scoring value equal to R_{ab} for amino acid a in columns r where nothing has been aligned to the query amino acid b, α is set to $m_r - 1$. A reasonable value for β is found to be 10.

6.4 HMM Profile

Hidden Markov models (HMMs) have been used in speech recognition since the early 1970s. In the early 1990s the first applications in bioinformatics appeared and since then they have become an important tool for sequence analysis and, in particular, protein family analysis. The most common usage of HMMs in protein family analysis is analogous to sequence profiles as described above. HMMs provide an alternative, statistically based tool for describing conserved parts of sequence families. Here we give a very brief and informal introduction to HMMs focusing on HMM profiles.

It is customary and natural to describe an HMM as a device for generating sequences of symbols, in our case sequences of amino acids. For our purpose, an HMM will be used to distinguish between sequences that are members of a family and nonmembers. The parameters of the HMM will be set so that the probability of the HMM generating member sequences is relatively high and higher than the probability of it generating nonmembers. In this way we can set a cut-off on the probability of the HMM generating a (query) sequence and predict it as a member if the probability of the HMM generating the sequence is above this cut-off. Note, however, that the HMM is *not* used to generate sequences, but to calculate the probability of a query sequence being generated by the model, where the resulting probability is used to predict family membership of the query sequence.

6.4.1 Definitions for an HMM

An HMM consists of a set of states including a start state T_0 and a stop state T_m. For each pair of states (T_i, T_j) there is a probability of moving from state T_i to state T_j; let us call this $P(i, j)$. The probabilities are parameters that can be adjusted in the HMM, but the following must hold:

$$0 \leqslant P(i, j) \leqslant 1 \text{ for all } (i, j),$$
$$\sum_j P(i, j) = 1.$$

For a model, one can follow many different paths through it, let one be $\Pi = \pi_1 \cdots \pi_n$, where $\pi_1 = T_0$ and $\pi_n = T_m$. We assume that the probabilities for the state transitions are independent of each other, hence the probability that the particular path Π is chosen in a passage is

$$P(\Pi) = \prod_{i=2,\ldots,n} P(\pi_{i-1}, \pi_i).$$

Each state can emit a symbol, in our case an amino acid, and each state has its own probability distribution over the symbol alphabet (the set of amino acids). The probability of state T_j generating symbol a can be denoted $P(a \mid j)$. These probabilities are also parameters that can be adjusted, but the following must hold:

$$0 \leqslant P(a \mid j) \leqslant 1 \text{ for all } a \text{ and } j,$$

$$\sum_a P(a \mid j) = 1 \text{ for all } j.$$

The HMM generates sequences by following a path Π, each state emits a symbol, and the concatenation of the emitted symbols is the generated sequence. The probability of following one specific path Π and that the model generates the sequence $q = q_1 \ldots q_n$ is then

$$P(q, \Pi) = P(\Pi)P(q \mid \Pi) = \prod_{i=2,\ldots,n} P(\pi_{i-1}, \pi_i) \prod_i P(q_i \mid \pi_i). \qquad (6.6)$$

Some minor changes are needed when one include states not emitting amino acid symbols. Also, when implementing an HMM one should calculate the logarithms of the probabilities instead of the probabilities themselves in order to achieve higher numerical stability.

6.4.2 Constructing a profile HMM for a protein family

Usually, one starts with a multiple alignment of the set of sequences (of a family) for which one wants to construct an HMM, and a state is defined for each column one wants to include. A state corresponding to a conserved position can then have a very high probability of emitting the conserved amino acid and the other probabilities are low. 'Hydrophobic columns' can have states with a high probability of emitting hydrophobic amino acids.

An HMM for a family shall be compared with query sequences, and be able to distinguish between sequences that are members of the family and those which are not. It is the value of the parameters that make this possible, and these values are determined in an HMM *training algorithm* that takes as input a set of (normally aligned) member sequences. In order to be able to compare (align) a sequence with a profile HMM, the HMM should have a set of three states per alignment column: one *match* state, one *insertion* state and one *deletion* state. The match state is used to match an amino acid from the query. The deletion state is used to bypass this

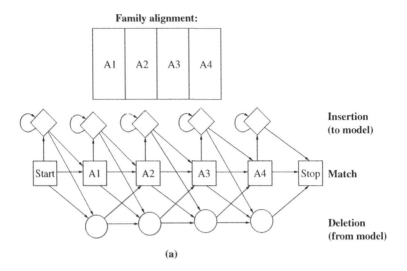

(a)

Query sequence: q1 q2 q3 q4 q5 q6

HMM and alignment with query sequence (Viterbi):

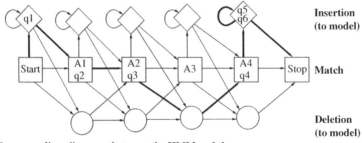

Corresponding alignment between the HMM and the query:

```
q1 q2 q3 - q4 q5 q6
 - A1 A2 A3 A4 - -
```

(b)

Figure 6.5 (a) An alignment of four columns is given, and an HMM is constructed with match states (and start- and stop-states) as squares, insertion states as diamonds, and deletion states as circles. Only state transitions that can have nonzero probabilities are shown as arcs. All the probabilities are learned in a training phase, such that the model can be used to predict family membership and construct an alignment between a query sequence and the HMM (corresponding to an alignment between the query and the original multiple alignment). (b) Shows a path for the query sequence *q* giving optimum alignment (thicker arcs), and the corresponding alignment.

alignment column and emit a blank. The insertion state allows the insertion of one or more amino acids (more than one is permitted by having a state transition from the inserted state to itself), the emission probabilities of this state are calculated either from a background distribution of amino acids or from the observed insertions in the alignment. See Figure 6.5 for a schematic showing a linear profile HMM.

6.4.3 Comparing a sequence with an HMM

When a sequence is to be compared with an HMM (to find whether the sequence is member of the family described by the HMM), one normally finds one path that has the maximum probability of the HMM generating the sequence. This can be done by an algorithm known as Viterbi, which is a dynamic programming algorithm. It finds for each prefix $q_{1...i}$ of the query and for each state T_j the number $D_{i,j}$ which is the maximum probability of a path from the start state ending at state T_j and generating $q_{1...i}$. $D_{i+1,j}$ is found by the recurrence:

$$D_{i+1,j} = \max_{0 \leqslant k \leqslant j-1} (D_{i,k} P(k,j) P(q_{i+1} \mid j)).$$

Note that for profile HMMs most of the probabilities $P(i, j)$ for state transitions are zero, and efficient implementations of the Viterbi algorithm exist. Sometimes one also uses the sum of probabilities over all paths that can generate the query sequence. Such a procedure does not directly produce a maximum scoring alignment.

6.4.4 Protein family databases

There are protein family databases with signatures for each family including the PROSITE, PFam, PRINTS, BLOCKS, Prodom and SMART databases. Some use regular expressions as family signatures, others use profiles or HMMs. The PFam database defines an HMM for each family and when searching a sequence against PFam the sequence is compared (aligned) with each HMM and family membership predicted depending on the probability being above a defined threshold for each family (in their implementation a score is calculated). The PROSITE database uses both profiles and sequence motifs (see Chapter 7). There is a very close relationship between profiles and linear profile HMMs (as they are described here), Figure 6.6 indicates the relationships between the different entries of an HMM with quantities in a profile.

6.5 Exercises

1. Suppose that a column in a multiple alignment contain five A, three C and two E. Calculate the ratio V_{rA}/V_{rC} for the different equations for calculating V_{ra} from T_{ra}.

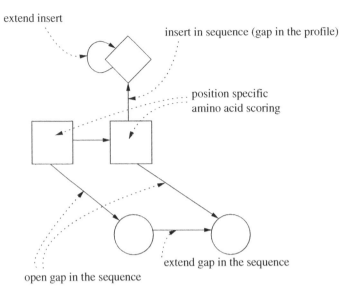

extend insert

insert in sequence (gap in the profile)

position specific
amino acid scoring

extend gap in the sequence

open gap in the sequence

Figure 6.6 The figure indicates which elements of a profile HMM are related to different parts of a profile. For example, the probability of going from a match state to a delete state is related to the penalty of opening a gap in the sequence in profile analysis, and the probability of moving from a delete state to a delete state is correspondingly related to the penalty of extending a gap in the sequence.

2. A scoring matrix and a multiple alignment is given:

	A	C	D	E		
A	1.0	0.2	−0.4	−0.2	d_1:	ACA−DE
C		1.2	−0.2	0.0	d_2:	ADCDDE
D			1.0	0.2	d_3:	CDADEE
E				0.8	d_4:	A−AD−C

   (a) Make a profile based on this when the weighting is similar to the one used in the first method in Subsection 6.1.3. Do the calculations for the first column.

   (b) Do the calculations for the second column of the profile, given that the sequences have weights 2, 1, 2, 2. Use the middle-score method (Equation (6.2)) for the position weights.

3. In the caption of Figure 6.2 it is mentioned that two of the alignments do not correspond to the underlying multiple alignment. Show how the alignments should be if they were consistent with Figure 4.1.

4. We have a sequence $q =$ MPCKART. Find the best local alignment when aligning q and the first seven rows of the profile in Table 6.1. Use a linear gap penalty with $g = 20$ for gaps in the sequence and $g = 40$ for gaps in the profile.

5. When using PSI-BLAST the sequences in Q do not necessarily have to be in Q after the next cycle. Explain why.

6. Show graphically the cells of the matrix D that are used for calculating the value of $D_{i,j}$ when the Viterbi algorithm is used for finding the probability when using an HMM for comparison.

7. Suppose we have an alphabet of five amino acids: A, D, E, L, V, and that the following multiple alignment is given:

```
ADLDL
AEV-A
L-V-A
VEL-A
```

We shall construct an HMM for this alignment, using the following rules.

- There shall be a match state for every column with at least 40% nonblanks. In our case it will be four: $A1$, $A2$, $A3$, $A5$.

- In order to calculate the emission probabilities, a pseudo-count of 1 should be used. That means that before calculating the probabilities, 1 is added to the number of occurrences for each amino acid. For the first column, A gets probability $\frac{3}{9}$, L and V get $\frac{2}{9}$, and D and E get $\frac{1}{9}$. When nothing can be derived for an insert state, use equal emit probabilities for all amino acids.

- The state transition probabilities also use a pseudo-count of 1. That means that the state transition probabilities from the match state with $A1$ is $\frac{4}{7}$ to match state for $A2$, $\frac{2}{7}$ to the delete state, and $\frac{1}{7}$ to the insertion state. Choose some reasonable probabilities if the transition probabilities cannot be derived.

(a) Construct an HMM model for the alignment, complete with the probabilities.

(b) Show a (reasonable) path through the model for the sequence $q = $ EVDAL, and show the alignment to the model.

(c) Calculate the probability that the path is chosen and q generated.

6.6   Bibliographic notes

Profiles were described early on by Gribskov et al. (1987), Gribskov (1990) and Gribskov and Veretnik (1996). ProfileWeight is described in Thompson et al. (1994a). Position-based scoring is in Henikoff and Henikoff (1996, 1994). Iterative scanning with alignment blocks is described in Tatusov et al. (1994) and PSI-BLAST in Altschul et al. (1997).

Pseudo-count is introduced in Tatusov et al. (1994). Dirichlet mixtures are described in Henikoff and Henikoff (1996) and Sjølander et al. (1996).

An introduction to HMM and HMM profiles is given in Durbin et al. (1998) and Eddy (1998). Pfam is described in Sonnhammer et al. (1997, 1998).

7

Sequence Patterns

When a collection of diverse proteins shares a common function or structure, sometimes all that is conserved between them is a few common residues that are critical for their structure and function. If the proteins are enzymes, these residues are typically those that are involved in the chemical catalysis in the active site. These can also involve nearby residues that play a structural role in maintaining the relative positions of the catalytic residues. The same applies to proteins that are involved in specific (noncatalytic) interactions, such as binding a nucleic acid. Again, a small number of residues may be highly conserved.

These patterns of conservation are not necessarily all found in one place in the sequence of the protein. An active site can (and usually is) formed from distant parts of the sequence. This appears as clusters of conserved residues. Sometimes these will have limited insertions and deletions within one cluster (sometimes called a motif), but between motifs there can often be very large insertions and deletions of sequences. This situation makes it difficult to recognize the relationship between these sequences using the alignment tools that we have considered so far. A global alignment method will have difficulty dealing with the large insertions and deletions between motifs (which may have almost no similarity), while a local alignment method might find one motif but will neglect the others. This has led to a class of analysis tools based on pattern matching which we will consider in this chapter.

In Chapter 6 we considered models for describing similarities among sequences (of a family), such as profile and profile HMM. Comparing such models with a sequence results in scores reflecting how well the sequence matches the models, thus the models are *statistical*. In this chapter we describe a *deterministic* model: *sequence pattern*. A sequence pattern is either matched or not matched by a sequence. The part of a sequence that actually matches a pattern is called an *occurrence* of the pattern.

Example

An example of a deterministic pattern is [RK]-x(2,3)-D, which is matched by a (sub)sequence beginning with R or K, then two or three arbitrary amino acids, followed by a D. The sequence ACRLTDEF matches the pattern with the occurrence RLTD. △

Protein bioinformatics: an algorithmic approach to sequence and structure analysis
I. Eidhammer, I. Jonassen and W. R. Taylor   © 2004 John Wiley & Sons, Ltd   ISBN: 0-470-84839-1

Sequence patterns are often used for characterizing proteins with structural similarity, for example, the PROSITE entry (see the section of PROSITE below) for the SH3 domain, which is a small protein domain of about 60 amino acid residues. The PROSITE entry gives a pattern for this family:

```
T-[ST]-R-x(2)-[KR]-x(2)-[DE]-x(2)-G-x(2)-Y-x-[FY]-[LIVMK]
```

If one finds that a new protein matches the pattern, one can be reasonably sure that the protein contains an SH3 domain. Patterns can also be used in relation to functional properties. For example, PROSITE gives a pattern characteristic for the sequences of a certain class of aminoacyl-tRNA syntethetases (a group of enzymes which activate amino acids and transfer them to specific tRNA molecules as the first step in protein biosynthesis). If a new sequence matches this pattern, one can hypothesize that it is an enzyme having this function.

Sets of proteins having the same structure and/or function can be analysed and one can find which residues are most informative about the structure/function, which are allowed to vary more freely, etc. Generally, one can define sequence patterns that are characteristic of each family. Then comparing a new sequence with a collection of patterns (one for each structure family) is more efficient and also provides a more sensitive and specific test of family membership than pure sequence comparison.

Patterns describing biologically meaningful similarities are called *motifs*. Motifs can thus be used to predict functional or structural properties of the protein, or to describe (nontrivial) features common to biologically (structurally or functionally) related proteins. Having found a pattern, one wants to evaluate whether it is a motif, i.e. if it has a biological meaning. Since in automatic methods there is no way of assessing the biological meaning of a pattern directly, mathematical methods for pattern evaluation have been developed as an alternative. We call them *fitness* or *scoring* functions.

Different *languages* (or formalisms) for describing patterns exist, and for any language there is a mechanism for deciding whether or not a sequence matches a pattern. (Note that it is usually enough that part of a sequence actually matches a pattern.) How specific a motif can be described depends on the language. We first describe the PROSITE language for deterministic patterns (Hofmann et al. 1999). (The PROSITE database also includes profiles for many of the protein families.)

7.1   The PROSITE Language

PROSITE is a database of protein families and domains. For each family the most conserved regions are searched, and by analysing the constant and variable properties of the groups of similar subsequences of such regions, one attempts to derive a signature for a protein family or domain which distinguishes its members from all other unrelated proteins. The PROSITE language for sequential patterns is described by the following.

- The standard one-letter codes for the amino acids are used.

- The symbol 'x' is used for an arbitrary amino acid.

- Ambiguities are listed between square parentheses '[ ]'. For example, [AGL] stands for Ala or Gly or Leu.

- Amino acids that are not accepted at a given position are listed between '{ }'. For example, {CH} stands for any amino acid except Cys and His.

- '-' is used for separating the elements.

- Repetition of an element is specified with a numerical value or a numerical range between parentheses, such that x(3) corresponds to x-x-x and x(1,3) corresponds to x or x-x or x-x-x.

- When a pattern is restricted to either the N- or C-terminal of a sequence, that pattern either starts with a '<' symbol or ends with a '>' symbol.

- A period ends the pattern.

Example

```
[RK]-x(2,3)-[DE]-x(2,3)-Y.

D-x-[DNS]-{ILVFYW}-[DENSTG]-[DNQGHRK]-{GP}-[LIVMC]-
[DENQSTAGC]-x(2)-[DE]-[LIVMFYW].
```

KLRACEDEEYRE matches the first pattern, and MADANADDDCTAADWST matches the second.   △

7.2   Exact/Approximate Matching

Exact matching means that the patterns must occur exactly in the sequences, approximate means that the occurrences might have one or several mismatches or insertions/deletions from the pattern. With exact matching, the variation allowed is described in the pattern, while this need not be the case when approximate matching is used.

Example

Assume that we have the following sequences:

ACDELT   RRCCENR   CEESV   SCDFRT

It is not difficult to see that the pattern CxE is common to the first three (meaning that they all exactly match the pattern). On the other hand, we also see that they approximately (to an edit distance of 1) match the more specialized pattern CDE, which is also approximately matched by the last sequence. We cannot find a pattern with two fixed positions common to all four sequences. Summing up, CDE matches all sequences approximately (to edit distance 1), and CxE matches three of them exactly.

△

When patterns are used in protein bioinformatics, exact matching is most commonly used. In the following we consider exact matching.

7.3   Defining Pattern Classes by Imposing Constraints

The patterns can be grouped into (overlapping) pattern *classes*, depending on the languages in which they are described. A method for discovering common patterns can then discover patterns of specific classes.

We consider a deterministic sequence pattern which consists of *components* and *wildcard regions*, and it is the restrictions on these which define the classes. Note that a pattern always starts and ends with a component.

- A component refers to occurrences of fixed length. Each position in a component can either be

    unique—only a specified amino acid is allowed—if all positions are unique, the component describes a unique segment, or

    ambiguous—amino acids from a given set are allowed—a component with an ambiguous position is itself ambiguous.

- A wildcard region consists of positions where all amino acids are allowed. A wildcard region is *fixed* if it has fixed length (is matched by a fixed number of amino acids in a sequence), otherwise it is *flexible*.

A pattern then consists of consecutive alternating components and wildcard regions (where wildcard regions can be of length 0).

The language of PROSITE patterns also allows for repetitions of components, e.g. [DE](2,3). This will not be considered further.

Example

The PROSITE pattern

$$[RK]-x(1,2)-[DE]-x(1,3)-Y-C.$$

has four components, all of fixed length (1). The first two are ambiguous, the last two unique. There are two flexible wildcard regions, and one fixed of length 0, which is implicit between Y and C. It could also be regarded as consisting of three components, the last of length 2.   △

A pattern with one or more flexible wildcard regions is itself called flexible. The language of PROSITE patterns allows flexible patterns. The *flexibility* of a wildcard region is 1 plus the difference between the two values defining the wildcard region; thus $x(2,4)$ has flexibility 3. (We see that a fixed wildcard region has flexibility 1.)

The methods for pattern discovery can be classified based on what types of patterns can be discovered, e.g.

- what type of components (fixed, ambiguous),

- the number of components (1, 2 or more),

- what type of wildcard regions (fixed, flexible).

A finer classification can be made by giving maximum (and sometimes minimum) values for a set of quantities. For example, we have the following.

- Maximum length of a pattern, the length of a pattern is defined as the maximum length of an occurrence of the pattern. Thus, the length of the pattern A-[ST]-x(1,3)-C is 6.

- Maximum (and minimum) number of components.

- Maximum (and minimum) length of each component.

- Maximum length of a wildcard region.

- Maximum number of flexible wildcard regions.

- Maximum flexibility of the wildcard regions.

- Maximum (total) flexibility of a pattern, which is defined as the product of the flexibility of its wildcard regions.

- Maximum number of ambiguous positions.

Example

An example of pattern class is that the patterns consist of two unique components, each of length 2 or 3, and a wildcard region of maximum length 4 and flexibility of at most 2. Each of the components has $20^2 + 20^3$ possibilities, and there are nine alternatives for the wildcard region. Hence, the number of patterns in the class is $9 \cdot 20^4 \cdot 21^2$. For the sequences ACRLHWRLKL, HCRRWRLA and CLCRAVWRLVK we can discover the common pattern CR-x(1,2)-WRL in this class.   △

7.4   Pattern Scoring: Information Theory

Information theory can be used for scoring a pattern, and we first give a brief introduction to this theory.

7.4.1   Information theory

Let $\mathcal{C}$ in this section be a set of symbols $\{a\}$, where each symbol has a background probability $\{p_a\}$. Assume a symbol is drawn using the background probability distribution. The *information* $I(a)$ associated with a is measured as how much information is gained if we are told that a is drawn. Constraints on $I(a)$ are that if $p_a = 1$, then

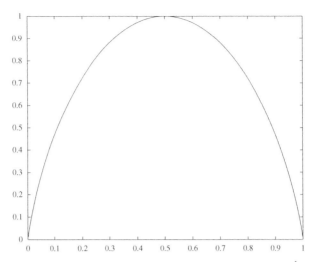

Figure 7.1   The entropy for a binary alphabet, with probability $\frac{1}{2}$ each.

$I(a) = 0$ (no new knowledge), and that $I(a)$ should increase for decreasing p_a. This is obtained by the following equation:

$$I(a) = -\log p_a. \tag{7.1}$$

We see that $I(a) = \infty$ when $p_a = 0$. When the logarithmic base is 2, the information is measured in *bits*.

Suppose we now draw m symbols using the distribution. If m is large, a will be drawn mp_a times (the law of large numbers). Middle information per symbol occurrences, uncertainty or *entropy*, is defined as (in bits per symbol if the base is 2)

$$H(\boldsymbol{p}) = -\frac{1}{m}\sum_{a \in \mathcal{C}} mp_a \log p_a = -\sum_{a \in \mathcal{C}} p_a \log p_a. \tag{7.2}$$

Example

Suppose we have a binary alphabet:

- $\boldsymbol{p} = \{1/2, 1/2\}$, then $H(\boldsymbol{p}) = 1$ bit;

- $\boldsymbol{p} = \{1/4, 3/4\}$, then $H(\boldsymbol{p}) = 0.8$ bits;

- $\boldsymbol{p} = \{1, 0\}$, then $H(\boldsymbol{p}) = 0$ bits.

This is illustrated in Figure 7.1.   △

For k symbols the entropy always has values in the number interval $[0, \log k]$.

7.4.2 Scoring patterns

Let $\{p_a\}$ be the background probabilities (distribution) of the amino acids. Scoring a pattern can be done by giving a score to each position of the components, and to each wildcard region, and then taking the sum of these.

To score a position i, its *information content* can be used. The information content of a position is defined as *the decrease in uncertainty when it becomes known that only residues from a set of amino acids K_i can be in that position*. The uncertainty when nothing is known is the entropy as defined in Equation (7.2).

When it becomes known that only amino acids from a set K_i can be in a position i, revised probabilities must be used for calculating the uncertainty. The revised probability is $p'_a = p_a/p_{K_i}$, where p_{K_i} is the accumulated background probabilities of the amino acids in K_i. The uncertainty then becomes

$$-\sum_{a \in K_i} \left(\frac{p_a}{p_{K_i}} \log \frac{p_a}{p_{K_i}} \right). \tag{7.3}$$

Then the information content of position i (decrease in uncertainty) can be formulated by use of Equations (7.2) and (7.3):

$$I'(K_i) = -\sum_{a \in \mathcal{M}} p_a \log p_a - \left(-\sum_{a \in K_i} \left(\frac{p_a}{p_{K_i}} \log \frac{p_a}{p_{K_i}} \right) \right). \tag{7.4}$$

We see that if $K_i = \mathcal{M}$, then $I'(K_i) = 0$ (no reduction in uncertainty), and for $K_i = A$ (a set of one amino acid), $I'(K_i)$ becomes equal to Equation (7.2) with $\mathcal{C} = \mathcal{M}$ (no uncertainty left).

The score of a pattern should decrease with increasing flexibility of the wildcard regions (less specific), and one way of doing this is to score a wildcard region k (where the wildcard region is specified as $x(i_k, j_k)$) as

$$-c(j_k - i_k), \tag{7.5}$$

where c is a constant. We see that fixed wildcard regions score 0.

Putting Equations (7.4) and (7.5) together gives the score of a pattern P as

$$I(P) = \sum_i I'(K_i) - c \sum_k (j_k - i_k), \tag{7.6}$$

where i runs over all positions in the components, and k over all (flexible) wildcard regions.

Example

Let (in this example) $\mathcal{M} = \{A,C,D\}$, and $p_A = p_C = p_D = \frac{1}{3}$ and a pattern $P = [\text{AC}]-\text{x}(1,3)-\text{D}$.

Then

$$I(P) = H_{\mathcal{M}} - H_{[AC]} + I(x(1,3)) + H_{\mathcal{M}} - H_D$$
$$= -3(\tfrac{1}{3}\log\tfrac{1}{3}) + 2(\tfrac{1}{2}\log\tfrac{1}{2}) - 2c - 3(\tfrac{1}{3}\log\tfrac{1}{3}) + 1\log 1$$
$$= -2\log\tfrac{1}{3} + \log\tfrac{1}{2} - 2c$$
$$= \log 4.5 - 2c.$$

$\triangle$

7.5   Generalization and Specialization

Generalization of a pattern means weakening it. It implies that if a sequence *s* matches a pattern *P*, *s* will also match every generalization of *P*. That is, if *U* is a set of sequences matching *P*, and *V* is a set of sequences matching a generalization of *P*, then *V* is a superset of *U* (if *U* and *V* are taken from the same superset). A pattern can be generalized in different ways.

- Increase the allowed set of amino acids in a position:

  A-x(2,3)-[CD] is a generalization of A-x(2,3)-C.

- Increase the flexibility:

  A-x(1,3)-[CD] is a generalization of A-x(2,3)-C.

Several special cases of these exist, for example, to remove the last position:
A-x(2,3)-C is a generalization of A-x(2,3)-C-x-[ARD].
This is the same as allowing all amino acids in the last position, or that the pattern can occur at the end of a sequence (the end of the sequence . . . . ACCD matches the generalized pattern but not the other). Another special case is to replace a component with (part of) a wildcard region: A-x(1,2)-C-D can be generalized to A-x(2,3)-D.

  Specialization is the inverse of generalization, hence A-x(2,3)-C-x-[ARD] is a specialization of A-x(2,3)-C.

7.6   Pattern Discovery: Introduction

There is a close connection between constructing *local* multiple alignments and pattern discovery. If a multiple alignment is given, it is easy to derive a pattern describing a given block. However, the pattern need not be reasonable, for example, with a lot of x's (but then the block may not be biologically informative).

```
BXA1_BOMMO   ..----------------------KR----GIVDECCLRPCSVDVLLSYC---...
BXD1_BOMMO   ..----------------------KR----GIADECCLQPCTNDVLLSYC---...
INS_HUMAN    ..----------------------KR----GIVEQCCTSICSLYQLENYCN--...
INS_SHEEP    ..----------------------KR----GIVEQCCAGVCSLYQLENYCN--...
INS_LOPPI    ..QME---------------MMVKR----GIVEQCCHRPCNIFDLQNYCN--...
IGF1_CANFA   ..-------------------------IVDECCFRSCDLRRLEMYCAPL...
IGF2_HORSE   ..------------------------R----GIVEECCFRSCDLALLETYCATP...
MPI2_LYMST   ..SSESA------------LTYLTKRQRTTNLVCECCFNYCTPDVVRKYCY--...
INL5_HUMAN   ..LWKCY.-----------KHSVMSRQ---DLQTLCCTDGCSMTDLSALC---...
RELX_PIG     ..QNEAEDKSLLELKNLGLDKHSRKKRLFRMTLSERCCQVGCTRKDTARLC---...
                                            :  **    *    :    *
```

Figure 7.2   The PROSITE motif for the insulin family is C-C-{P}-x(2)-C[STDNEKPI]-x(3)-[LIVMFS]-x(3)-C. In SWISS-PROT release number 40.7 there are 181 proteins containing this motif, of which 180 belong to the insulin family. Hence there is one false positive; in addition, six of the insulin family do not contain the motif (false negatives). Above is a block of the multiple alignment for 10 of the sequences (chosen randomly), and the conserved cysteines are marked with *. (Alignment found by CLUSTAL.)

Example

We have the following multiple alignment:

```
ADLSQWAR
TD--QA--
LDL-TVS-
WDR-QLMN
```

A pattern derived from columns 2–5 is D-x(0,2)-[QT].
   For the alignment

```
ARK
NWG
GLQ
```

we get the pattern xxx if the AACH defined in Figure 5.1 is used.   △

Figure 7.2 illustrates motif and multiple alignment for the insulin family.

In the opposite direction, given a pattern and its occurrences, a local multiple alignment can be derived, using the pattern as a 'guide'. However, the alignment in the wildcard regions is not necessarily unique.

Example

Let a pattern be D-x(0,2)-[QT], and the occurrences

```
DAXQ
DST
DT
DQQ
```

An alignment consistent with the pattern is

```
DAXQ
DS-T
D--T
DQ-Q
```

△

There is also a third related problem: find segments from each of the sequences where the pairwise mutual similarities are higher than a given threshold (or that the similarities to one of the segments are higher). Although we are mainly interested in *patterns*, all three problem variants will be included in the following. All are special cases of the more general problem: find common similarities between a set of objects.

Describing methods for pattern (similarity) discovery can be done along different axes.

1. What type of algorithm is used. There is a close connection between *local* multiple alignment and *global* multiple alignment, hence between pattern discovery and global multiple alignment. Indeed, on a coarse level the algorithmic approaches used for solving the two problems can be classified in the same way. However, what approaches are most popular differ.

   We will here use a two-way classification of the approaches: *comparison based* and *pattern driven*.

   Comparison based. First, patterns ('similarities') are found using one (or a few) sequence(s). One then investigates whether or not these patterns (or generalizations of them) also occur in other sequences.

   Pattern driven. In the simplest algorithms conceptually all patterns of a class are generated, and for each pattern one finds how many sequences match the pattern. Patterns with overlapping or neighbouring occurrences might be combined.

2. Does one want to discover patterns with occurrences in *all* the sequences, or is it enough that there are occurrences in at least k (k being a constant) of the sequences? The latter is a harder problem than if occurrences in all sequences are required.

3. How is the fitness of the patterns scored?

4. Is a method guaranteed to find the best patterns (requiring that the goodness of the patterns can be measured)?

5. What class of patterns can be discovered?

6. Is exact or approximate matching used?

7.7   Comparison-Based Methods

This approach is based on a general approach for finding common similarities for a set of objects. Pairwise comparison is the basic operation, either pairwise between objects, between an object and a similarity description (e.g. alignment), or between two similarity descriptions. A pairwise comparison method can be extended to the multiple case using one of the following four ways.

PI (pivot). Use one object as pivot, and compare it successively to all the other objects. The results are then compared so that one finds the part(s) of the pivot that has similarities in all the other objects, or in at least $k - 1$ of them.

LP (linear progressive). Start with one object, and compare successively the other objects to the result. The results are descriptions of the similarities of the objects involved.

TP (tree progressive). Compare (results) using a (possibly implicit) tree where the leaves are objects.

AA (all_by_all). All $m(m-1)/2$ pairwise comparisons are performed to find common pairwise similarities. Comparisons (e.g. intersections) are done on these results to find similarities occurring in, for example, k of the objects (a clique of similarity relations).

The different approaches are illustrated in Figure 7.3, where the objects are sequences and the similarities are patterns.

All pairwise methods can be extended to the multiple case by the PI approach. Note, however, that a description (pattern) of the similarities is not necessarily given, only the part of the pivot which has similarities with all (or $k - 1$) other objects, and the location of the similar regions in the other sequences.

If a pairwise comparison method can express each of the (one or several) best local similarities as a pattern, and if these patterns can be compared with other objects, then the pairwise method can be extended to the multiple case by using the LP approach. If patterns can also be compared with patterns to find common generalizations, the TP approach can be used.

Using AA, the result of all pairwise comparisons can be used to identify similarities between all objects (or some minimum number of the objects). It is used in Vingron and Argos (1991), where the sequence similarities are represented by dot matrices (reflecting residue similarities) and they are combined using matrix multiplication to filter out similarities not shared by all objects.

In principle, PI and AA are able to give the same results if the similarity relation used is transitive. (Transitivity exists if similarity between A and B and between B and C implies that A is similar to C.) However, the results of the PI and AA approaches depend on the more detailed aspects of the algorithm, for example, how similarities are scored and how many similarities are included in each step.

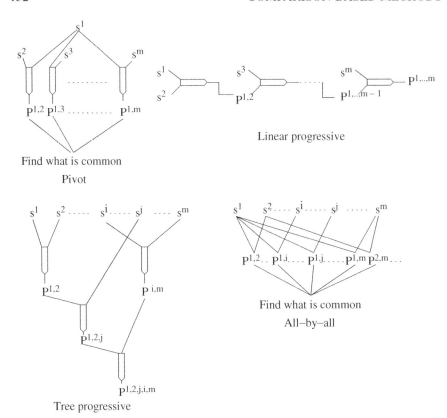

Figure 7.3   Illustrations of four principles for comparison-based pattern discovery. s^i is sequence and P^i pattern or similarity specification.

7.7.1   Pivot-based methods

One sequence is used as pivot, and all others compared with this in order to find similarities. In a naive implementation, every segment of all possible lengths from the pivot sequence is compared with each of the other sequences, and this takes an unsatisfactorily long time. Heuristics are therefore used, as in Roytberg (1992). Only segments of a certain length are used, and those in the pivot for which there exist similar segments in the other sequences are found (similarity defined by edit distance). Next the combination of such segments to form longer segments is attempted.

Example

We have the four sequences in the leftmost column below. Let the length of the segments we use in the pivot be 6, and the requirement for similarity be an edit distance less than or equal to 1. By using the first sequence as pivot we find similar segments in all the other sequences, shown in the second column. Using the last sequence as pivot we only find similar segments in the first sequence (* indicates pivot):

```
RARTGHSLKACRSWE        *  RTGHSL              RTGHSL
VCRGHSLTAADWAT            RGHSL
CRTGGHSLRTL              RTGGHSL
LLSTGHSLLGVT            STGHSL            *  STGHSL
```

In this case it is easy to generate a pattern describing the segments found with sequence one as pivot:

$$[RS]-x(0,1)-G-x(0,1)-H-S-L \text{ or } [RS]-x(0,2)-G-H-S-L \qquad \triangle$$

In general, it may not be possible to find nontrivial patterns describing the results using regular expressions. We also see that the result depends on the pivot; hence every sequence should be used as pivot.

A stronger requirement is if all the segments found for a pivot segment should also mutually satisfy the similarity constraint. This is not the case for the example above; RGHSL is, for example, not similar to an edit distance of 1 to RTGGHSL. Note that the result when using this stronger constraint is not dependent on the choice of pivot.

The problem of finding all similarities between at least k sequences can be done by using $m - k + 1$ of the sequences as pivot, one at a time, and each time finding those segments in the pivot which have similar segments in at least $k - 1$ of the other sequences.

7.7.2   Tree progressive methods

These methods start with m trees of one node each, each representing a single sequence. Then progressively two trees are joined in each cycle. The joining is done by creating a new root, and making the two trees subtrees of the root. The root is labelled with pattern(s) generalizing the patterns/sequences of its two children, if the pattern scores high enough. The iteration stops when there is only one tree, covering all sequences. The approach is therefore similar to the method described in Algorithms 4.2 and 4.3 in Chapter 4. The main task is to decide which trees to join in each cycle.

We illustrate the approach with an example from a method by Smith and Smith (1990). The scoring scheme is as defined in Figure 5.1 in Chapter 5. Each pattern will be a string over the alphabet $\mathcal{M} \cup \{a, b, c, h, i, j, k, l, p, x, g\}$, and its score is the sum of the match scores of the positions different from g (gap), minus an affine gap penalty.

Dynamic programming is used for pairwise comparison, and special considerations are taken for gap penalty (during the DP) when there are gaps in one or both of the patterns. This has as a consequence that a pattern will have identical match score when aligned with any of the 'parental' sequences from which the pattern was derived. Especially, the final pattern will have equal score when aligned to any of the original sequences.

Example

Assume three sequences $s^1 = \text{FADEFHH}$, $s^2 = \text{SAENHHL}$, $s^3 = \text{GREKFHN}$, and use the gap penalty $g_o = g_e = 1$.

- Locally aligning (s^1, s^2) results in

    ```
    ADEFHH
    A-ENHH
    ```

    and the pattern $P^{1,2} = \text{AgExHH}$, with score $(3 + (-2) + 3 + 0 + 3 + 3) = 10$ (an alternative pattern with score 10 can also be found).

- Aligning (s^1, s^3) gives $P^{1,3} = \text{plEgFH}$, with score $(2 + 1 + 3 + (-2) + 3 + 3) = 10$.

- Aligning (s^2, s^3) gives $P^{2,3} = \text{pgElxH}$, with score 7.

A node for $P^{1,2}$ is then created (we could also have chosen $P^{1,3}$), and $P^{1,2}$ must be aligned with s^3. This results in $P^{(1,2),3} = \text{pgExxH}$, with score $(2 + 0 + 3 + 0 + 0 + 3) = 8$. Aligning $P^{(1,2),3}$ with each of s^1, s^2, s^3 results for all in $P^{(1,2),3}$, which means the same score. $\triangle$

Only one pattern (the highest scoring) is kept at each internal node. Note also that g here is the same as a flexible gap $(x(0, 1))$, so the identified pattern written in PROSITE-like language is $\text{[AG]-x(0,1)-E-x(2)-H}$.

7.8   Pattern-Driven Methods: Pratt

The principle for the pattern-driven methods is

- define a pattern class,

- for each pattern in the class investigate if there are occurrences of it in all (or at least k) of the sequences.

We will illustrate these methods by describing the program Pratt (Jonassen 1997; Jonassen et al. 1995).

The class of patterns that Pratt can discover is described by

- the components can be ambiguous or unique,

- the component length is 1,

- the wildcard regions can be flexible (or of length 0, effectively giving arbitrarily long components).

Thus the patterns are of the form

$$C_1 - x(i_1, j_1) - C_2 - x(i_2, j_2) - \cdots - C_{p-1} - x(i_{p-1}, j_{p-1}) - C_p.$$

Example

Find the 'best' pattern matched by at least three of the sequences:

LKACRARVLST, LQRACVH, ACVRTLST, LACKSQLP

A pattern matched by the first and the two last is A-x(0,0)-C-x(1,2)-[RS]-x(1,1)-L, in short form A-C-x(1,2)-l-x-L. Wildcard regions of length 0 are removed, and the AACH codes of Figure 5.1 are used for the ambiguous components.

△

Maximum values can be set for several of the quantities given in the list of Section 7.3.

7.8.1 The main procedure

The main procedure of Pratt consists of three steps.

Preprocessing. Collect information from the sequences and save in special data structures.

Searching. Search for flexible patterns with unique components.

Specialization (refinement). Specialize the patterns to allow ambiguous components and more flexibility.

Ambiguous components can also be found in the search (see later), but the standard procedure is to wait until the specialization. Also, the explanation of the method will be easier if we separate the two processes.

For the following explanation we first present some definitions.

Definitions

- An *l-segment* is a segment of length l which occurs in one of the sequences $S^{1,m} = \{s^1, \ldots, s^m\}$, when the sequences are extended by $l - 1$ end symbols (+). Note therefore that an l-segment need not be a real segment of a sequence, it might have end symbols at the end. The reason for this is to be able to discover occurrences at the end of the sequences. (The standard value for l is 50.)

- An *h-pattern* is a pattern of length $\leqslant l$ with h ($1 \leqslant h \leqslant l$) components (of length 1). Recall that the length of a pattern is the maximum length that an occurrence of the pattern can have.

- B_P^l is the set of all l-segments which match the h-pattern P, with the additional requirement that the first position in the segment must match the first component in the pattern.

- $B_\emptyset^l$ is the set of l-segments matching the empty pattern, that is, all l-segments from the sequences.

△

Example

Let the sequences be MGRSTAR and LTSRRU. $B_\emptyset^4$ then contains

> MGRS, GRST, STAR, TAR+, AR++, R+++ from the first sequence, and
>
> LTSR, TSRR, SRRU, RRU+, RU++, U+++ from the second sequence.

For $P = $ TxR we get $B_P^4 = \{$TAR+,TSRR$\}$. $\triangle$

7.8.2 Preprocessing

The sequences $\{s^1, \ldots, s^n\}$ are preprocessed. One value for l is chosen (which is the maximum length of a pattern that can be discovered), and $B_\emptyset^l$ is found. Thus it is $B_\emptyset^l$ which is used in the search. *Note that $B_\emptyset^l$ is found for only one l*; therefore, we will often omit l in the further explanation if it is otherwise clear what is meant. In addition, some other data structures are filled.

7.8.3 The pattern space

The pattern space has a state for every pattern in the defined class, that means every h-pattern ($1 \leqslant h \leqslant l$), and the empty pattern. Each state contains a quadruple (Node, P, B_P, H_P), where

- Node is identifying the state,

- P is the pattern,

- B_P is the set of all l-segments matching P, and

- H_P is the set of all sequences containing a segment in B_P.

These values are not computed before they are needed.

7.8.4 Searching

A depth-first search is done in the space of all patterns of the pattern class. A node in the search tree corresponds to a state in the pattern space. The search tree is defined as

- the root contains the empty pattern (which is matched by all segments);

- a node on level k contains a pattern with k components (k-pattern);

- a child node of a node N with pattern P, contains a pattern which is an extension of P to the right with a wildcard region and one component (the root's children have empty wildcard regions);

- there is a child node of N for every allowed extension of P with a wildcard region and a component to the right.

Example

Let the pattern of node N be P, maximum flexibility 2, and maximum length of a wildcard region 3. Then child nodes of N will be created for the following patterns:

```
P-x(0,0)-A which is the same as P-A
P-x(0,1)-A
P-x(1,1)-A which is the same as P-x-A
P-x(1,2)-A
P-x(2,2)-A which is the same as P-x(2)-A
P-x(2,3)-A
P-x(3,3)-A which is the same as P-x(3)-A
P-x(0,0)-C
..............
```

For this example 7×20 nodes will be created, thus the branching factor of the tree is 140, except for the root, for which it is 20, one child for each amino acid. Note, however, that limits on the maximum length (l) and maximum total flexibility, can reduce the branching factor for nodes at higher levels. △

When traversing the search tree, for each node N the associated pattern P's occurrences (B_P) are found.

Example

Let a pattern P be A-x(1,2)-C, $l = 5$, and one of the sequences be ATACCCRS, then the segments from this sequence in B_P are

```
ATACC - containing the occurrence ATAC
ACCCR - containing the occurrence ACC
ACCCR - containing the occurrence ACCC
```

△

Algorithm 7.1 shows the searching procedure.

Algorithm 7.1. Pratt.

The search subprocedure for Pratt. DFS is first called by the main program.
All children to Node are created.

const

wm	maximum length of wildcard regions
fm	maximum flexibility of a wildcard region

var

a	amino acid
Q	a child of P

procedures

$\mathcal{F}$ a procedure for finding all occurrences of the pattern Q (B_Q)

$\mathcal{G}$ a procedure finding all sequences with occurrences of Q (H_Q)

                 ($\mathcal{F}$ and $\mathcal{G}$ is in the implementation a common procedure)

create_node   create a new node

DFS(Node, P, B_P, H_P)

in-parameters

Node   the node

P, B_P, H_P the values of Node

begin

    for *each amino acid* $a \in \mathcal{M}$ **do**

        for $i := 0$ **to** wm **do**

            for $j := i$ **to** $\max(wm, i + fm - 1)$ **do**

                $Q := P$-x(i,j)-a

                $B_Q := F(B_P, Q)$ find the occurrences of Q

                $H_Q := G(B_Q, H_P)$

                if Q *is to be extended* **then**

                    create_node(Nod1, P, Q, B_Q, H_Q)

                    DFS(Nod1, Q, B_Q, H_Q)

                elseif Q *has score high enough* **then**

                    report Q

                end

            end

        end

    end

end

The main procedure will generate the nodes for the patterns
consisting of one amino acid '

begin

    for *each amino acid* $a \in \mathcal{M}$ **do**

        $Q = a$; $B_Q := F(B_\emptyset, Q)$; $H_Q := G(B_Q, \{1, 2, \ldots, n\})$

        if Q *is to be extended* **then**

            create_node(Nod1, Q, B_Q, H_Q)

            DFS(Nod1, Q, B_Q, H_Q)

        end

    end

end

Finding B_Q, $Q(P)$ a child-pattern of P

Since $Q(P)$ is an extension of P, then B_Q is a subset of B_P. Let first $Q(P)$ be a *fixed* extension of P, $Q_a^i(P) = P$-x(i)-a. Then for each segment s in B_P we examine whether s also matches $Q_a^i(P)$. This is done by using the occurrence of P in s (always a prefix), and see if the $(i + 1)$th symbol after the occurrence is equal to a.

Example

Let

- $l = 7$,

- $P = \text{A-x(0,1)-E}$,

  $B_P = \{A E\text{RAEVA}, A E\text{ERVVH}, A C\text{ELEVH}, A E\text{ERVVR}\}$,

  where the occurrences (prefixes) of P are marked, and

- $Q_V^2(P) = \text{P-x(2)-V}$.

Then $B_{Q_V^2(P)}$ becomes $\{A E E R V\text{VH}, A C E L E V\text{H}, A E E R V V\text{R}\}$. $\triangle$

If $L_{i,a}$ is the set of all l-segments with amino acid a in position i, $B_{Q_V^2(P)}$ can be efficiently found by set operations on (subsets of) B_P and (subsets of) $L_{i,a}$. $L_{i,a}$ is found in the preprocessing.

It is now easy to see that B_Q for a *flexible* extension $Q_a^{(i,j)}(P) = \text{P-x(i,j)-a}$ ($i < j$) can be found by using the sets for the fixed extensions ($i \ldots j$), and the following recursion can be derived:

$$B_{Q_a^{(i,j)}(P)} = \bigcup_{i \leqslant k \leqslant j} B_{Q_a^k(P)} = B_{Q_a^{(i,j-1)}(P)} \cup B_{Q_a^j(P)}. \tag{7.7}$$

The sets for all extensions can therefore be calculated by dynamic programming (see Exercise 5).

Pruning the search tree

One way of pruning the search tree is to demand that each pattern should occur in at least k sequences. Therefore, each occurrence is labelled with the sequence in which it occurs. Using that H_P is the set of sequences with segments matching P, we get the following recursion:

$$H_{Q_a^{(i,j)}(P)} = H_{Q_a^{(i,j-1)}(P)} \cup H_{Q_a^j(P)}.$$

Letting H be represented by bit vectors this can be efficiently calculated using bitwise *or* operations.

If $|H_a^{(i,j)}| \geqslant k$ is not satisfied for any (i, j) (P cannot be extended to a sequence matching at least k sequences) and if P has a score above a defined threshold, then P is saved for possible later specializations.

Another way of pruning is to restrict the tree level to a constant depth.

7.8.5   Ambiguous components

In reality it is easy to implement ambiguity by extending the alphabet, for example, by using the lowercase letters for amino acid groups as in the AACH in Figure 5.1.

This can be implemented as ordinary child nodes, e.g. P-x(i,j)-b, which is P-x(i,j)-[FWY], and using

$$B_{Q_b^{(i,j)}(P)} = B_{Q_F^{(i,j)}(P)} \cup B_{Q_W^{(i,j)}(P)} \cup B_{Q_Y^{(i,j)}(P)}.$$

The standard method is, however, not to treat ambiguity in the search, but to postpone it to the specialization, by trying to replace positions in wildcard regions by ambiguous components.

7.8.6 Specialization

Two means of specialization are tried:

- replacing wildcard positions with ambiguous components,
- extending the patterns to the right (but not to the left).

The same search procedure as before is used, but now it is only searched in B_P (and not in $B_{\emptyset}$).

Example

Assume that we have a pattern P = A-x(1,2)-C-x(1)-D and B_P = {AKCQDQQ, ARVCNDN, AHTCADQ}.

If AACH is used (see Figure 5.1), the following specialized pattern is found: A-i-x(0,1)-C-x(1)-D-j, where i=[HKR] and j=[NQ]. △

Heuristics are used in the specialization step; hence it is not guaranteed that all possible ambiguous positions are found.

If a pattern P is specialized to Q, it is searched further for more specializations with both P and Q.

Pattern graph

A variant of the search program uses a *pattern graph* to restrict the number of nodes in the search tree, and thus speed up the searching time.

7.8.7 Pattern scoring

The patterns matching at least k sequences are ranked according to their information contents as defined in Section 7.4.2.

7.9 Exercises

1. An alphabet of three symbols is given, {A, C, D}, and a class of patterns that satisfies the following conditions: the components are ambiguous, the length

of each component is 2, and maximum number of components is 3. There is a maximum of two alternatives for each ambiguous position, the maximum flexibility is 2, and the maximum size of a wildcard region is 2.

  (a) How many patterns does this class contain?

  (b) CD-x(1,2)-DA is a pattern in the class. Find some generalizations and specializations of this pattern which are also in the class.

2. Let an alphabet of three symbols be {A, C, D}, the background probabilities be $p_A = p_D = \frac{1}{4}$, $p_C = \frac{1}{2}$, and the score be as defined in Section 7.4.2.

  (a) Find the score of the pattern C-x(1,3)-[AC].

  (b) The patterns C-x(1,3)-[AC] and A-x(1,3)-[AC] score equally. Some would argue that the pattern A-x(1,3)-[AC] should score higher. Explain why.

3. This problem treats the comparison-based method in Section 7.7.2. Four sequences are given:

$$s^1 = \text{ARSHIKLMRY}, \qquad s^2 = \text{WLTGHILYG},$$
$$s^3 = \text{QLSGIRLPP}, \qquad s^4 = \text{PRTAHILC}.$$

The pairwise local alignments found for (s^i, s^j) are

s^1:	RSHIKLM	s^1:	SHIKL	s^1:	RS-HIKLM
s^2:	TGHI-LY	s^3:	SGIRL	s^4:	RTAHI-LC

s^2:	LTGHIL	s^2:	TGHILY	s^3:	SGIRL
s^3:	LSGIRL	s^4:	TAHILC	s^4:	TAHIL

  (a) Write for each alignment a pattern and calculate its score by using AACH (Figure 5.1) and $g_l = 1.5 + 0.5l$. The first and the last positions of the pattern should have score > 0.

  (b) Choose the highest-scoring pattern and align it to each of the two other sequences. Score each alignment. Use gap cost as in (a) and let the score of aligning a gap against a blank be zero.

  (c) Based on the scores, decide which patterns/sequences to align next, and perform the alignment.

  (d) Do the last aligning and find the resulting pattern. Remove any subpatterns with score 0 from the start or the end. Draw the evolutionary tree corresponding to the alignment procedure.

  (e) Align the pattern that you have found to each of the original sequences and calculate the scores.

  (f) Write the pattern as a PROSITE pattern.

4. This problem treats the pattern-driven approach (Pratt). We have the alphabet {A,C,D}, $l = 5$, $d_{max} = 2$ (the maximum number of levels in the tree), the maximum length of a gap is 1 and the maximum flexibility is 2. Three sequences are given: s^1: AACDAC, s^2: CAACDA, s^3: ADACC.

    (a) Draw the search tree and find B_P for the patterns P which are on the first level of the tree and the nodes in level two with the patterns AC, AD, AxC, AxD, CxA, A-x(0,1)-C and A-x(0,1)-D.

    (b) Use the result from (a) to find the segments which match the nonunique pattern A-x(0,1)-[CD].

    (c) Two patterns (P_1, P_2) can be combined if l-segments matching P_1 have at least one residue in common with l-segments matching P_2, and the common residues are inside the pattern occurrences. Examine whether patterns AxD and CxA can be combined, when we claim that the combined pattern must be matched by at least two of the sequences, and the difference of the start positions of the occurrences must be $\leqslant 2$ in the sequences. What is the resulting pattern, and the occurrences?

5. Explain why Equation (7.7) is correct. Draw the DP matrix, showing which cells are used for calculating the value of a cell. Also make an algorithm for calculating $B_{Q_a^{(i,j)}(P)}$ and $H_{Q_a^{(i,j)}(P)}$.

7.10   Bibliographic notes

PROSITE is described in Hofmann et al. (1999). More about the exact and approximate matching of strings and sequences are in Gusfield (1997) and Stephen (1994). Scoring of patterns is discussed in Brazma et al. (1998), and examples can be found in Lawrence et al. (1993), Neuwald and Green (1994), Jonassen et al. (1995), Sewell and Durbin (1995) and Brazma et al. (1996).

Pratt, including use of pattern graphs, is described in Jonassen et al. (1995) and Jonassen (1997). A survey of deterministic patterns discovery is given in Brazma et al. (1998). Examples of comparison-based methods are to be found in Smith and Smith (1990), Vingron and Argos (1991) and Roytberg (1992). Examples of pattern-driven (or combined) methods are in Neuwald and Green (1994), Jonassen et al. (1995), Jonassen (1997) and Wolferstetter et al. (1996) (for nucleic acids). A method using minimal description language is in Brazma et al. (1996). Other methods for finding motifs are in Blanchette et al. (2000) (for orthologous sequences) and Pevzner and Sze (2000) and Pavesi et al. (2001) (for DNA sequences).

A probabilistic method for motif discovery is the Gibbs motif sampler:
        http://bayesweb.wadsworth.org/gibbs/gibbs.html
described in Lawrence et al. (1993) and Liu et al. (1995).

A method using suffix-trees for pattern discovery is in Vilo (1998).

Part II

STRUCTURE ANALYSIS

8

Structures and Structure Descriptions

Proteins perform a wide variety of functions in the cell: some of these require an active role for the protein (mobility and catalysis) while others are more passive structural roles. Proteins involved in the former tend to be soluble and globular in shape while the latter can be very long and fibrous in nature. A third class of protein is found embedded in the lipid *membrane*, where they control import and export of substances across the membrane of transmitted cellular messages. In this chapter we focus on the globular proteins which generally exist in an aqueous environment in the cells. They fold into fairly compact units and, typically, the number of amino acids are in the range from 100 to 1000.

Achieving a stable fold is important for the function of a protein, and this fold is achieved at a state with near minimum energy, called the *native state*. This stable state is marginal, in that stability is only achieved in restricted values for solvent and temperature conditions. Within this operating window, a variety of conformations can be adopted. These can range from small random (thermal) fluctuations to large movements associated with function. On folding, proteins lose energy through the formation of *hydrogen bonds* (H-bonds); therefore, the number of bonds is maximized, either between atoms in the protein, between atoms in the protein and the surrounding aqueous, or to other proteins or ligands. Only *polar* side chains can form hydrogen bonds with water; those amino acids are referred to as *hydrophilic* (a polar molecule means roughly that the molecule has a separated charge, but a polar molecule can as a whole be charged, positive or negative, or partially charged). The amino acids with nonpolar side chains are called *hydrophobic*. Put simply, the hydrophilic amino acids are *soluble* in water while the hydrophobic ones are not. Therefore, in order to maximize the number of H-bonds, it is reasonable to think this can be done by placing the polar amino acids in contact with water (aqueous environment), i.e. on the surface of the protein. The result of the hydrogen bonding is that the (globular) proteins consist of a hydrophilic surface around an hydrophobic core, and thus as a whole protein is soluble in water. The *solvent exposure* of a protein or part of a protein

Protein bioinformatics: an algorithmic approach to sequence and structure analysis
I. Eidhammer, I. Jonassen and W. R. Taylor © 2004 John Wiley & Sons, Ltd ISBN: 0-470-84839-1

(down to individual residues or atoms) is a property reflecting how large a part of the protein is exposed to the solvent. The rest are *buried*.

Another property having impact on the folding is that all amino acids have polar atoms in the backbone, the amino group (N–H) (positive) and carboxyl group (C=O) (negative). H-bonds can be formed between different residues along the backbone chain, and the way this is done is often remarkably regular. H-bonds are formed in two different ways involving links that are local or more remote in the sequence. This results in the two dominant structures called the α-helix and the β-strand, together referred to as secondary structures. There are also some minor variants.

In addition to these bonds there exist other chemical forces contributing to the folding and stability of proteins. These factors, which tend to have a secondary effect on the structure, include the formation of disulphide bridges (bonds) between the sulphur atoms found in cysteine residues. Bonds also occur when metal ions are integral parts of the structure (cofactors), such as the zinc atoms in Figure 8.3.

The folding to the native state is a global activity, in that the whole protein (or at least a large part of it) is necessary. Forces both on individual residues and between secondary structure elements (SSEs) contribute. These forces reinforce each other on different levels, resulting in the formation of compact globular structures.

As the structure of a protein is much more informative than the sequence for understanding the function and evolutionary history of the protein, comparison of structures becomes an even more important task than comparison of sequences. Hence, structure patterns and structure motifs play an important role in the analysis of proteins. While the comparison of sequences is almost always on a residue level, the comparison of structures is either on a *fine level* (mostly residue) or on a *coarse level* (mostly SSE). Consequently, structure patterns can be divided into two types, depending on the description level. These two levels serve different purposes. The fine level is used for finding important common local (spatial) similarities, such as active sites or binding sites. Fine level structure patterns can be used to make a hypothesis of the function of a protein, for example, the 'coordinate templates' for finding Ser-His-Asp catalytic triads in the serine proteinases and lipases. The coarse level is mostly used for comparing proteins on a global level, and is used for classifying the proteins into structural classes or hierarchies.

Thus there is a need for methods describing and revealing common functionally important units in (related) structures, and storing descriptions of such in databases. Figure 8.1 illustrates the different tasks and the connections between them.

Note that in structure comparison, it is not always necessary that the residues or SSEs in a pattern occur in the same sequential order in every position. As with many comparison problems, complications arise as there is neither one best way to make the comparison nor to evaluate the answer. This situation pertains in sequence comparison where, although there is an optimal alignment algorithm for pairs of sequences, its results depend on a model of sequence relatedness that is based on much less certain ground. In structure comparison, we do not even have an algorithm that guarantees an optimal answer for pairs of structures and, with the added complexity of structure, relative to sequence, the models of relatedness are correspondingly more varied.

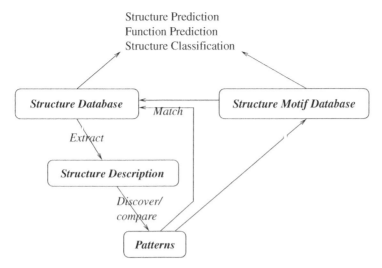

Structure Prediction
Function Prediction
Structure Classification

Figure 8.1 Protein comparison and motif discovery in a context with databases. Reproduced from Eidhammer et al. (2000) by permission of Mary Ann Liebert.

The **PDB** (Protein Data Bank, http://www.rcsb.org/index.html) at Research Collaboratory for Structural Bioinformatics (RCSB)) is the main public database for protein structures, while a number of other databases exist with related information. The PDB is an archive of experimentally determined three-dimensional structures of biological macromolecules, together with an extensive annotation, and in December 2002 it had 19 623 entries.

In this and the following chapters, descriptions and representation of structures and patterns are investigated, as well as scoring and algorithms for comparison and the problem of automatic discovery of patterns common in a set of structures.

8.1 Units of Structure Descriptions

Briefly, we can say that a protein structure consists of *elements*: atoms, residues, fragments or secondary structure elements (SSEs). A fragment is the structure of a sequence segment. A structure description can consist of *architecture*, *topology* and *properties*.

Architecture is the position of the elements, coordinates or relative positions. When the elements are atoms or residues, the architecture is sometimes called *geometry*.

Topology is the elements' order along the backbone. Usually, by the topology of a structure one means the architecture and the order of the elements.

Properties of the elements, e.g. physio-chemical properties of the residues and types or exposure of the SSEs.

The most common description is *element based*, meaning that the description has reference to each element. Further, a description can be on a *fine* (*low*) level or a *coarse* (*high*) level.

Fine: the elements are atoms or residues.

Coarse: the elements are fragments or SSEs.

Several descriptions exist for specifying the architecture and topology of proteins. For the fine level it is done by specifying coordinates, distances or torsion angles.

8.2   Coordinates

The fundamental three-dimensional (3D) structure description consists of the specification of the coordinates of each atom, as given in the PDB (see Figure 8.2). The coordinates are determined by either X-ray crystallography or by nuclear magnetic resonance (NMR). Note that the structures are 'vibrating', and that there might also be uncertainty in the determinations. Figure 8.3 shows visualizations of all the atoms of the chain in Figure 8.2.

In structure comparison it is common to let one or two atoms represent each residue, often the C_α atom. The coordinates of C_β are sometimes used, in order to include some information on the orientation of the side chains (see Figure 8.4). The side chain is alternatively represented by a 'mean' side chain atom.

8.3   Distance Matrices

A distance matrix for a structure shows the pairwise distances between elements. In this way it is a 2D representation of the 3D structure. Figure 8.5 shows a distance matrix for 1chc, where the distances between the C_α atoms are used. The distances are in angstroms and rounded to integers, and distances larger than nine are represented by a dot.

A distance matrix contains more than enough information to reconstruct the 3D structure, except for *handedness* or *chirality* (mirror images) (see Figure 8.6).

Example

In order to indicate that a distance matrix contains (more than) enough information to reconstruct the structure (except for handedness), think of a set of points, $a, b, c, d, e, \ldots$ in two dimensions, where all pairwise (Euclidean) distances are known. We want to calculate the coordinates of each point, e.g. (a_x, a_y). First consider the points a, b, c. There are six unknowns (two for each point), and three equations can be defined on these (one for each distance). However, any reference (coordinate) system can be used. We can choose the one with origin at a and the y-axis along the line $a - b$. Then only the coordinates of c are unknown. Two equations can be

```
..........
FORMUL    2    ZN     2(ZN1 ++)
HELIX    1 1    VAL     31    ARG     38  1 RIGHT-HANDED
SHEET    1 S1  3 HIS    26    CYS     29  0
SHEET    2 S1  3 TYR    18    LEU     22 -1        SER     19        PHE     28
SHEET    3 S1  3 SER    52    THR     56 -1        VAL     54        MET     20
SITE     1 S1  4 CYS     8    CYS     11 CYS    29 CYS    32
SITE     1 TWO 4 CYS    24    HIS     26 CYS    43 CYS    46
..........
                                                        Occup.  B-value
ATOM     1    N   MET    1      66.104  56.583 -35.505  1.00    0.00
ATOM     2    CA  MET    1      66.953  57.259 -36.531  1.00    0.00
ATOM     3    C   MET    1      67.370  56.262 -37.627  1.00    0.00
ATOM     4    O   MET    1      67.105  55.079 -37.531  1.00    0.00
ATOM     5    CB  MET    1      68.227  57.852 -35.867  1.00    0.00
ATOM     6    CG  MET    1      67.848  58.995 -34.899  1.00    0.00
ATOM     7    SD  MET    1      66.880  58.593 -33.421  1.00    0.00
ATOM     8    CE  MET    1      68.253  58.332 -32.269  1.00    0.00
ATOM     9   1H   MET    1      66.566  56.636 -34.574  1.00    0.00
ATOM    10   2H   MET    1      65.969  55.585 -35.765  1.00    0.00
ATOM    11   3H   MET    1      65.179  57.056 -35.460  1.00    0.00
ATOM    12    HA  MET    1      66.373  58.046 -36.989  1.00    0.00
ATOM    13   1HB  MET    1      68.894  58.236 -36.625  1.00    0.00
ATOM    14   2HB  MET    1      68.743  57.078 -35.317  1.00    0.00
ATOM    15   1HG  MET    1      67.283  59.729 -35.455  1.00    0.00
ATOM    16   2HG  MET    1      68.760  59.479 -34.578  1.00    0.00
ATOM    17   1HE  MET    1      68.880  57.524 -32.617  1.00    0.00
ATOM    18   2HE  MET    1      67.847  58.062 -31.306  1.00    0.00
ATOM    19   3HE  MET    1      68.822  59.245 -32.159  1.00    0.00
ATOM    20    N   ALA    2      68.018  56.788 -38.637  1.00    0.00
ATOM    21    CA  ALA    2      68.498  55.965 -39.793  1.00    0.00
ATOM    22    C   ALA    2      70.028  56.064 -39.893  1.00    0.00
ATOM    23    O   ALA    2      70.561  56.995 -40.466  1.00    0.00
ATOM    24    CB  ALA    2      67.833  56.491 -41.076  1.00    0.00
```

Figure 8.2 The coordinates of the atoms of the first residue (Methionine), and some of the second (Alanine) for the PDB entry 1chc, a RING-finger domain. Occupancy and B-value indicate some of the uncertainties in the results of NMR. Occupancy can be used to represent alternative conformations of side chains, usually 1.00, which represent no alternative conformation. B-value is a temperature factor: an indication of uncertainty in this atom's position due to disorder or thermal vibrations. A high B-value means high disorder. For this structure all B-values are set to 0 (means not given).

defined on these unknowns (using the distances dist(a, c) and dist(b, c)). Solving the equations will result in c_y being unique, and $c_x = \pm k$, where k is a constant. The coordinates of d can be found in the same way, and by use of the distance dist(c, d) it can be found if c and d are on the same side of the y-axis or on the opposite side. Then e's coordinates can be found, and also its side relative to c. We then see that the distance dist(d, e) is superfluous. It follows that for each following atom it is enough with three distances. Hence, for a description with n elements in two dimensions it is enough to have $(1 + 2) + 3(n - 3) = 3n - 6$ distances. $\triangle$

For three dimensions, the minimum number of distances required is $4n - 10$ (Exercise 1). When going in the opposite direction, from distances to coordinates, the number of coordinates we have to calculate is $3n - 6$: for n atoms there are $3n$ coordinates, but three of the first atom, two of the second and one of the third can be set

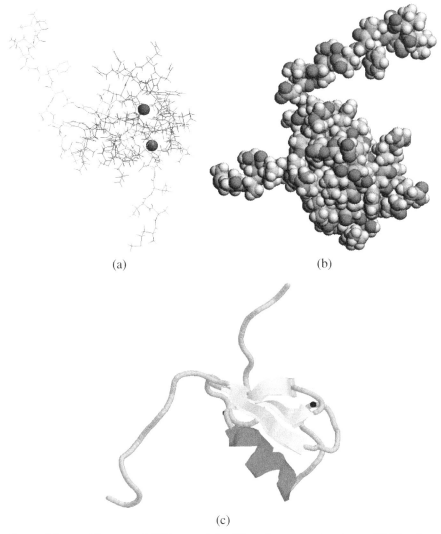

Figure 8.3 (a) All atoms of PDB entry 1chc. The spheres are zinc atoms. (b) The atoms as spheres. (c) The SSEs are drawn, one helix and a sheet of three strands. (Drawn using RASMOL.)

arbitrarily, so there are $3n - 6$ coordinates to decide. We see that the required number of distances is not the same as the required number of coordinates (remember that six coordinates are required for calculating the distance between two points).

The reconstruction of coordinates from distances can be done by *distance geometry methods*. We here give a very brief overview of the method. When all the distances between the n atoms (points) are known, we can calculate the vectors from an origin

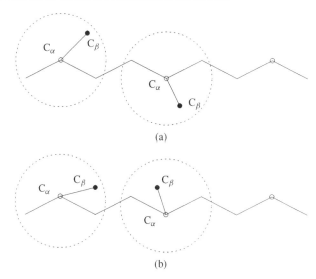

(a)

(b)

Figure 8.4 The figure illustrates the use of C_α and C_β atoms for comparison. Two structures (A, B) which are similar when only the C_α atoms are used might not be similar when the C_β atoms are taken into account. Using C_β distances is more sensitive to differences in orientations of the side chains.

to all points in the following way: first calculate

$$g_{ij} = \tfrac{1}{2}[d_{io}^2 + d_{jo}^2 - d_{ij}^2] \quad \text{for all } (ij),$$

where d_{ij} is the distance between points i and j, and d_{io} is the distance between point i and the origin (which can be found by use of Lagrange's theorem). Then define the $(N \times N)$ matrix G of the elements $\{g_{ij}\}$, and the eigenvectors of G are the required atom points.

The most common source for distance information is NMR experiments, which measure distances between H atoms. For protein structures there are many constraints between the distances. By use of these structural constraints and the measured values, more distances can be found. (It can, for example, be formulated as a type of informatics problems called the constraint satisfaction problem (CSP).) The constraints that hold for real structures can also be used in structure prediction. Some distances are proposed, and by using the constraints (as for NMR experiments), further distances can be deduced. It can then be investigated whether the resulting set of distances can represent a real structure, and additional knowledge about the protein can be used to assess the proposed structure. In this way the initially proposed distances can be iteratively refined.

For illustration, the possible distance values in a distance matrix are often divided into intervals, each interval represented by grey levels or colours. Let v be the distance value in angstroms, then the cells satisfying $0 \leqslant v < 4$ can be coloured red, those satisfying $4 \leqslant v < 8$ light red, etc.

```
    123456789A123456789B123456789C123456789D123456789E123456789
 1  0469.....................................................
 2  40469....................................................
 3  640479...................................................
 4  9640479..................................................
 5  .9740469.....778.........9...............................
 6  ..9740468..8.869.9......8779.............................
 7  ...9640469857669.........989............................
 8  ....964045547769.........9857..88.........99............
 9  .....8640466..9..........7...98..9.....967..............
10  ......9540469.............89..76.97.......79............
11  ......856404688..........98..89.........................
12  .....8546640456..........99.............................
13  ......77.9640468.........9..............................
14  ....7867..8540469........8...............................
15  ...76669.866404679.......8658.9.........................
16  ...8999....864047........968..................9..
17  ...........964047.......98559.................8868.
18  .....9........774046...9564579...............98676..
19  .........9.740469..8855645788...............66578..
20  ..........64046985546888.99................85578...
21  ............6404765678..9.78..9.....8....9.676579....
22  ............964034579....................9966568.....
23  ............9730469.....................897779......
24  ............86440469................9.98.7.89........
25  ............8555640469.....................988.....
26  .....8..9....98567964046...9.........9.89..........
27  ...9798......8.955479.96404789.9.............999....
28  .....7857899..6986568...9640477669..................
29  ....997.98998565468......7404657................9....
30  ...........885548......87404689...............99698..
31  ..........97589......9764046579...............798....
32  ......8978...9..97.......6564046679.................
33  ...8869......897....9967864045569....88.......9.....
34  ...........898......9.9564045569.9........9.97.....
35  ........9.................7654046679...............
36  .....97..................97555046667.88............
37  ...........9...........9656404544768....88.........
38  .......................966640466..................
39  .......................97654046...................
40  ......................9646404788..................
41  ......................9.746640468..8867.........
42  .......................7..74047874457...........
43  ........9.......8..9.9......8..86..86404554577.........
44  ....967...............8..88..887404669...........
45  .....979.........9.8..........8540469...........
46  ..............8.9..........75640469...........
47  ...........................8446640479...........
48  ............9987..........84599640469..........
49  ...........99........8...657..974047.......
50  ............6678.........9..8...777..9640468.....
51  ............7679................974047.....
52  ........8657.9........9..........64047.....
53  ......965569.8.9..97.97..........874047....
54  ........86578..8.9..99...........740469..
55  ........86579.....9.968...........740469.
56  ........8778.........9.............64047.
57  ........9668.........8.............964047
58  ........8.........................97404
59  ...................................740
```

Figure 8.5 A distance matrix for the first 59 residues of PDB entry 1chc, where each residue is represented by its C_α atom. The distances are rounded to integers, and distances larger than nine are represented by dots.

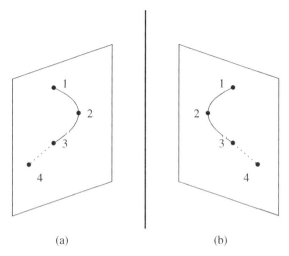

Figure 8.6 A structure A and its mirror image B. They will get equal distance matrices, but trying to superpose B onto A will fail. Assume the plane drawn goes through points 1, 2, 3. Placing points 1, 2, 3 from B onto those from A will result in the two residues 4 being on opposite sides of the plane.

8.4 Torsion Angles

Geometrically, the backbone chain of a protein is a succession of points (atoms) in space:

$$\cdots C_{i-1}=N_i-C_i^\alpha-C_i=N_{i+1}-C_{i+1}^\alpha-C_{i+1},\cdots$$

where '−' means a single bond and '=' a double bond.

The distance between the successive atoms on the backbone is approximately constant. They are specified by Schulz and Schirmer (1979) as 1.47 for N–C^α, 1.53 for C^α–C and 1.32 for C–N (in angstroms). The angles between the two bonds of each atom are also approximately equal. The only freedom the proteins have in folding is to rotate around the bonds (in the backbone and side chains). Generally, three points define a plane, and due to the constant values the position of the fourth can be defined by an angle relative to this plane. This angle is the rotational angle between the second and third of the three points. Consider atom C_i in Figure 8.7. The angle δ and the distance C_i^α–C_i are constants, hence the only freedom C_i has relative to the plane of the points $(C_{i-1}, N_i, C_i^\alpha)$ is the angle ϕ_i 'around' the bond (N_i, C_i^α) (the angle of the line (C_i^α, C_i) to the plane of $(C_{i-1}, N_i, C_i^\alpha)$).

Positive values of the angles are defined to be in the clockwise direction. We can then define the angles showing the conformational freedom as in Figure 8.7.

- The only freedom C_i has to be placed relative to the plane of $(C_{i-1}, N_i, C_i^\alpha)$ is to rotate 'around' the bond (N_i, C_i^α). This angle is denoted ϕ_i.

- The freedom N_{i+1} has relative its three preceding atoms is to rotate around the bond (C_i^α, C_i). This angle is denoted ψ_i.

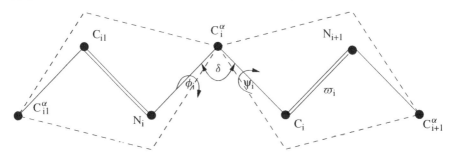

Figure 8.7   Figure illustrating the torsion angles.

- The freedom C_{i+1}^{α} has relative to its three preceding atoms is to rotate around the bond (C_i, N_{i+1}). This is the peptide bond, which is effectively a double bound and not free to rotate. The two possibilities are (approximately) 0, called *cis* and (approximately) 180, called *trans*. Almost all such bonds are trans; the only cis bonds appear between a proline and the preceding residue.

We now see how we can successively calculate the coordinates of the backbone chain by using these angles. The 3D structure of a protein is therefore completely specified by the torsion angles and the rotational angles of all the side chains. Note also that we can 'go in the opposite direction': the torsion angles can be calculated from the coordinates.

The rotation around the single bonds is only restricted by possible steric collisions in the conformation. A Ramachandran plot (named after the Indian biophysicist, G. N. Ramachandran) is a plot where the angles (ϕ, ψ) are plotted. In Figure 8.8 the sterically 'allowed' values of (ϕ, ψ) are plotted along with the distribution of the angles for the PDB structure 1chc. The preferred rotational angles for the side chains are called *rotamers*.

8.5   Coarse Level Description

The overall 3D shape of a protein, the *fold*, remains essentially stable during evolution, even if the sequence undergoes large changes. This makes it useful to classify proteins on the basis of their structures. Global classification on the residue level would involve a large number of parameters in comparing structures. However, the overall shape can be described by the organization (*architecture* and *topology*) of the SSEs, thus comparison of the overall structures is mainly done on the level of SSEs. In Figure 8.3 one of the models is shown to emphasize the SSEs.

8.5.1   Line segments (sticks)

Before comparison of the structures, the SSEs must be found and represented. A common way to represent SSEs is by *line segments* (also called *sticks*). The most

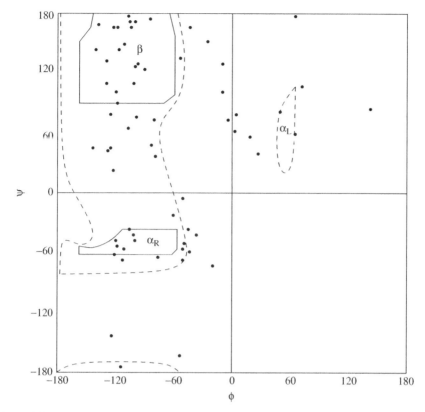

Figure 8.8 A Ramachandran plot indicating the sterically 'allowed' values of (ϕ, ψ) (dashed lines) and the areas for β-strands, right-handed α-helices and left-handed α-helices. The distribution of the angles for the PDB structure 1chc is also plotted. Note that not all residues have angles inside the 'allowed' regions.

straightforward method to determine a stick is to fit a line to the C_α atom of each residue of the SSE (or of a fragment) by a least-squares method (see Figure 8.9).

8.5.2 Ellipsoid

An alternative description to the sticks is to describe the SSEs (or any other fragments) as ellipsoids. An ellipsoid has three inertial axes, and the width of the elements can also be described explicitly. This description can be used to capture the SSEs of a structure (see Section 8.6), where the longest axis corresponds to the secondary structure line segment (stick) representation. The ellipsoid description can also be used for proteins as a whole or for domains. A method for determining the axes and examples is given in Taylor et al. (1983).

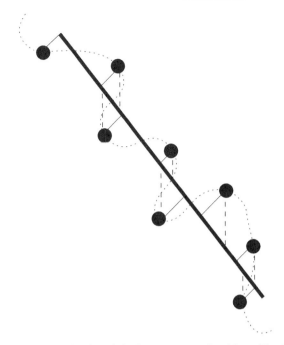

Figure 8.9 A least-squares fit of a stick (line) to a set of residues. The fitting is over the solid lines, which have to make angles of 90° to the stick. Note that this is a variant of the more common problem where one has to minimize the sum of the squares of the lengths of the vertical dashed lines. (Both problems can be solved by constructing two equations by use of partial derivation.)

8.5.3   Helices

The main SSEs, helices and strands, are formed by hydrogen bonds. Let Hbond(i, j) mean that there is a hydrogen bond between the C=O group of residue i and the N–H group of residue j. Hbond is thus a logical function which is *true* when there is an H-bond between the residues given as its parameters.

Helices are formed by hydrogen bonds between residues in the same helix, as shown in Figure B.6. Three different types of helices exist.

α-**helix** is made by successive hydrogen bonds:

Hbond($i, i + 4$), Hbond($i + 1, i + 5$), ...

The average length is 10 residues. This is by far the most common helix.

3_{10}-**helix** is made by successive hydrogen bonds:

Hbond($i, i + 3$), Hbond($i + 1, i + 4$), ...

π-**helix** is made by successive hydrogen bonds:

Hbond($i, i + 5$), Hbond($i + 1, i + 6$), ...

They are very rare in proteins.

The bonds forming helices restrict the torsion angles, and the idealized angles for 'geometrically correct' α-helix are $\phi = -57.8$ and $\psi = -47.0$. However, the real angles usually deviate from these; the regions for these are illustrated in the Ramachandran diagram (Figure 8.8).

8.5.4 Strands and sheets

Strands and sheets are formed by successive hydrogen bonds between residues which can be far apart in sequence, as shown in Figure B.7. The backbone hydrogen-bonding groups (N–H and O=C) are in the plane of the sheet, with the bonding groups from successive residues pointing in opposite directions.

Let residue i be in one strand, and residue j in another. Then the bonding of the two strands can be either parallel or antiparallel.

- Parallel bonding is formed by each residue forming hydrogen bonds to two residues on the other strand, separated by a residue in the sequence. This means successive hydrogen bonds:

[Hbond((i, j), Hbond($(j, i+2)$)], [Hbond($(i+2, j+2)$), Hbond($(j+2, i+4)$)], ...

- Antiparallel bonding is formed by each residue forming two hydrogen bonds with a single residue on the other strand. This means successive hydrogen bonds:

[Hbond((j, i), Hbond((i, j))], [Hbond($(j-2, i+2)$), Hbond($(i+2, j-2)$)], ...

Sheets can be parallel, antiparallel or mixed (with both parallel and antiparallel bondings).

The idealized strand satisfying these constraints can be thought of as a helix with two residues per turn, with torsion angles of approximately $\phi = -120$ and $\psi = +120$ (see Figure 8.8).

8.5.5 Topology of Protein Structure (TOPS)

TOPS *cartoons* are a simplified way of describing protein structures in two dimensions (Flores et al. 1994). Only the SSEs are shown, with each alpha-helix as a circle and each beta-strand as a triangle. Loops are shown by lines connecting the two SSEs. The direction of the SSEs relative to the protein fold are shown through the connecting loops: if the connection is drawn to the centre of an SSE, it points downwards into the plane; if the connection is drawn to the boundary of an SSE, it points upwards, out of the plane. For beta-strands, the direction is also shown more directly: an upper triangle points upwards and vice versa. Figure 8.10 shows the structure 1chc using TOPS cartoons.

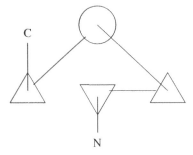

Figure 8.10   TOPS cartoons for the structure 1chc.

A TOPS *diagram* is a formalization of a TOPS cartoon, containing more information, it also describes the hydrogen bonds and the chirality between the SSEs. TOPS
diagrams are used for both comparison of structures and for finding common patterns
in a set of structures.

8.6   Identifying the SSEs

Irregularities in real structures mean that the identification of the SSEs is not an easy
problem. The constraints above are not precise enough for a unique definition. It is
especially at the ends of the elements that the identification is problematic, i.e. to
exactly identify where an SSE starts and stops.

However, there does not exist a precise universal definition for SSEs; the few
definitions that exist are biased according to the author's view. On the other hand,
one wants a consistent automatic and 'correct' method to identify SSEs, and some
automatic methods for identification do exist. Three different tools are mainly used:
angle plots, distance matrices and hydrogen bonds. Plots of the torsion angles are least
used, mostly because the helices and strands can have angles that differ considerably
from the typical values.

8.6.1   Use of distance matrices

Distance matrices can be useful, either manually or automatically, to indicate where
there can be SSEs. For idealized α-helices, the distances between the C_α atoms from
the start of the helix can be roughly calculated to be 3.8, 5.4, 5.1, 6.3, 8.7, 9.9, 10.6,
12.5, ... These distances are found by using an idealized angle pair for α-helices
and the distances between the backbone atoms (Section 8.4). Real helices usually
deviates from these due to irregularities, as shown in Table 8.1(a).

In a distance matrix a helix will turn up as an area of small distances along the main
diagonal as shown in Figure 8.5, where the area of distances less than 10 is widest
where the helix is (residues 31–38).

Table 8.1 (a) Some distances in the helix (residues 31–38) of PDB entry 1chc.
(b) Some distances in a strand (residues 18–22).

(a)

	31	32	33	34	35	36	37
31	0.0	3.9	5.5	5.1	7.1	9.3	10.4
32	3.9	0.0	3.8	5.5	5.8	7.0	9.4
33	5.5	3.8	0.0	3.8	5.3	4.8	6.3
34	5.1	5.5	3.8	0.0	3.8	5.4	5.4

(b)

	18	19	20	21	22
18	0.0	3.8	6.2	9.5	12.2
19	3.8	0.0	3.8	6.3	9.2
20	6.2	3.8	0.0	3.8	6.1
21	9.5	6.3	3.8	0.0	3.8
22	12.2	9.2	6.1	3.8	0.0

For an idealized β-strand the successive distances from a residue i can be calculated to be 3.8, 6.6, 10.3, 13.5, 16.9, ... Again, real strands deviate from these, as shown in Table 8.1(b).

It is sometimes possible to recognize adjacent strands by looking around the diagonal of a distance matrix, but as the distances grow, the strand interactions appear further away from the diagonal in the distance matrix (Figure 8.5). However, it is often possible to detect the connections between the strands in a sheet. These occur as areas of small distances around local subdiagonals and anti-subdiagonals. Parallel sheets appear as areas around subdiagonals, antiparallel as areas around anti-subdiagonals. The structure in Figure 8.5 has strands in residues 18–22 (β1), 26–29 (β2) and 52–56 (β3), with antiparallel connections between β1 and β2 and between β1 and β3. These appear clearly in the distance matrix. The other areas around anti-subdiagonals, crossing the main diagonal, are loops (see Figure 8.3).

8.6.2 Define Secondary Structure of Proteins (DSSP)

The most commonly used program to identify (define) SSEs from structures is probably Define Secondary Structure of Proteins (DSSP) by Kabsch and Sander (1983), which is mainly based on H-bonding patterns.

DSSP identifies both the SSEs and solvent exposure of proteins. Since it is mainly based on H-bond patterns, there must exist a method for identifying the H-bonds.

Determining H-bonds

Hydrogen bonds in proteins can be described by electrostatic models, and a good H-bond has a binding energy of about -3 kcal mol^{-1}. However, there is no generally correct H-bond definition, and any such definition is empirically tailored to a particular

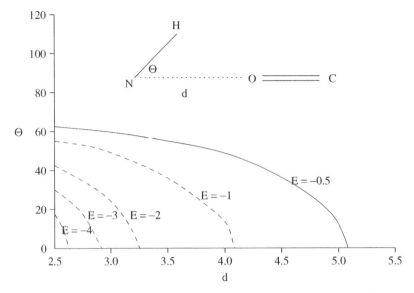

Figure 8.11   The H-bond between peptide units is described here by the dominant electrostatic part E of the H-bond energy, drawn in contours for different values of E as a function of the distance d and the alignment angle θ. An ideal H-bond has $d = 2.9$ Å, $\theta = 0$, and $E = -3.0$ kcal mol^{-1}. We assume an H-bond for E up to -0.5 kcal mol^{-1} (solid line). Thus, a misalignment of up to 63° is allowed at the ideal length; an N–O distance of up to $d = 5.2$ Å is allowed for perfect alignment. This definition of the H-bond is particularly simple and physically meaningful. It is more general than the historical definition of hydrogen 'bond' and could be called the polar interaction. Reproduced from Kabsch and Sander (1983) by the permission of Wiley-VCH.

purpose. DSSP allows for a rather large deviation from the ideal case, in that it lets binding energies of less than -0.5 kcal mol^{-1} define a bond. But instead of using this one-parameter model, DSSP uses a geometrical description of the bonding in terms of one distance and one angle, and finds the correspondence between the values of these parameters and the values of the one-parameter model. The model and the correspondence to the one-parameter energy model are shown in Figure 8.11.

Defining SSEs

A minimal helix of length n ($n = 3, 4, 5$) from residue i to residue $i + n - 1$ is defined by Hbond($i - 1, i + n - 1$) and Hbond($i, i + n$) as illustrated in Figure 8.12. Longer helices are defined by overlaps of minimal helices (see Figure 8.12). Note that nothing is, for example, required about the H-bond state of residue $i + 1$. We see that fragments can be defined as helices without all the involved residues satisfying the H-bond criteria.

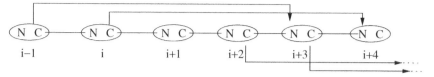

Figure 8.12   Illustration of the definition of the minimal helix for $n = 4$ (α-helix).

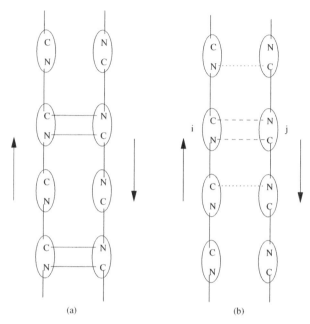

Figure 8.13   Illustration of the definition of antiparallel bridge used in DSSP. (a) The general definition of antiparallel sheet; H-bonds are shown by solid lines. (b) The definition of antiparallel bridge used in DSSP; the two alternative hydrogen bonds for residues (i, j) are drawn with dotted and dashed lines, respectively.

To determine the strands the concept of the *bridge* is defined:

$$\text{Parallel\_bridge}(i, j) := [\text{Hbond}(i - 1, j) \text{ and Hbond}(j, i + 1)]$$
$$\text{or } [\text{Hbond}(j - 1, i) \text{ and Hbond}(i, j + 1)],$$
$$\text{Antiparallel\_bridge}(i, j) := [\text{Hbond}(i, j) \text{ and Hbond}(j, i)]$$
$$\text{or } [\text{Hbond}(i - 1, j + 1) \text{ and Hbond}(j - 1, i + 1)].$$

Figure 8.13 illustrates the definition for the antiparallel bridge. Consecutive bridges form strands, one strand involving residue i and one involving residue j. Strands then form sheets. Irregularities in the strands are allowed by explicit definition of strands with bulges. Two strands satisfying Antiparallel_bridge$(i - 2, j + 5)$ and Antiparallel_bridge(i, j) will, for example, form a bulge in strand 'j'.

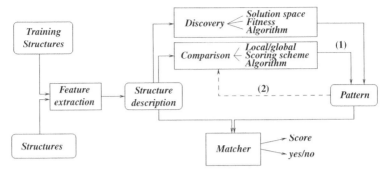

Figure 8.14 Overview of the process of discovering patterns, either by the direct discovery method or using a comparison algorithm and pattern matching. The edge marked (1) indicates that the result of a comparison method may need some further processing in order to be expressed as one or several patterns. Furthermore, if the result of comparing two structure descriptions can be represented as a pattern and then compared with another structure (or another pattern), the edge (2) can be used to find patterns matching progressively larger sets of structures. Note that 'Matcher' and 'Comparison' are very closely related and may in fact be implemented using the same method. Reproduced from Eidhammer et al. (2000) by permission of Mary Ann Liebert.

8.7 Structure Comparison

Structures are compared for the same reasons as sequences are compared: to find homologous proteins, for classification reasons, and for the discovery of motifs (common substructures), in this case *structure motifs*. Comparing structures can reveal relations that are not possible to identify using sequences alone. Also, as explained at the beginning of this chapter, comparing structures is a more complicated problem than comparing sequences, and a larger variety of methods exist. An overview of the main tasks involved in structure comparison, especially for discovering structure motifs and matching new structures against them, is given in Figure 8.14.

The different steps involve the following.

Feature extraction. In this step the features to be used in the comparison of the structures or in the pattern discovery method are extracted. This might include comprehensive computing, e.g. assigning secondary structures to the residues. Not all methods perform this step explicitly, but all take into account a well-defined set of structural features and conceptually this can be seen as if they first extract the relevant features and then subsequently work on the resulting structure descriptions.

Comparison. This takes as input a pair of structure descriptions (or a pair of description/pattern) and finds (local or global) similarities between the two, optimizing a similarity measure and outputting a score. The similarity may also be represented as a pattern.

Discovery. Patterns matching many or all of the input structures are found. The patterns are chosen from a solution space so that their fitness with respect to the input structures are as high as possible.

Matcher. This takes as input one pattern and one structure and evaluates the match between the two; the output is 'yes' or 'no' if the pattern is deterministic or a score if the pattern is probabilistic. Note the similarity between the matching and comparison operations; they may sometimes be implemented using the same method.

The steps involved in these operations will be discussed in more detail in the following chapters. The extension from pairwise alignment to multiple alignment can be done using the same approaches as for sequences, described in Chapter 4.

8.7.1 Structure descriptions for comparison

When performing structure comparison one must first decide on which structure level similarities are sought (e.g. atom group, residue, secondary structures). Also, one needs to decide whether the similarities should require sequence order to be preserved, a reasonable requirement if we assume that the proteins are evolutionarily related, but not if common features have arisen independently. The structure description to be used as input to the comparison or pattern discovery algorithms should contain only the features which we would like to compare and/or to describe as patterns. As already mentioned, a structure description will consist of architecture (geometry), topology and properties, and for each of these we need to decide which features to include.

In the framework used, patterns are to be found from structure descriptions so that they represent features common to a set of such structure descriptions. Patterns will therefore be generalizations of structure descriptions, and are limited to features included in these.

In order to provide the comparison (pattern discovery) algorithms with a good starting point, the structure descriptions should ideally satisfy the following properties.

1. *Invariant to trivial changes*, such as translation and rotation.

2. *Robust*—the description should not change drastically due to minor changes in the structure. This is important, since the determination of structures can contain errors. Small errors should result in similar structure descriptions, for example, the SSEs found at the same positions.

3. *Similar* structures should get similar descriptions. Note that this is different from the point above, since it refers to different proteins (or chains). This point is important for classification.

4. *Different* structures should get different descriptions.

Loosely, the definition of 'similar' and 'different' will depend on what aspects of protein structure one wants to compare or capture in a pattern. For example, if the comparison is to be done at the level of packing of secondary structure element, the descriptions of structures with similar packings should be similar.

A natural way to describe a complex object like a protein structure is to break it into pieces (units) and to describe each unit separately and (most often) the relationship between the units. As already noted, the natural structure of elements such as *atoms, residues, fragments, SSEs* are used as the basic units. These descriptions make use of the *element class, property* and *relation*.

Element class. The level of the description varies: atom (group), residue, backbone fragment and secondary structure element.

Property. This is used for specifying the properties of each element, such as three-dimensional coordinates, physico-chemical properties, amino acid type, secondary structure type, curvature and torsion.

Relation. This is used for describing the relation between the elements. In practice, the relations are binary, such as geometrical distances, difference in orientation and bonding.

Space-based description

We call the description above, where it is based on elements, the *element-based* description. Another type is occasionally used, which we call *space based*. In this type of description the space in which the structure is located is divided into (possibly overlapping) geometrically defined cells, e.g. by using a grid or shells around centre points. Figure 8.15 illustrates a space-based description using two dimensions, where the space is divided into squares (in three dimensions it could be cubes). For each square some values are calculated, for example, the number of residues in the square, the number of hydrophobic residues, how many belong to each SSE type, etc. Comparing structures using space-based descriptions involves comparing values in two sets of cells. The space-based description is sometimes called statistical description. In the following discussion we will mainly concentrate on the element-based description.

The distinction between element-based and space-based descriptions may seem a bit artificial. One difference is that relation between elements (distance, direction, etc.) is often explicitly given in element-based descriptions, and that the sequential order might be important. The sequential order is always disregarded in space-based descriptions.

Substructure descriptions

Some methods divide the structures into substructures, describe each substructure, and then find (local) patterns common to substructures from different structures. This initial step is often followed by a step where the identified local patterns are combined to form larger patterns. For example, a substructure can be defined to consist of elements where the distance between any two elements is below a given threshold. Figure 8.16 illustrates substructures.

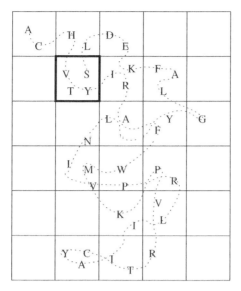

Figure 8.15 Illustration in two dimensions of space-based structure description. The space is divided into squares. The cell marked with thick lines contains three polar residues.

8.7.2 Structure representation

For each protein, the selected features are represented in a data structure. Typically, each unit (element or cell) is represented separately and combined to form a representation of the whole structure. For example, each residue can be represented by its coordinates and amino acid, and the whole protein structure can then be represented as a list (ordered) or set (unordered) of residue representations.

We can group the different representations into five groups.

Strings. It is possible to represent structural features as one or several strings, so that methods for discovering patterns in sequences can be used. One technique is to let residue i be represented by a letter reflecting the C_α's position relative to the positions of the C_α of residue $i - 2, i - 1, i + 1$. Since the letters in the positions are not independent, these representations are appropriate only for matching (consecutive) substrings. Let a string representation be $...ACDGHI...$. If the residue represented by G is deleted, then the surrounding residues letters might have to be changed.

List of unit descriptions. For example, extract one or several coordinates per residue: the coordinates of the C_α atom, the mean coordinates of the side chain, two pseudo-atoms, or unit vectors (the vectors between succeeding C_α atoms). In addition to the coordinates, each unit can have associated additional properties, such as amino acid type, physio-chemical properties, degree of burial, SSE status, etc.

Set of unit descriptions. Same as list, but the units are unordered.

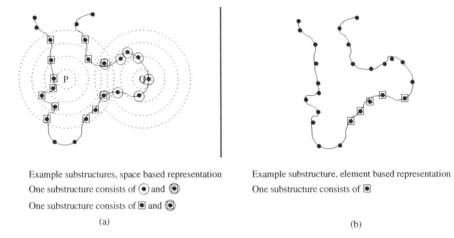

Example substructures, space based representation
One substructure consists of (•) and (⊡)
One substructure consists of ⊡ and (⊡)

(a)

Example substructure, element based representation
One substructure consists of ⊡

(b)

Figure 8.16 Substructures illustrated in two dimensions. The structures are shown by solid curves, the dots represent elements. (a) Two substructures in a space-based description, one is the part of the structure falling in the shells around P, the other is the part falling in the shells around Q. (b) A substructure in an element-based description, where a substructure here is a fragment (succeeding residues).

Graphs. A labelled graph can be used to represent all element-based feature descriptions, nodes representing the elements and the edges represent the relations. For instance, identify SSEs as nodes, and label the edges with the distance and angular relationships between the SSEs of the nodes, or label them with the type of parallelism.

Feature arrays. Features of the structure can be represented by fixed-length arrays. This can be used for both space- and element-based description, for example, in a space-based description let each cell be described by an array summing up the properties of its residues.

External/internal representations

For reason of efficiency most of the methods use *internal* representation of the structures, hidden from the user (who only sees the external representation). Internal representations can, for example, facilitate more direct comparison of spatial similarity. For instance, in the 'double dynamic programming' method, a local coordinate system is constructed for each residue into which the remaining residues' coordinates are transformed. Then, assuming a particular residue pair is aligned, the spatial similarity of the remaining residues can be assessed and scores assigned to be used in a (lower level) dynamic programming alignment step (see Chapter 9.3).

The same approach can be used in geometric hashing; a local coordinate system is constructed for each residue, but normally in geometric hashing the coordinates of only the neighbouring residues are transformed. Then for a pair of residues (one from each structure), their spatial neighbourhoods can be compared simply by counting the number of neighbouring residues with similar positions (in the respective local coordinate systems). Special hash tables are used to speed up the computations (see Chapter 10.1).

Another approach, used for example in the SPratt method (Chapter 13.5), is to represent the spatial neighbourhood of each residue by a string of the spatially close residues (having spatial distance below some threshold). The strings do not contain information about the exact geometry of the neighbours, but do give the sequence order (along the backbone) of the neighbouring residues. Patterns shared by sets of these strings can be discovered using efficient sequence pattern discovery methods.

The external representations can be generalized to be useful for pattern descriptions, which is not the case for the internal representations. When algorithms using an internal representation are to present the result to the user as a pattern, it must be described in another (external) form.

8.8   Framework for Pairwise Structure Comparison

The fundamental operation for structure comparison is the pairwise comparison. The most natural way to compare two objects, each of which is represented by a collection of elements, is to try to find element correspondences between the two. More formally, for two objects A and B having elements $A_1, A_2, \ldots, A_m$ and $B_1, B_2, \ldots, B_n$, respectively, an *equivalence* is defined as a set of pairs $E(A, B) = (A_{i_1}, B_{j_1}), (A_{i_2}, B_{j_2}), \ldots, (A_{i_r}, B_{j_r})$. The equivalence is called an *alignment* if the elements of A and B are ordered and if the pairs in $E(A, B)$ are collinear, i.e. if $i_1 < i_2 < \cdots < i_r$ and $j_1 < j_2 < \cdots < j_r$.

Many different algorithms exist which, given two structure descriptions and a scoring scheme, aim to find the equivalences giving the highest score. A schematic overview of the steps involved is given in Figure 8.17. The general problem is NP-hard; therefore, some simplifications have to be made, either in the search or in the scoring.

Even if we restrict the comparison to finding the best alignment, dynamic programming (as used for sequences) cannot be used. Let $\mathcal{A}$ be the first columns in the best *sequence* alignment of (A, B), and let A_i, B_j be the last elements of A and B in $\mathcal{A}$. Dynamic programming can be used since the best sequence alignment of $A_1, \ldots, A_i$ and $B_1, \ldots, B_j$ is $\mathcal{A}$. This follows from the fact that the scoring of a matched pair of elements is independent of the other matched pairs, and inserted gaps can be penalized independently. This is not the case for structure alignment. For the example above, if $\mathcal{A}^*$ is the first column in the best *structure* alignment, and A_i, B_j are the last elements of $\mathcal{A}^*$, the best structure alignment of $A_1, \ldots, A_i$ and $B_1, \ldots, B_j$ is not necessarily $\mathcal{A}^*$. This follows from the fact that a score for aligning (structurally) an element A_k with an element B_l is not independent of how the other elements are aligned. We will illustrate this by measuring how well two structures are aligned: that is, how well they can be 'fitted' together when they are placed on top of each other, allowing for rotation and translation of the structures.

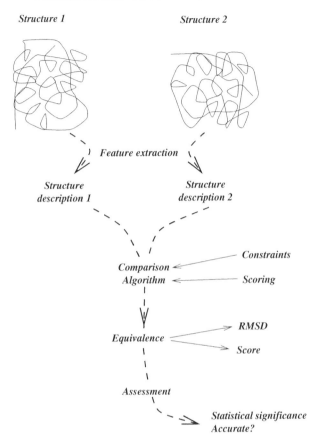

Figure 8.17 Framework for pairwise structure comparison. First the relevant features are extracted and represented in structure descriptions. These are input to a comparison algorithm which finds an equivalence obeying certain constraints with as high scores as possible. The equivalence is output together with the RMSD value. It can be assessed with respect to whether it is accurate (if a standard of truth is available) and to whether it is statistically significant, i.e. if the similarity is stronger than could be expected by chance. Reproduced from Eidhammer et al. (2000) by permission of Mary Ann Liebert.

Example

Assume that we have two structures *A*, *B* with sequences

ACSLDRTSIRV and ATLREKSSLIR.

If the best alignment is as shown on the left below, we know that the best alignment of the first five residues are as on the right below:

```
ACSL-DRTS-IRV        ACSL-D
A-TLREKSSLIR-        A-TLRE
```

Let the structures be as in the upper part of Figure 8.18, and the best structure

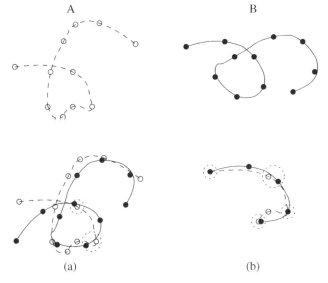

Figure 8.18   Illustration that dynamic programming cannot be used directly for structure alignment (see the text).

alignment be as in (a), and the best when only the first five residues are used be as in (b). We see that the paired residues (one from each structures) are not the same in (b) as in (a).   △

It follows from the above explanation that there does not exist any general scoring matrix which can be used to score pairs of elements. The best we can hope for is to calculate a *position-dependent scoring matrix* for every two structures which are to be aligned (which, for example, is done in the methods in Chapter 9).

Several methods exist for comparing pairs of structures, and also pairs of sequences and structures. A classification of the methods can be done along different lines; in this book a classification of the *approaches* is used. We consider the structures as rigid bodies, see the bibliographic notes for a comparison method allowing flexible structures.

8.9   Exercises

1. Show that for describing a structure using distance matrices, the number of required distances is $4n - 10$ when n is the number of atoms.

2. Show that when going from a distance matrix to coordinates, the number of coordinates that must be calculated is $3n - 6$.

3. A distance matrix in two dimensions is given:

	B	C	D	E
A	4	5	3	2.4
B		3	5	3.2
C			4	5.5
D				4.8

Calculate the coordinates of the points in a coordinate system where the origin is point A, and the y-axis along the line AB.

4. Explain why SSEs are so important in the analysis of protein structures.

5. Let the torsion angles of a structure be $(-70, -20)$, $(-72, 60)$, $(-70, 120)$, $(-60, 170)$, $(-65, 125)$, $(-100, 45)$, $(-100, -65)$, $(-105, -66)$, $(-100, 60)$. Use the Ramachandran plot in Figure 8.8 to find possible α-helices and β-strands.

6. n points are given in two dimensions: $(x_1, y_1), \ldots, (x_n, y_n)$. We want to approximate these with a straight line $y = ax + b$.

   (a) Write a formula for the sum of the quadratic error in the y values, when the line is used instead of the correct y values. One wants to minimize this expression.

   (b) Instead of minimizing the errors in the y-direction, one now wants to minimize the quadratic sum of the distances from the points to the line (see Figure 8.9). Write a formula for this error.

7. Two succeeding (in sequence) β-strands can be connected in three ways, depending on parallelism and how the loop between two of them goes, either over or under a surface of the two strands.

   (a) Draw a figure illustrating this.

   (b) The number of different sheet-topologies (how n strands can be combined) which can be constructed from n strands is $n!3^{(n-1)}/2$. Show this.

8. A structure description can contain both architecture and topology or only architecture. Explain what this means when comparing structures. Homologous proteins often have similar topologies; explain why.

9. Two sequences are given: FDKLCDTA and SYNVDRGP. Consider AACH in Figure 5.1.

   (a) Make an equivalence of at least six pairs, where each pair scores at least 2.

   (b) Find the highest-scoring alignment when the gap score is 0 (try without using dynamic programming).

8.10   Bibliographic notes

A book describing protein architectures and topologies is Lesk (2001). Taylor et al. (2001) describe several of the topics discussed.

More information about PDB can be found in Berman et al. (2000). For other structure databases, see Baxevanis (2003).

Survey articles discussing structure comparison include Brown et al. (1996) and Eidhammer et al. (2000). Some reviews are also in Gibrat et al. (1996) and Godzik (1996).

Distance geometry methods are described in Aszódi and Taylor (1994). The Ramachandran plot is defined in Ramachandran et al. (1963), and its use for judging the quality of protein structure is described in Hooft et al. (1997).

A method for least-squares fitting of SSEs is Thomas (1994). Ellipsoid representation of SSEs is described in Taylor et al. (1983). TOPS was first described in Flores et al. (1994), and used for automatic structure comparisons in Gilbert et al. (1999a).

DSSP is described in Kabsch and Sander (1983), and a method for determining SSEs in which only geometrical values are used is in Taylor et al. (2001).

Examples of space-based descriptions are in Bagley and Altman (1995) and Kastenmüller et al. (1998) and examples of substructure description in Escalier et al. (1998) and Bagley and Altman (1995). Different representations are described in Matsuda et al. (1997) (strings); Taylor and Orengo (1989), Jonassen et al. (1999), Artymiuk et al. (1994), Chew et al. (1999) (lists); Alexandrov et al. (1992) (sets); Grindley et al. (1993), Koch et al. (1996) (graphs); Bagley and Altman (1995) (features).

Matching is not discussed separately; for an example of a matching algorithm, see Gilbert et al. (1999a).

A method for aligning flexible proteins (containing a hinge) is described in Shatsky et al. (2002a).

9

Superposition and Dynamic Programming

Most methods for comparing structures include some use of superposition and dynamic programming.

9.1 Superposition

Superposition can be used to find and score equivalences, by measuring how close the equivalent pairs can come together. One way of thinking of it is to put the structures on top of each other so that the equivalenced elements from the two structures lie as close as possible. If the geometry of the structures is not changed in this process, it is referred to as *rigid-body* superposition. The score can then be a function of the distances between the elements of each equivalent pair in the equivalence. Commonly, the root of the mean of the squares of the distances is used, and is called the *root mean square deviation* (RMSD). Low RMSD values are best, zero indicates exact equality.

Note that superposition can be used to measure (score) equivalences, not necessarily alignments directly. Two different measures are mainly used.

9.1.1 Coordinate RMSD

Superposition can be done by a *transformation* of structure A over B such that the equivalent pairs come as close as possible.

Let $(\alpha_1, \beta_1), \ldots, (\alpha_r, \beta_r)$ be the coordinate sets of the equivalenced elements of the equivalence E (α_i from A and β_i from B, for three dimensions a coordinate set consisting of three values). The problem is then to find a transformation T for A which minimizes the *coordinate* root mean square deviation, that is,

$$\text{RMSD}_C(E) = \min_T \sqrt{\frac{1}{\sum_{i=1}^{r} w_i} \sum_{i=1}^{r} w_i (T\alpha_i - \beta_i)^2}, \qquad (9.1)$$

Protein bioinformatics: an algorithmic approach to sequence and structure analysis
I. Eidhammer, I. Jonassen and W. R. Taylor © 2004 John Wiley & Sons, Ltd ISBN: 0-470-84839-1

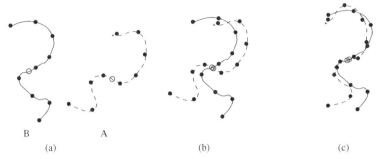

B   A
(a)   (b)   (c)

Figure 9.1   Figure illustrating the superposition in two dimensions. (a) The structures; o shows the centroids. (b) A moved so the centroids coincide. (c) A rotated to the minimum RMSD.

where w_i are weights corresponding to each pair (α_i, β_i) (and often set to 1). For residue level structure descriptions, there is usually one coordinate set per residue (e.g. C_α atom).

A transformation can be performed as a *translation* (three distances), and a *rotation* (three angles, around each of the x-, y- and z-axes). (The rotation can also be performed in one operation around a line, the direction of the line has to be calculated for each rotation; cf. Euler's theorem (Gelbert et al. 1977).)

It has been shown that a transformation for the minimum RMSD can be found by first shifting the centroids (geometrical centres) of each structure to the origin of a common coordinate system, and then finding the rotation of A which minimizes the RMSD$_C$, as shown in Figure 9.1.

A rotation around the origin can be described by an *orthogonal* matrix $R_{3,3}$ (3D space) with determinant equal to 1. (There exist equations describing the connections between the angles (3) and the values of the matrix (9) (Gelbert et al. 1977, pp. 530–535).) A matrix is orthogonal if the scalar product of any two different columns is 0, and the result of taking the scalar product of any column with itself is 1. The matrix must be orthogonal to assure that the distances between the points of the same structure are not changed (cf. rigid-body superposition).

The formula can therefore be described by a rotation matrix R and a translation vector t, and we search for a pair (R, t) which minimizes the expression (assuming $w_i = 1$ for all i):

$$\sum_{i=1}^{r}(R\alpha_i + t - \beta_i)^2. \tag{9.2}$$

Example

Let the matrix R be

$$
\begin{Bmatrix}
\overset{C_1}{\dfrac{1}{\sqrt{3}}} & \overset{C_2}{\dfrac{1}{\sqrt{3}}} & \overset{C_3}{\dfrac{1}{\sqrt{3}}} \\[2ex]
-\dfrac{1}{\sqrt{2}} & 0 & \dfrac{1}{\sqrt{2}} \\[2ex]
\dfrac{1}{\sqrt{6}} & -\dfrac{2}{\sqrt{6}} & \dfrac{1}{\sqrt{6}}
\end{Bmatrix}
$$

We can show that R is orthogonal, for example,

$$C_1 \cdot C_2 = \frac{1}{\sqrt{3}}\frac{1}{\sqrt{3}} + \frac{1}{\sqrt{2}}0 + \frac{1}{\sqrt{6}}(-\frac{2}{\sqrt{6}}) = 0$$

and

$$C_1 \cdot C_1 = \frac{1}{\sqrt{3}}\frac{1}{\sqrt{3}} + \frac{1}{\sqrt{2}}\frac{1}{\sqrt{2}} + \frac{1}{\sqrt{6}}\frac{1}{\sqrt{6}} = 1.$$

A point $(1, -1, 1)$ will then be transformed as

$$\left\{ \begin{array}{ccc} \frac{1}{\sqrt{3}} & \frac{1}{\sqrt{3}} & \frac{1}{\sqrt{3}} \\ \frac{1}{\sqrt{2}} & 0 & -\frac{1}{\sqrt{2}} \\ \frac{1}{\sqrt{6}} & -\frac{2}{\sqrt{6}} & \frac{1}{\sqrt{6}} \end{array} \right\} \left\{ \begin{array}{c} 1 \\ -1 \\ 1 \end{array} \right\}$$

to $(\frac{1}{\sqrt{3}}, 0, \frac{2\sqrt{2}}{\sqrt{3}})$. $\triangle$

The superposition problem can therefore be formulated as finding the orthogonal matrix $R_{3,3}$ which minimizes the function

$$\sum_{i=1}^{r}(R\alpha_i - \beta_i)^2, \tag{9.3}$$

where $(\alpha_1, \beta_1), \ldots, (\alpha_r, \beta_r)$ are now the coordinates after the structures are moved (translated) to a common origin. Algorithms exist for finding such a matrix that either use iterative or direct methods.

In Equation (9.1) a weight is specified. For example, one could let w_k be a measure of how similar the amino acids in the equivalent pair are, and/or how similar the *environments* around the residues of the kth pair are, the environment meaning the spatial position of the neighbouring residues.

9.1.2 Distance RMSD

The distance score method *Distance RMSD* (RMSD$_D$) alleviates the need for finding a translation and rotation of one of the structures and is given by

$$\text{RMSD}_D(E) = \frac{1}{r}\sqrt{\sum_{i=1}^{r}\sum_{j=1}^{r}(\delta_{ij}^A - \delta_{ij}^B)^2}, \tag{9.4}$$

where δ_{ij}^A is the spatial distance between the elements of A in pairs i and j of the equivalence. Since there is no need to calculate a transformation, it is a faster calculation. However, it has a (sometimes serious) weakness: it is invariant under reflection. This means that if structure B is the mirror image of structure A, then $\text{RMSD}_D(A, B) = 0$ and $\text{RMSD}_D(C, A) = \text{RMSD}_D(C, B)$ for all structures C.

The two measures are experimentally shown to have a close to linear relation (when all weights are equal to 1) as $RMSD_D = \rho_1 \times RMSD_C + \rho_2$, where $\rho_1 \approx 0.75$ and ρ_2 is between 0 and 0.2 (Cohen and Sternberg 1980).

9.1.3   Using RMSD as scoring of structure similarities

The problem of pairwise structure comparison is often the problem of finding equivalences with low RMSD value(s). However, several quite different equivalences with similar scores might be found and which of these equivalences represent the 'correct' solution is not an easy task to decide. However, *one always needs to consider how many elements were equivalenced, since for random comparisons the expected RMSD value seems to be proportional to the square root of the number of equivalenced residues.* When taking this into consideration, different measures can be used for evaluating how well two structures can be superposed.

1. Find the equivalence that minimizes the RMSD divided by the square root of the length of the equivalence: $\min_E RMSD(E(A, B))/\sqrt{n_E}$, where n_E is the number of pairs in the equivalence E.

2. Define a threshold L. Find the maximum number of elements that can be superposed such that RMSD is less than or equal to L.

3. Define a threshold l. Find the maximum number of elements that can be superposed such that the distance between each equivalenced element is less than or equal to l.

The two last methods are mostly used to improve detection of regions of similar topology, excluding structurally unrelated regions.

Most scoring schemes for evaluating equivalences between structure descriptions contain factors related to the RMSD. Many structure comparison programs give as output an equivalence and a resulting RMSD even if they do not use RMSD internally to score equivalences. See Chapter 12 for more discussion of scoring structure comparison.

9.2   Alternating Superposition and Alignment

The methods using alternating superposition and alignment operate on residue level, and the goal is to find a 'best' alignment for the two structures. An initial equivalence (a *seed*) of atoms from each structure is first given, $E_0 = (a_{i_1}, b_{j_1}), (a_{i_2}, b_{j_2}), \ldots, (a_{i_r}, b_{j_r})$. Note that the A superposition is then performed with respect to the equivalence E_0 using the transformation T_0 which minimizes an RMSD measure. Then the whole structures are superposed using T_0. The distances between *all* pairs of atoms (residues) from the two structures (after superposition), can then be used to define a new scoring matrix R_0, which is used to obtain an alignment A_0 by dynamic programming. Note that usually a scoring matrix gives highest values to 'similar' elements, hence a score could be the reciprocal of the distance. Note also that R_0 is a *position-dependent scoring matrix*: there is a score for every pair of elements (residues) from A and B.

A new equivalence E_1 of the r equivalences from the alignment with least distances (using R_0) can then be found. From this T_1 is computed, and thereafter R_1, as explained above. This iteration is performed until convergence ($E_{i+1} = E_i$) or some maximum number of cycles is done. Algorithm 9.1 describes the method and Figure 9.7(a) illustrates the approach. Note that in an implementation it is not necessary to use a new variable for each E_p.

Algorithm 9.1. Alternating superposition and dynamic programming.

Comparing structures A and B by finding a 'best' equivalence for them.
const

p_{max}	maximum number of cycles
E_{init}	initial equivalence
r	the number of pairs in the equivalences

var

p	cycle number
R	scoring matrix
E_p	the equivalence of cycle p
T	the (minimum) transformation for E_p

proc

$dist(a_i, b_j)$	the distance between residues
$score(d)$	calculate a score from a distance

begin

    $E_0 := E_{init}; \; p := 0$
    repeat
        $T :=$ *the transformation for* $\text{RMSD}_C(E_p)$
        $A^* := T(A)$ Superpose A onto B giving A^*
                       Calculate the new scoring matrix:
        forall *pairs* (i, j) **do** $R_{ij} := score(dist(a_i^*, b_j))$ **end**
        $(s, P) := \mathcal{A}_R(A, B)$ DP on (A, B) using R (find path P with score s)
        $p := p + 1$
                       Pick the r pairs from P with lowest distances in R
        $E_p := \{(a_{i_1}, b_{j_1}), \ldots, (a_{i_r}, b_{j_r})\}$
    while $E_p \neq E_{p-1}$ **and** $p < p_{max}$
end

Example

Let $r = 4$ and the initial equivalence be

$$E_0 = \{(a_1, b_2), (a_2, b_4), (a_5, b_7), (a_9, b_5)\}.$$

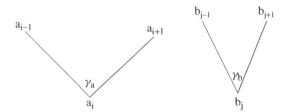

Figure 9.2   Example of a spatial component of a score of (a_i, b_j),
the difference of the angles γ_a and γ_b (remember that three points define a plane).

The transformation for best superposition of (a_1, a_2, a_5, a_9) on (b_2, b_4, b_7, b_5) is found, and then used on the whole structure A. The $m{\times}n$ matrix R of *all* pairwise distances (after the superposition) is calculated, and used to define the scoring matrix for a dynamic programming algorithm. Let this result in the best alignment be

$$a_1\ a_2\ a_3\ a_4\ \text{---}\ \text{---}\ a_5\ a_6\ a_7\ a_8\ a_9$$
$$\text{---}\ b_1\ b_2\ b_3\ b_4\ b_5\ b_6\ \text{---}\ b_7\ \text{---}\ b_8$$

An equivalence with six pairs is defined by this alignment, and taking the four (r) of them with smallest distances (using R_0) might result in the equivalence

$$E_1 = \{(a_3, b_2), (a_5, b_6), (a_7, b_7), (a_9, b_8)\}.$$

This is the equivalence used in the next cycle. $\triangle$

The initial seed equivalence is critical to the final result, therefore several initial seeds should be tried. As explained in Section 9.1.3, the final results can be quite different while still giving similar quality measures (RMSD and/or number of aligned residues).

In the presentation we have used r as a constant, instead it could be, for example, the number of pairs $(a_{i_p}^*, b_{j_p})$ on P for which the distances R_{ij} are below a given limit ($a_{i_p}^*$ are the positions of the atoms of A after superposition). In this way only pairs which superpose well are used in the next cycle, hopefully giving faster convergence. However, there is no reason for saying that this is always for the best, so both options should be available.

The scoring of (a_i, b_j) for making the scoring matrix can be changed to include several components, e.g. a sequence component and a local structure component. The sequence component might reflect the similarity between the amino acid types of the two residues (e.g. using a PAM matrix) and the local structure component might reflect, for example, the spatial similarity between (a_{i-1}, a_i, a_{i+1}) and (b_{j-1}, b_j, b_{j+1}). The spatial similarity can, for example, be measured by the difference between two angles, as shown in Figure 9.2.

The method of performing alternating superposition and alignment is also used for refining (postprocessing) the results found by other methods, mostly methods using coarse level description.

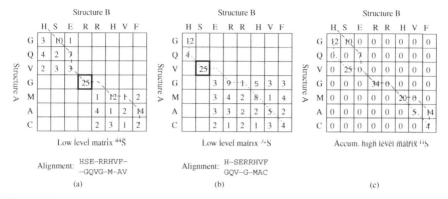

Figure 9.3 Application of the double dynamic programming method. (a), (b) Two of the 56 low-level scoring matrices are shown. The expression used for low-level scores gives the value of 25 to $^{ij}S_{ij}$. The best (DP) paths using the scoring matrices and a linear gap penalty with $g = 4$ are shown by broken and dotted lines, respectively. (c) The low-level scores for the two optimal paths are summed up in the high-level scoring matrix.

9.3   Double Dynamic Programming

As explained earlier, traditional dynamic programming cannot alone be used for aligning structures. Since any choice to align two substructures will affect the scoring of the alignment of the complete structures, the independence requirement is violated and DP can no longer guarantee an optimal solution. But if one assumes that the structures are already superposed, reasonable scoring schemes can be devised which allow the structures to be aligned optimally using DP (Section 9.2). However, ideally, one might wish to *simultaneously* align and superpose the structures to optimize a score depending on how well-aligned substructures superpose. Here we describe one method using this approach, the structure sequence alignment program (SSAP) program (developed by Taylor and Orengo (1989)). This is based on a method called double dynamic programming (DDP). The main idea is to use two levels of dynamic programming, constructing a final scoring matrix that can be used on the high level for finding the (best) alignment between the two structures by an ordinary DP algorithm.

Conceptually, the method looks at each residue pair (a_i, b_j), and for each it tries to find how likely it is that this pair is in an optimal alignment. A heuristic for this is to find an optimal alignment under the constraint that (a_i, b_j) is part of the alignment, and to define a (low-level) *position-dependent scoring matrix* under this constraint. In the DDP context this is called a low-level DP, and the scoring matrix uses a *low-level scoring matrix*. For each pair (i, j) there will be defined a separate (low-level) scoring matrix, denoted ^{ij}S. The matrix element $^{ij}S_{kl}$ will get a number (score) showing how well the residue a_k fits to b_l under the constraint that a_i is aligned with b_j. Note that these scoring matrices define scores between all pairs of *residues*, not between amino acids.

To force the (low-level) alignment to go through (a_i, b_j), the ordinary DP algorithm can then be used, either by finding an optimal path from (a_i, b_j) to (a_1, b_1) and from (a_i, b_j) to (a_m, b_n), or by giving the score of (a_i, b_j) so high a value that the optimal path is forced to go through it. Since we assume that a_i matches b_j, it is only necessary to calculate the scores of $^{ij}S_{kl}$ for $(1 \leqslant k < i, \ 1 \leqslant l < j)$ and $(i < k \leqslant m, \ j < l \leqslant n)$ (that is, the top left and bottom right of $^{ij}S_{ij}$).

The results from all low-level computations are 'summed up' in a *high-level scoring matrix* ^{H}S, and final (high-level) dynamic programming is done using ^{H}S. The summing is done by letting the contribution from the low-level matrix ^{ij}S be all $^{ij}S_{pq}$ such that (a_p, b_q) lies on the optimal (low-level) path when ^{ij}S is used as the scoring matrix. In this way the highest-scoring path from each low-level DP matrix is propagated to the high-level 'summary' scoring matrix. The procedure is illustrated in Figure 9.3, and given in Algorithm 9.2. Note that Figure 9.3 shows scoring matrices, not the 'DP matrices'.

Algorithm 9.2. Double dynamic programming.

Aligning the structures A and B using DDP
begin
   $^{H}S := \{0\}_{m \times n}$ Set high-level scoring matrix to zero
    for *each pair* (a_i, b_j) **do**
      Create the low-level scoring matrix ^{ij}S
      $(s, P) := \mathbf{DP}^*_{ij\,S}(A, B)$ Low-level DP, forced through (a_i, b_j)
      forall $(a_p, b_q) \in P$ **do**   Accumulate from path P
        $^{H}S_{pq}(P) :=^{H} S_{pq}(P) +^{ij} S_{pq}(P)$
    end
   end
    $(s, P) := \mathbf{DP}_{H\,S}(A, B)$ High-level DP using ^{H}S, best path in P
end

As ^{H}S contains a sum of values from low-level matrices, it might contain large values, and before the high-level computation is performed, the values are normalized or their logarithmic values are used.

^{H}S is constructed as the sum of values from mn low-level matrices, most of them representing pairs which are not in the final alignment. There is a general background or random score associated with the comparison with every dissimilar pair, and to damp this noise it should be removed from the accumulation. One way is to define a cut-off on the score of the low-level alignment and accumulate the scores only from alignments above this cut-off.

Different versions of DDP can be developed, depending, for example, on how the low-level scoring matrices are calculated and which sequence and structure properties are used. Examples are given in the following sections.

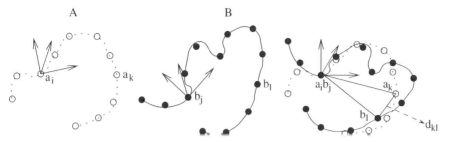

Figure 9.4 Calculation of a low-level scoring matrix ^{ij}S in double dynamic programming. Two structures A and B are shown schematically, and local coordinate frames are derived from local geometry at a_i and b_j, respectively. The structures are then translated and rotated such that the coordinate frames coincide (right). The scoring $^{ij}S_{kl}$ can then be found simply as a function of the distance between a_k and b_l shown as d_{kl} or as a more complex function.

9.3.1 Low-level scoring matrices

There are mn low-level scoring matrices ^{ij}S. All methods for calculating $^{ij}S_{kl}$ (showing how good a_k fits b_l when a_i is aligned to b_j) should calculate a superposition of the two structures A and B bond on the choice of i and j. One way of defining the scores is to first define local reference systems at a_i and at b_j, e.g. by using (the C$_\alpha$ of) a_{i-1}, a_i, a_{i+1} and b_{j-1}, b_j, b_{j+1}. With three points one can define a unique reference (coordinate) system (as long as they do not lie on a straight line). The coordinates of the remaining residues are transformed into the respective coordinate systems. The score of aligning a_k, b_l depends on the distance between a_k and b_l in the respective coordinate systems defined at a_i and b_j, as shown in Figure 9.4. One simple scoring scheme is to use a function of the distance between a_k and b_l.

A more comprehensive scoring scheme could also take into account.

- A *direction* component, the difference in the direction of the vectors (a_i, a_k) and (b_j, b_l) (see Figure 9.4).

- An *orientation* component o_{kl}. A local reference frame can also be constructed for a_k and b_l (using the neighbouring residues, as above), after the structures have been transformed to coincident reference frames at a_i and b_j. This component reflects the difference in the reference frames at a_k and b_l, and is measured by the angle between corresponding axes at a_k and b_l.

- A *sequence* distance component g_{kl}, calculated as an increasing function of $|k - i| + |l - j|$. This component should damp the contribution from near neighbours in the sequence (as the matching of local secondary structures). The reason for this component is that the structural similarity at a_k and b_l tends to be higher when they are near a_i and b_j in sequence, and we want all pairs to have 'equal sequential significance' (having a global view).

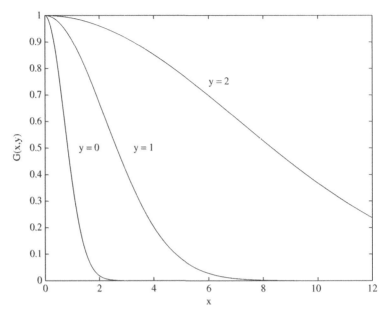

Figure 9.5   The Gaussian function for different values of x and y.
We see that the decrease of $\mathcal{G}(x, y)$ is highest for low y.

- A *spatial* distance component h_{kl}, calculated as a decreasing function of $d_{ik} + d_{jl}$, where d_{ik} is the distance between the (the C$_\alpha$ of) a_i and a_k. This will damp the contribution from combined large distances in space.

The two last components together give high weight to residues which are near in space but not near in sequence (i.e. candidates for being in a site around $a_i b_j$).

Combining the components: Gaussian transform function

The low-level scoring matrices are constructed as a combination of several components. Often the score of a component should be the inverse of a measured value, and also some normalization should be done, e.g. in some cases where the measured value is a distance, and we want to use similarity. One such decreasing and normalizing function (all values being in the interval [0,1] for nonnegative x and y) is the Gaussian transform function:

$$\mathcal{G}(x, y) = \exp(-x^2/10^y), \tag{9.5}$$

where x is the measured value which is to be used in a scoring function and y is a (different) constant for each component, defining the slope of $\mathcal{G}$ (see Figure 9.5). The final scoring value $^{ij}S_{kl}$ is the product of all the components after transformations.

9.3.2 High-level scoring matrix

The accumulated sums in the high-level matrix could be large, and the logarithms of the sums are used as the scoring values. A linear gap penalty is used. The variation of the values of the scoring matrix is high, with a large difference between 'good' and 'bad' pairs. A typical alignment will not have gaps in the 'good' regions.

9.3.3 Iterated double dynamic programming

As DDP is a heuristic, one might achieve better results by iteration—doing a DDP in each cycle. This, together with an idea of how to limit the work for each DDP cycle, is developed into a program called the structure alignment program (SAP).

The DDP algorithm described above requires a computation time proportional to the fourth power of the sequence length (for two proteins of equal length) as it performs an alignment for all residue pairs. To circumvent this severe requirement, some simple heuristics are devised based on the principle that comparing the environment of all residue pairs is not necessary. Thus some pairs, together referred to as the *seed*, are chosen before the first cycle. Low-level matrices are only constructed for the pairs in the seed, and used for the DDP algorithm in the first cycle. In each cycle the high-level scoring matrix is updated, and the pairs are also selected anew for the next cycle. The overall algorithm is shown in Algorithm 9.3. Two high-level matrices are used: $^H S$ as before, and a *bias* matrix Q. The reason for this will be explained below. In the algorithm a high-level DP is performed in each cycle; this is for the termination criterion.

Algorithm 9.3. Iterated double dynamic programming.
Aligning the structures A and B by iterated DDP
var
Q the high-level bias scoring matrix
E the equivalence used in each cycle
begin
    initialize the bias matrix Q
    $E :=$ *the seed*
    repeat
        $^H S := \{0\}_{m \times n}$ Set high-level scoring matrix to zero
                                           Calculate the high-level scoring matrix $^H S$:
        for each pair $(ij) \in E$ **do**
            $(s, P) := \mathbf{DP}^*_{ij\,S}(A, B)$ Low-level DP forced through (a_i, b_j)
            accumulate the low-level result in $^H S$
        end
        update Q using old Q and $^H S$
        $(s, P) := \mathbf{DP}_Q(A, B)$ High-level DP
        $E :=$ *select new pairs based on Q*
    until *termination criterion is satisfied*
end

Several of the statements in Algorithm 9.3 have to be described in more detail.

- How is Q initialized and the seed selected?

- How many residue pairs should be selected in each cycle, and how?

- How should Q be updated?

- Should the high-level scoring matrix contribute to the low-level matrices?

- What is the termination criterion?

Initialization

Based on local structure and environment, many residue pairs (indeed most) can be neglected. This selection can be made on secondary structure state (one would not normally want to compare an alpha-helix with a beta-strand and burial (those with a similar degree of burial are most similar) but a component based on the amino acid identity can also be used, giving any sequence similarity a chance to contribute. Contributions from all three components are combined for each pair of positions as a product (to ensure that all three have a reasonable value) giving a matrix Q, which is referred to as the '*bias matrix*' (Q is thus determined by use of the three components). No specific weights are introduced, instead the $\mathcal{G}$ transform is used to give a roughly equal contribution and taking their product makes the size of the component ranges less critical.

The seed can be selected by taking the pairs with highest values in Q, or, for example, using methods for discovering common sites in the structures, e.g. SPratt (see Chapter 13.5).

Selecting pairs and updating Q

The highest values in Q for each cycle are used for selecting the pairs to be used in the low-level computation. First a relatively small number of pairs is selected (10–20), but the number is gradually increased with each cycle, as hopefully more 'true pairs' are found. The initial sparse sampling can be unrepresentative of the truly equivalent pairs. To maintain a continuity through the early sparse cycles (hopefully towards the true equivalence) Q is used as a base for incremental revision (the *bias* matrix). If Q^p is the bias matrix on cycle p, then the next revision is calculated as

$$Q^{p+1} = Q^p/2 + \log(1 + {}^H S^{p+1}/20),\qquad(9.6)$$

where ${}^H S$ is the matrix as defined for SSAP formed from the summed traces from all the low-level comparisons. The scores in the high-level matrix (${}^H S$) are generally large but the logarithmic damping reduces these into a range with an effective maximum of 1, which is more commensurate with the range of values found in the bias matrix (Q).

To further ensure that the bias matrix does not become dominated by extreme values, its elements are normalized on each cycle.

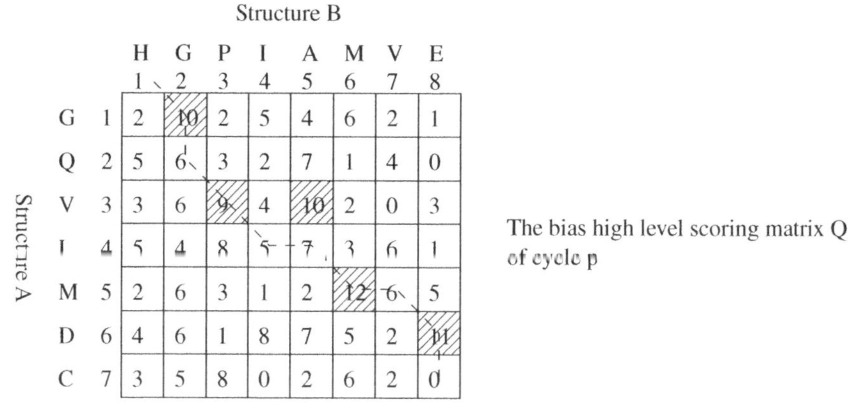

Structure B

		H 1	G 2	P 3	I 4	A 5	M 6	V 7	E 8
G	1	2	10	2	5	4	6	2	1
Q	2	5	6	3	2	7	1	4	0
V	3	3	6	9	4	10	2	0	3
I	4	5	4	8	5	7	3	6	1
M	5	2	6	3	1	2	12	6	5
D	6	4	6	1	8	7	5	2	11
C	7	3	5	8	0	2	6	2	0

(Structure A — row labels)

The bias high level scoring matrix Q of cycle p

Figure 9.6 Illustration of the selection procedure and stop criterion for the iterated DDP method. The figure shows the selected pairs in cycle p, when the number of pairs is five (shaded). It also shows the best alignment (by the broken line) for cycle $p + 1$, hence four of the five selected pairs are on this path ($r = 5$, $u = 4$). If all five lay on the alignment the iteration would have converged (and could be terminated).

This form of updating weakens the contribution from the initial Q for each iteration cycle, making the selection of pairs become increasingly determined by the dominant alignment, approaching (or attaining) by the final cycle a self-consistent state in which the alignment has been calculated predominantly (or completely) from pairs of residues that lie on the alignment.

The increasing number of pairs selected can be determined by the following function of the size of the two proteins (m and n) and the cycle number (p):

$$K = 10 + \frac{p+1}{20}\sqrt{mn} \tag{9.7}$$

($p = 0$ for the initial cycle). We see that the number of selected pairs can be larger than the length of the sequences, but this is taken care of in the stop criterion. See Figure 9.6 for an illustration of the selection.

Achieving coherence between the levels

The initial selection of residue pairs might be quite random with respect to the final set of 'true' equivalences and, as a consequence, the comparison of their environments might provide little coherent direction towards the final solution. Although the bias matrix provides a platform from which the selection of pairs can be refined, it has no effect on the scores derived from the low-level matrices. A contribution from the bias matrix as described above can therefore be introduced at this level to provide stability into the early cycles. This is done by adding a contribution from the bias matrix to the low-level matrix that decreases with increasing cycle number. Therefore, instead of using ^{ik}S as explained in Section 9.3.1, a revised matrix $^{ik}S^*$ is used:

$$^{ik}S^* = {}^{ik}S + Q^p \cdot \mathcal{G}(p, 1), \tag{9.8}$$

were p is the cycle number and $\mathcal{G}$ is the Gaussian transform. On the initial cycle the bias matrix has a full contribution ($\mathcal{G}(0, 1) = 1$), which decreases until after the fifth cycle there is effectively no contribution from the bias matrix ($\mathcal{G}(5, 1) = 0.082$). This provides a smooth transition from the, initially, local information in the bias matrix into the full global view provided by the comparison of the residue environments.

Termination criterion

The iteration should stop when the selected pairs in cycle p coincide with the best (high-level) path in cycle $p + 1$. Let r be the number of selected pairs at cycle p. Then a new Q is calculated, and the best path found. Let u of the selected pairs lay on this path. Then

$$k = \frac{u}{r}$$

can be used as stop criterion, as illustrated in Figure 9.6. If k becomes 1, the iteration stops. However, if the number of selected pairs is more than the length of the true alignment, then $k = 1$ can never be attained so the iteration can also terminate if k stops to increase. In addition, an upper limit for the number of cycles should exist (typically 5–10).

9.4   Similarity of the Methods

SAP and the methods using alternating superposition and alignment can both be looked upon as *two-level* methods. While the latter methods find the DP matrix using an optimal superposition for the current equivalence, SAP does not need to decide (or assume) one exact alignment to calculate the higher-level scoring matrix. Instead the residues pairs that participate in high-scoring low-level alignments receive high values in the high-level scoring matrix and are likely to be included in the final alignment (see Figure 9.7(b)).

9.5   Exercises

1. Regard two 'structures' (A, B) of three atoms in 2D space, defining an equivalence of three pairs:

       A: (1,4) (4,1) (4,4)
       B: (0,0) (2,0) (3,2)

   (a) Calculate $RMSD_D$.

   (b) Calculate the geometrical centres of the 'structures'. Then move the 'structures' so that the geometrical centres are at the origin. Find the new coordinates of the points, and plot them in a diagram.

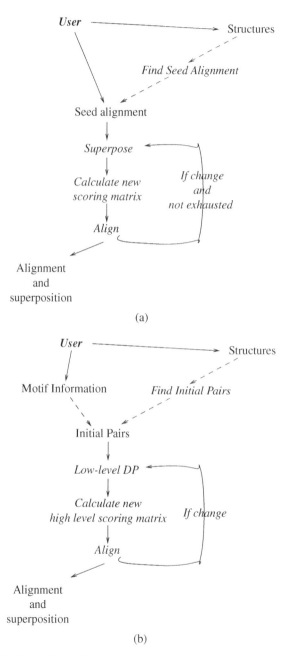

Figure 9.7 (a) Outline of algorithm alternating between alignment and superposition. (b) Outline of the SAP method. Reproduced from Eidhammer et al. (2000) by permission of Mary Ann Liebert.

(c) Calculate $RMSD_C$ (weight 1) for the 'structures' as they are, without rotation.

(d) Look at the 'structures'. Do you think you can find a lower $RMSD_C$ value by rotating one of the 'structures'?

2. This exercise is to illustrate the method of alternating superposition and alignment for finding the best alignment. For simplicity, we consider 'structures' in 2D space. Consider the structures

$$A = (a_1, a_2, a_3, a_4) \quad \text{and} \quad B = (b_1, b_2, b_3, b_4, b_5).$$

Let the position of (atoms representing) the residues be

a_1:	$(2, -0.414)$,	b_1:	$(1.2, -2)$,
a_2:	$(-0.121, 0.293)$,	b_2:	$(1.8, 1.2)$,
a_3:	$(1.293, 3.121)$,	b_3:	$(0.2, -0.4)$,
a_4:	$(4.828, 1)$,	b_4:	$(-1.4, -0.6)$,
		b_5:	$(-1.8, 1.8)$.

(a) Calculate the geometrical centres, and find the coordinates of the atoms when both are moved so that the geometrical centres are at the origin. (Control: the coordinates for a_1 should be $(0, -1.414)$.)

(b) Assume an initial equivalence (subalignment): $(a_2, b_1)(a_3, b_2)(a_4, b_4)$. Assume further that the rotation matrix for minimum $RMSD_C$ for this is

$$\begin{matrix} 1/1.414 & 1/1.414 \\ -1/1.414 & 1/1.414 \end{matrix}$$

(i) Show that the matrix is orthogonal.

(ii) Rotate all the coordinates of A, and calculate the new coordinates. (Control: the new coordinates of a_1 should be $(-1, -1)$.)

(c) Define a distance matrix D for all pairs (a_i^*, b_j), where a_i^* is the coordinates after rotation.

(d) Define a scoring matrix R by dividing 1 by the distances in D; use one decimal. Use R to find the best alignment for A, B when R is used. Use a linear gap penalty, where the penalty for one blank is the average of the values in R which are less than 1.

(e) Now find the highest-scoring equivalence (subalignment) of length 3.

3. This exercise is to illustrate double dynamic programming. Consider the 'structures' $A = (a_1, a_2, a_3, a_4)$ and $B = (b_1, b_2, b_3, b_4, b_5)$ in 2D space. To define a coordinate system for a_i, choose origin in a_i and the x-axis through a_{i+1}. First choose the pair (a_2, b_3) for low-level dynamic programming. The coordinate in this system are found to be

```
A:  (-1,2),  (0,0),  (2,0),  (1,1)
B:  (-2,1),  (-1,1),  (0,0),  (1,0),  (0,2)
```

(a) Fill in a distance matrix D for all pairs from A, B, and define a low-level scoring matrix $^{2,3}S$, by dividing 1 by the distances in D (to one decimal). Let the score for (a_2, b_3) be the double of the highest score. Use a linear gap penalty, and let the gap penalty for one blank be the average of the values in $^{2,3}S$, when (a_2, b_3) is not counted. Perform low-level dynamic programming.

(b) Choose (a_3, b_3) for the new low-level matrix. B will have the same coordinates as in (a). To find the new coordinates for (A), you can use the following procedure. Let

- X and Y be the x- and y-axes of the original coordinate system,
- X' and Y' be the x- and y-axes of the coordinate system with origin at a_3 (and x-axis along a_4),
- ϕ be the angle between X and X',
- $g_1 = a_3 x$ (the x-coordinate of a_3),
- $g_2 = a_3 y$ (the y-coordinate of a_3),
- $g_{11} = \cos(X, X') = \cos(\phi)$,
- $g_{12} = \cos(X, Y') = \cos(90 + \phi)$,
- $g_{21} = \cos(Y, X') = \cos(90 - \phi)$,
- $g_{22} = \cos(Y, Y') = \cos(\phi)$.

If (x, y) are the coordinates of a point in the original coordinate system, then the coordinates (x', y') in the new coordinate system become

- $x' = g_{11}(x - g_1) + g_{21}(y - g_2)$,
- $y' = g_{12}(x - g_1) + g_{22}(y - g_2)$.

You may want to draw a diagram with the coordinate systems for verifying the new coordinates you get. Then define a scoring matrix and perform dynamic programming as explained in (a).

(c) Make the high-level scoring matrix using the two low-level ones, and perform dynamic programming.

(d) Compare the three alignments, and comment the choice of pairs when only two are chosen.

4. When choosing the pairs for the next cycle in double dynamic programming, there is no test for inconsistencies among the pairs (one residue in A could be in two or more pairs). Discuss whether this should have been done.

5. Equation (9.6) shows how the bias matrix Q is updated. Now assume that a cell in the high-level scoring matrix $^H S$ in every cycle is calculated to a fixed value K, such that $K = \log(1 + {}^H S^{p+1}/20)$.

(a) Show that when p approaches infinity, the value of the corresponding cell of Q approaches $2K$. Hint: write the equation as $q^{p+1} = q^p/2 + K$ (where q^p is the value of the corresponding cell in cycle p). Then find an expression for q^{p+1} depending on K and q^0.

(b) Let $q^0 = 0.5$. Find the value of q^3 when (i) $K = 0.2$ and (ii) $K = 0.8$.

9.6　Bibliographic notes

Fundamental articles for superposition are McLachlan (1972, 1979), Kabsch (1978) and Cohen and Sternberg (1980).

More about the methods of alternating superposition and alignment can be found in Rao and Rossmann (1973), Rossmann and Argos (1975, 1976), Satow et al. (1986), Cohen (1997), Russel and Barton (1992), Ding et al. (1994), Holm and Sander (1995), Zu-Kang and Sippl (1996), Petitjean (1998) and Gerstein and Levitt (1998),

Double dynamic programming in the program SSAP is described in Taylor and Orengo (1989). An early iterative version is described in Orengo and Taylor (1990), while the algorithm above is described in Taylor (1997a, 1999a).

10

Geometric Techniques

A number of structure comparison methods use geometric techniques to find similar substructures between the structures. Such substructures can subsequently be used in a clustering method and combined into larger equivalences, as described in Chapter 11. In this chapter we focus on the first step, and we consider geometric hashing and distance-based techniques.

10.1 Geometric Hashing

Geometric hashing was originally developed as a computer vision technique for matching geometric features. In order to explain the principles, we first describe the basic ideas using two-dimensional geometric figures, and then show how geometric hashing is used for structure comparison.

10.1.1 Two-dimensional geometric hashing

Assume we have two two-dimensional geometric figures, a *model A*, and a *query B*, described by m and n points, respectively. The task is to discover common subfigures, invariant under both rotation and translation (rigid-body transformation, scale could also easily be included, but this will not be dealt with here since all structures are usually given in the same scale). One approach is to try to 'place the query on top of the model', and consider how many points coincide (ignoring the edges). This is illustrated in Figure 10.1, where six points coincide under a given threshold for point coincidence, meaning that the distance between points that coincide is less than the threshold.

Finding this maximal coincidence set is NP-hard. In addition we might not only want to find the maximal coincidence, but *all* coincidences with the number of points over a given threshold. Geometric hashing is a technique used for this problem. The geometric hashing technique defines coordinate systems, called *reference frames*, in both A and B, using, for example, two figure points for each frame (in the 2D case).

Protein bioinformatics: an algorithmic approach to sequence and structure analysis
I. Eidhammer, I. Jonassen and W. R. Taylor   © 2004 John Wiley & Sons, Ltd   ISBN: 0-470-84839-1

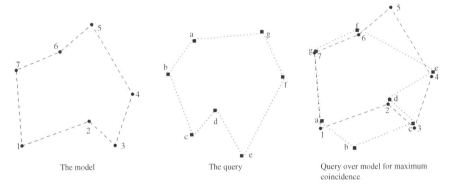

The model   The query   Query over model for maximum
                                                          coincidence

Figure 10.1 Two figures, model and query, are drawn. For reasons of illustration, the edges are also drawn, but only the nodes are used. The order of the points are irrelevant. The query is rotated and translated over the model to achieve a maximal coincidence set.

One way is to define the origin at the first point, and let one of the axes go through both points, the second axis defined to be orthogonal to this. The pair of points used to define the reference frame is called a *basis*, and the coordinates of all the other points are computed in the frame, constituting a *reference frame system*. Hence a reference frame system is the set of coordinates defined by a particular basis. We see that the reference frame system remains unchanged if the structure is rotated and/or translated. (If the distance between the basis points is defined to always be 1, then the reference frame is also invariant under scaling.) A frame system on A can then be compared with a frame system on B, and the number of point pairs (one from each figure) having equal or almost equal coordinates is the number of coincident points.

Example

In Figure 10.1 there are two figures. For reasons of illustration, the edges are drawn, but it is only the nodes which are used in the comparison. The order of the points is also irrelevant. Figure 10.2 illustrates the reference frame systems, using the figure points from Figure 10.1. In Figure 10.2(a) the points (1,3) (1 and 3) are chosen as a basis for the model, and the origin is defined at point 1. The frame system is divided into squares, indexed by horizontal and vertical indices. The position of each point can be described by the coordinates of the square into which it falls (the size of the squares reflects the tolerance in the comparison). We see that the points, including the basis, falls in squares:

$$(0, 0)(6, 2)(8, 0)(9, 4)(6, 10)(3, 8)(-1, 6).$$

Point 2 is on the border between square (5,2) and (6,2), and illustrates a complication which has to be handled: in which square to place points on or near the border. In reality the coordinates are floating point values, and very rarely on the borders.

In (b) (3,5) is selected as the basis, and the points are in squares:

$$(1, 8)(2, 2)(0, 0)(4, -2)(10, 0)(8, 3)(8, 7).$$

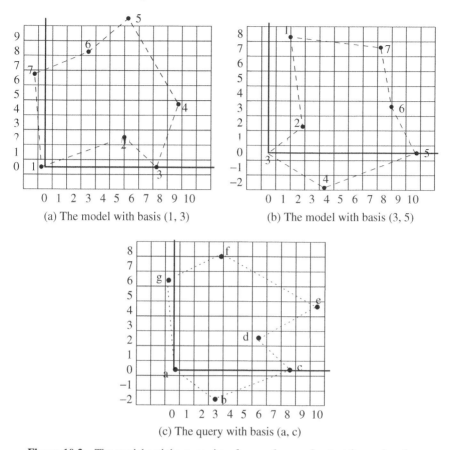

Figure 10.2   The model and the query in reference frames. See text for explanation.

In (c) the points (a,c) are selected as the basis for the query, and the points come in squares

$$(0, 0)(3, -2)(8, 0)(6, 2)(10, 4)(3, 8)(0, 6),$$

but some of the points are on borders. Comparing the query frame system with the two-model frame systems shows that four points coincide between (a) and (c) (including the origin), namely (1,a)(2,d)(3,c)(6,f). This is two fewer than what was found in Figure 10.1, namely (4,e) and (7,g). This demonstrates that the method is sensitive to the threshold, but this can be taken care of in the implementation of the method (by, for example, also considering neighbouring squares).

For the frame systems (b) and (c) it is only the origins that coincide.   △

Usually, one does not know beforehand which pairs to use as the basis in the model and query for finding maximal coincidences. Generally, all pairs should therefore be used as basis, and each frame system from the model compared with each frame system constructed from the query. However, one could have as a constraint that the basis should be in the coincidence set, hence the distances between the two points in the basis should be equal in the model and the query. This would reduce the number of frame systems without omitting good coinciding sets.

Using all reference frames may introduce redundancy. For example, let (a_i, a_k) be the basis for A, and (b_j, b_l) the basis for B, and let both (a_r, b_u) and (a_s, b_v) coincide in that (common) frame system. Then it is likely that we will get the same coincidence set if (a_r, a_s) and (b_u, b_v) are used as bases. Note, however, that similarity and not exact equality is used in the calculations and comparisons; hence getting the same set cannot be guaranteed.

In the naive method $(m(m - 1)/2)(n(n - 1)/2)$ frames are compared, where m and n are the number of points in the model and query. In order to perform all the comparisons efficiently, hashing is used, and therefore the approach is referred to as geometric hashing. It is especially efficient when several queries are to be compared with one or with several models.

Hashing

The local comparison problem can be formulated as follows. Given a query reference frame system, for each model reference frame system, find the number of coinciding points, or, in other words, in how many squares are there points from both the query and model frame system? The hashing technique makes it possible to simultaneously compare a query frame system to all model frame systems.

Hashing is a technique for storing and retrieving data in a (hash) table. A *hash function* is used to map the data (or more commonly the *key*) to (legal) indices into the table, and the data are stored in the indexed *bucket*. In that way several different items can be mapped to the same bucket, and special techniques must be used for handling this. For example, can each bucket have a linked list of elements.

Example

Assume that we want to store information about people in a one-dimensional hash table, and let the key be an identification number of 11 digits. Let a (simple) hash function be to take the three digits in positions five to seven. The hash table will then be of length 1000, and all records with the same digits in the three chosen positions will be mapped to the same bucket.   △

For simplicity, let us for our 2D figures define a 2D hash table H, and let it have a bucket for each square in the frame systems (see Table 10.3). Hence the hash function is very simple: it is the positions of the squares. In a *preprocessing* phase, the coordinates of all points in each model frame system are found. If there is a point in the square (p, q) in the frame system with basis (a_i, a_k), then the basis (a_i, a_k) is placed in the bucket $H(p, q)$. The table H for the two frame systems in Figure 10.2 is shown in Table 10.3. The bases (1,3) and (3,5) are both placed in six buckets of H. There might in total be $m(m - 1)/2$ or $m(m - 1)$ reference frame systems, depending

x \ y	−2	−1	0	1	2	3	4	5	6	7	8	9	10
−1									(1,3)				
0													
1											(3,5)		
2					(3,5)								
3											(1,3)		
4	(3,5)												
5													
6					(1,3)								(1,3)
7													
8			(1,3)		(3,5)				(3,5)				
9					(1,3)								
10			(3,5)										

Figure 10.3 A simple hash table for the model with the bases in Figure 10.2.

on whether there are one or two reference frames for each pair (using both (a_i, a_k) and (a_k, a_i) as basis). The number of pairs in H will therefore be either $m(m − 1)^2/2$ or $m(m − 1)^2$. Generally, a bucket will contain several bases when several reference systems are stored for the model, although this is not the case in the example.

After the preprocessing phase comes the *recognition phase*, where a query is compared with the model. To find coinciding points, a pair is chosen as basis in the query, and for each of the other points the coordinates in the frame system are computed, as for the model. These are then used as indices in H. For each bucket being indexed, a vote is given for each of the bases in the bucket. The number of votes for a model basis is the number of coinciding points between the model frame system and the query reference frame. For the example in Figure 10.2c), using Table 10.3, three votes are given for (1,3), from the points c,d,f (if the point at the origin is not included). There are no votes for (3,5). Hence there are four coinciding points between model (a) and the query (including origin), and only the point at the origin between model (b) and the query, using the specified basis.

Labels (e.g. colours and/or forms) assigned to the points might also be included, such that coinciding pairs must also have equal labels. This inclusion can be done in (mainly) two different ways.

1. The labels can be stored in the hash table buckets together with the basis. In the recognition phase, for a query point to match a model point, an explicit test for equality (or similarity) of labels must be done, e.g. whether the colour of the model point matches the colour of the query point.

2. The hash table can be indexed by the label(s) in addition to the coordinates; in this way the test for equality of labels is done implicitly in the hashing.

Example

Let the squares be numbered $1 \cdots m$ in one dimension, and $1 \cdots n$ in the other, and let the points be coloured by one of five different colours. Number the colours from one to five. A hash function to a one-dimensional table can then be such that a point in square (x, y) with colour c is hashed to the bucket $nm(c-1) + n(x-1) + (y-1)$ (the same as an implementation of a 3D table). $\triangle$

It is now straightforward to extend the hashing technique to include several models, so that a query is simultaneously compared with several models. The only extension in the hash table is to store the model identifications along with the basis.

10.1.2   Geometric hashing for structure comparison

In order to compare structures we need methods which are invariant to translation and rotation. Geometric hashing has these properties. Additionally, geometric hashing as explained here does not use the sequential order of the points and is therefore also appropriate for structure comparison in cases where the sequential order should be ignored. Note that finding a coincidence set therefore means finding an equivalence, and not an alignment.

As explained earlier, comparison of structures can be done at different levels (e.g. residue or secondary structure element) and geometric hashing can be used in all cases. We first discuss its use at residue level.

Structures can be looked upon as sets of points in three-dimensional space (e.g. taking the C_α of each residue, or several atoms from each residue). For using geometric hashing, 3D reference frames must be defined (the bases should consists of three points). Any three points (a_i, a_k, a_r) not falling in a straight line (noncollinear) can be used. These points define a plane. One way of defining a reference frame is to let a_i be the origin, let the x-axis lie along the vector (a_i, a_k), the y-axis on the plane and orthogonal to the x-axis in a counterclockwise direction. The z-axis is orthogonal to the plane, such that the x-, y- and z-axes defines a right-handed coordinate system. The coordinates of all residues (atoms) are then calculated for this system.

The residues might have assigned labels (e.g. amino acid type, secondary structure element, exposure), and as explained in Section 10.1.1 this can be implemented explicitly or in the hashing. The preprocessing in Algorithm 10.1 assumes that the labels are used in the hashing. For indexing the table for a point a_p, a hash function, F, of the coordinates and the label is used: $F(a_i, a_k, a_r, a_p, p_L)$, where p_L is the label of a_p. F must calculate the coordinates of a_p in the frame (a_i, a_k, a_r).

Algorithm 10.1. Preprocessing phase of geometric hashing.

var
M_u the model structures
H the hash table
F the hash function
p_L the label of atom a_p
begin
   for *each model* M_u **do**
        for *each (ordered, noncollinear) triple* $(a_i, a_k, a_r) \in M_u$ **do**
          calculate the reference frame R_{ikr}
          for *each atom* $p \in M_u$ **do**
            calculate $F = F(R_{ikr}, a_p, p_L)$; $H(F) := H(F) \cup (M_u, R_{ikr})$
          end
        end
   end
end

Any hashing function F could be used, but ideally the granularity should be so fine that no two atoms from the same structure are hashed into the same bucket.

As the reference frame with basis $(a_i a_k a_r)$ is different from one with, for example, basis $(a_k a_i a_r)$, there are $m_u(m_u - 1)(m_u - 2)$ different reference frames for each model; hence the time complexity of the preprocessing is $O(m_u^4)$ per model (where m_u is the number of atoms in model M_u). It is possible to reduce the complexity by not using all frames, but without any additional knowledge this might exclude some solutions. The space requirement of the hash table for a model is $O(m_u^4)$.

In the recognition phase, a reference frame (using three atoms) is defined for the query, and for each of the atoms the coordinates and indices into H are calculated. All (M_u, R_{ikr}) in the indexed bucket are then voted for. At the end all such pairs (structure, reference frame) with high votes are investigated further for the similarity of the model structure with the query structure. Algorithm 10.2 describes the procedure. The complexity depends on how many cycles are done in the repeat-until loop, but in the worst case it is of $O(n^4)$, where n is the number of atoms in the query structure. We assume that the processing of each hash table bucket takes constant time.

Algorithm 10.2. Recognition phase of geometric hashing.

Compare the query structure with each model structure in the hash table
var
V the two-dimensional vote table
H the hash table
F the hash function
q_L the label of atom a_q
M a model structure
R a reference frame
begin
   repeat
      initialize the vote table V to 0
      choose 3 atoms (b_j, b_l, b_s) in the query as basis,
      defining the reference frame R_{jls}
      for *each atom q in the query* **do**
         calculate $F = F(R_{jls}, b_q, q_L)$
         for *each pair $(M, R) \in H(F)$* **do** $V(M, R) := V(M, R) + 1$ **end**
      end
   until *satisfactory coincidence set is found or all query ref. frames are used*
end

The result of this step is a list of (M, R_{ikr}, R_{jls}), showing that there are coincidences between the model M and the query that have been found using the reference frames R_{ikr} (for M) and R_{jls} (for the query). The points constituting the coincidence set are also known, (the data structures used includes more than shown in this overview presentation). A superposition using the coincident points can be performed, thus defining a transformation.

Note that joining by using similar transformations requires that the superposition must be done relative to a 'global' reference system, the same for all superpositions. This could be the original coordinate system, in which the coordinates of the structures were given.

The procedure explained is efficient for establishing a database of hash tables for a set of model structures that are later to be compared with query structures. When the task is to compare two structures, not hashed before, we can (for reasons of efficiency) reduce the number of model reference frames. One can, for example, claim that for a triplet (a_i, a_k, a_r) to be a reference frame there must be a triplet (b_j, b_l, b_s) in the query, so that the triangles they form must be highly similar and the corresponding atoms must have matching labels.

In order to find local similarities, usually only the atoms lying within a sphere (of a defined radius) with centre at a chosen residue are included in the reference frame system.

10.1.3   Geometric hashing for SSE representation

Geometric hashing can also be used when the elements are SSEs. Typically, the elements are represented by *sticks*, for example, determined by the C_αs of the SSE, e.g. by using a least-squares adaption, and direction from the N-terminus to the C-terminus. The description below is based on a method by Holm and Sander (1995).

By the use of two sticks, a coordinate reference frame can be defined in many different ways, usually with one axis along one of the sticks. Depending on how the frames are defined, there are constraints on the relation between the sticks, e.g. that they have to be noncollinear. One example of a reference frame is shown in Figure 10.4.

Preprocessing

In the preprocessing step, a reference frame is defined for each pair of SSEs (or, to save time, one can restrict it to, for example, each pair with midpoint not more than h Å apart, h being a constant). For each reference frame a 3D grid is defined, with a resolution of h' Å (h' being another constant). The cells of the grid make up a hash table. The position of the midpoint and the direction in the reference frame is calculated for each of the other SSEs, and the midpoint is used for indexing the hash table. A bucket in the hash table contains a set of descriptions, each describing information for one SSE:

- an identification of the SSE;

- an identification of the basis;

- the type of SSE (α-helix, β-strand);

- the transformed midpoint of the SSE (i.e. the midpoint in the reference frame);

- the transformed direction of the SSE;

- possibly other information.

Recognition

In the recognition phase reference frames are defined for each (relevant) pair of SSEs of the query (B), and the hash table is used for finding reference systems with similar spatial architectures.

The procedure is described in Algorithm 10.3. Since only the midpoint is used for hashing, a comprehensive comparison for similarity must be performed in the recognition phase. The operations are inexact, hence several (neighbouring) buckets are investigated.

Algorithm 10.3. Recognition phase of geometric hashing, SSE-representation.

Compare the query structure B with each model structure
Comments to some of the statements:
**1) B_q might match an A_p in cells surrounding the one found by B_q (see *2),
therefore several neighbouring cells are investigated.
**2) match occurs if similar midpoints to less than h'' Å, less than $h'''°$
between the direction, and similar SSE type (some sequential constraints
could also be included).
**3) saving for further investigation.
var
B the query structure
H the hash table
b, b_q buckets of H
N a set of neighbouring buckets
begin
   for *each (relevant) pair of SSEs* $(B_j, B_l) \in B$ **do**
      define the reference frame R_{jl}
      for *each SSE* (B_q) *of* B **do**
         map B_q *to bucket* b_q *in* H
         $N := b_q \cup$ *all neighbour buckets to* b_q *in* H**1)*
         for *each bucket* $b \in H$ **do**
            for *each list in* b **do**
               let A_p *be the SSE of the list*
               if *the values of* B_q *are similar to the ones of* A_p **then****2)*
                  increase the vote for the reference (A_i, A_k) *of the list,*
                  and save (A_i, A_k, A_p, B_q)**3)*
               end
            end
         end
      end
   end
end

We now have equivalences of SSEs, and those can be further investigated. For example, can the residues of the SSEs of an equivalence be superposed and refined by an alternating superposition and alignment procedure?

10.1.4   Clustering

By geometric hashing similar substructures in A and B are found. The next step can then be to find sets of similar substructures which can be grouped, making larger similar substructures. This is described in Chapter 11.

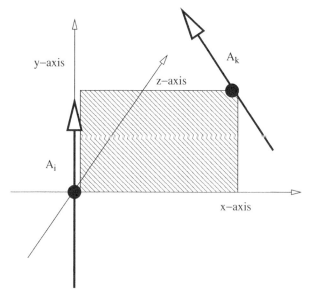

Figure 10.4 A reference frame defined by A_i and A_k. The origin is at the middle of A_i and the y-axis along A_i. The middle of A_k lies in the x-positive xy-half-plane. (The middle of A_k must not be along the y-axis.) The z-axis is orthogonal to the plane.

10.2 Distance Matrices

Geometric hashing is used for finding similar substructures, independent of sequential order. Distance matrices are mainly used to find similar substructures where the sequential order is the same (but this does not need to be a requirement in the clustering step). Comparison using distance matrices can be done in a *global* or in a *local* manner.

In a global manner we try to uncover common SSEs, by looking at the whole matrices. Figure 10.5 is part of the distance matrix for PDB entry 1chc, and Figure 10.6 is part of the distance matrix for PDB entry 1rmd.

Only by looking at the patterns (ignoring the values) the figures reveal some similarities between the submatrix for residues $\sim$(5–50) of 1chc and the submatrix for residues $\sim$(25–70) of 1rmd. Both show four areas around antidiagonals (possible connections between antiparallel strands), and one sheared area around the main diagonal (potential helix). This might indicate that the SSEs found in 1chc (see Figure 10.5) have corresponding SSEs in 1rmd. Indeed there are only two strands in the shown area of 1rmd, corresponding to the first two in 1chc, as can be seen in Figure 10.7. Also there exists a helix in 1rmd corresponding to the helix in 1chc.

A local comparison approach is used when one compares small fragments of two structures. Similar subareas around the main diagonals in two matrices represent similar substructures, and similar subareas off the main diagonal represent similar

```
                          bbbbb    bbbb aaaaaaaa              bbbbb
                          11111    2222 11111111              33333
             123456789A123456789B123456789C123456789D123456789E123456789
        1  046.......................................................
        2  4046......................................................
        3  64047.....................................................
        4  .64047....................................................
        5  ..74046......77...........................................
        6  ...74046.....6...........77...............................
        7  ...64046..5766............................................
        8  ....6404554776...........57..............................
        9  ......640466.............7...............67...............
       10  ......54046.................76..7......7.................
       11  ......564046..............................................
       12  .....546640456............................................
       13  ...77..64046..............................................
       14  ...7.67...54046...........................................
       15  ....7666...6640467.........65.............................
       16  ............64047.........6...............................
       17  ............64047.........55.........................6.
   b1  18  ...........774046......56457.....................676..
   b1  19  ...........74046.....556457.....................6657...
   b1  20  ..........64046..5546..........................557....
   b1  21  .........640476567.....7...................67657......
   b1  22  .........6403457...........................6656......
       23  ..........73046............................777........
       24  .........644046...........................7..........
       25  .........55564046.........................
   b2  26  .........567.64046........................
   b2  27  .....7.....5547..64047....................
   b2  28  ...7.57.....6..656....64047766...........
   b2  29  .......7.....56546......7404657...........
       30  ...........554........74046..............6....
   a1  31  ...........75......76404657.............7......
   a1  32  ........7......7.....656404667.................
   a1  33  ........6.......7.....67.6404556...............
   a1  34  ...................56404556...........7........
   a1  35  .................765404667.....................
   a1  36  ........7........7554046667....................
   a1  37  ................65640454476....................
   a1  38  ................66640466.......................
       39  ................7654046........................
       40  ................6464047........................
       41  ................74664046.....67................
       42  ...............7..74047.74457.........
       43  ..............6...6404554577.........
       44  .......67.............740466..........
       45  .......7...............54046..........
       46  ......................7564046.........
       47  ......................44664047........
       48  ...............7.......45..64046......
       49  ......................657...74047......
       50  ...........667........777...64046......
       51  ...........767...............74047.....
   b3  52  ...........657...............64047.....
   b3  53  ...........6556........7..7......74047....
   b3  54  ...........657...............74046...
   b3  55  ...........657........6.......74046..
   b3  56  ...........77.................64047.
       57  ..........66.................64047
       58  .............................7404
       59  .............................740
```

Figure 10.5   A distance matrix for the first 59 residues of PDB entry 1chc, where each residue is represented by its C$_\alpha$ atom. The distances are rounded to integers, and distances larger than 7 Å are represented by dots. The SSEs (as given in PDB) are shown, 'a' for helix and 'b' for strand.

```
    123456789A123456789B123456789C123456789D123456789E123456789F1234567
 1  0456.........................................................................
 2  404556.......................................................................
 3  5404557......................................................................
 4  654046.......................................................................
 5  .554047......................................................................
 6  .6564047..................................67.7...............................
 7  ..7.74047.7...............................66................................
 8  .....740456...............................56................................
 9  .....740456...............................66................................
10  ......54045.................................................................
11  ......7654046...............................................................
12  ........6540467.............................................................
13  .........64047..............................................................
14  ..........64046.............................................................
15  .........774047.............................................................
16  ..........04040.70..........................................................
17  ...........7404666..........................................................
18  ............6404556.........................................................
19  ............6404556.........................................................
20  ...........765404557........................................................
21  ...........6655404567.......................................................
22  .............65540465.......................................................
23  ..............6554046.......................................................
24  ............7664047.....66..................................................
25  ............7564046..5655...................................................
26  ...............7404565665..........646......................................
27  ..............640466...............66...................7..................
28  ..............54046...............7...77...............7...................
29  ..............664046...............7.......................................
30  ...........55664047.........................................................
31  .....6656.......66..64046....................................................
32  ..7666.........656...74046...................................................
33  ..............655...640466........75.......................................
34  ....7...............64046.........57.......................................
35  ..................64047.......56...........................................
36  ..................664047.....6656..........................................
37  ...................74047....6666...........................................
38  ...................74047.6657..............................................
39  ...................74045557................................................
40  ...................74046...................................................
41  ...................54046..................................................7.
42  ...................6564047.................................................
43  ...................65.64047.............................7..................
44  .............66.......6657..74047..........................................
45  .............467...7..667....74046.66......................................
46  .............6......55556.....7404656......................................
47  ...................7666......6404656.......................................
48  ...........................6404656.........................................
49  .................77........656404667.......................................
50  .................7.........6666404556......................................
51  ..........................656404656........................................
52  ..........................665404556........................................
53  ..........................7564045557.......................................
54  ..........................65540466557......................................
55  ..........................655404667........................................
56  ...........................6564046.........................................
57  ...........................5664047.........................................
58  ...........................75664046........................................
59  ...........................57.74047.....6..................................
60  ...........................7..64046..554...................................
61  ...........................740456456.......................................
62  ...............7...........640466..........................................
63  .............7.............7.............54046..............................
64  ..........................................664046.............................
65  ..........................................54664046...........................
66  ............................7.............55..6404...........................
67  ..........................................646...640...........................
68  ..........................77............7..556...64...........................
69  ...........................................7..6......7.......................
70  ...........................................6..7..............................
71  ............................................................................
72  ..............................66............................................
73  .............................6657...........................................
74  .............................557............................................
75  .............................756............................................
76  .............................766............................................
77  ............................................................................
78  ............................................................................
79  ............................................................................
```

Figure 10.6 A distance matrix for the first residues of PDB entry 1rmd, where each residue is represented by its C_α atom. The distances are rounded to integers, and distances larger than 7 Å are represented by dots.

```
     1rmd   1chc           1rmd   1chc           1rmd   1chc
a     A 18   1 M      **T 39   21 A** b       *Y 60   42 T*
a     H 19   2 A*       S 40   22 L** b      **C 61   43 C**
a     F 20   3 T       *C 41   24 C           *P 62   44 P*
a   **V 21   4 V        K 42   25 L           *S 63   45 L*
a     S 23   5 A**     *H 43   26 H*  b        C 64   46 C
    **I 24   6 E*  b  b **L 44  27 A*  b        R 65   47 K
    **S 25   7 R*     b **F 45  28 F** b        Y 66   48 V
    **C 26   8 C**    b  *C 46  29 C*  b        P 67   49 P*
    *Q 27    9 P*     a  *R 47  30 Y*         **C 68   50 V*
    *I 28   10 I*     a   I 48  31 V*  a       *F 69   51 E
    *C 29   11 C      a  *C 49  32 C*  a   a  *D 72    52 S   b
    *E 30   12 L*     a  *I 50  33 I** a      *L 73    53 V*  b
    **H 31  13 E      a  *L 51  34 T*  a       E 74    54 V*  b
    *I 32   14 D*     a  *R 52  35 R   a      *S 75    55 H*  b
    *L 33   15 P**    a  *C 53  36 W*  a      *P 76    56 T   b
    *A 34   16 S      a **L 54  37 I** a       V 77    57 I*
     D 35   17 N*     a   K 55  38 R   a   a   K 78    58 E*
b   *P 36   18 Y*  b  a   V 56  39 Q       a   S 79    59 S
b  **V 37   19 S** b      G 58  40 N*
b  **E 38   20 M*  b     *S 59  41 P*
```

Figure 10.7   The first part of an alignment of 1chc and 1rmd using SAP. The SSEs as given in PDB are specified by 'a' for α-helix, and 'b' for β-strand.

(spatial) relations (connections) between substructures. Note, however, that similar relations do not imply that the corresponding substructures are similar.

Example

In order to illustrate the relational patterns we consider a structure of elements in two dimensions. Assume two substructures of elements $A_i = \{a, b, c\}$ and $A_k = \{v, x, y, z\}$, respectively, as shown in Figure 10.8. The distance matrix is partly shown in Table 10.1.

The distance matrix is shown in part in Table 10.1.

Now suppose we have another structure, with substructures $B_j = \{a', b', c'\}$ and $B_l = \{v', x', y', z'\}$. Suppose further that the relation pattern (distance submatrix) between B_j, B_l is similar to the relational pattern between A_i, A_k. This does not imply that the distance submatrix for B_j is equal to the one for A_i. $\triangle$

10.2.1   Measuring the similarity of distance (sub)matrices

When searching for common substructures, or common relations, in two structures one at the same time wants to find as many corresponding elements as possible, and that the set of corresponding elements 'fit as well as possible' to each other. When using distance matrices for the representation, one must therefore have a scoring function which at the same time takes into account the number of corresponding

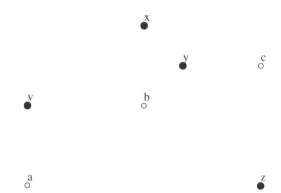

Figure 10.8 Two substructures used for illustrating distance matrices and relation patterns, where only the elements are shown. See the text in the example for explanation.

Table 10.1 Distance matrix for the (2D) structures in Figure 10.8.

\	a	b	c	. . . . .	v	x	y	z	. . . . . .
.									
.									
a	0	3.7	6.8		2	5	5	6	
b		0	3.2		3	2	1.5	3.6	
c					6.2	3.2	2	3	
.									
.									
v					0	3.6	4.2	6.5	
x						0	1.5	5.1	
y							0	3.7	
z								0	
.									

cells, and the difference between the corresponding cells. Such a scoring is presented in Holm and Sander (1993b) (for use in the DALI program). Consider (without loss of generality) two submatrices (of equal size) $D_A(i_1 \cdots i_2, k_1 \cdots k_2)$ of structure A, and $D_B(j_1 \cdots j_2, l_1 \cdots l_2)$ of structure B. A formula for the similarity score is

$$S(D_A(i_1 \cdots i_2, k_1 \cdots k_2), D_B(j_1 \cdots j_2, l_1 \cdots l_2))$$

$$= \sum_{p=0}^{i_2-i_1} \sum_{q=0}^{k_2-k_1} \phi(D_A(i_1 + p, k_1 + q), D_B(j_1 + p, l_1 + q)), \quad (10.1)$$

where ϕ is a similarity measure.

When the submatrices are around the main diagonal, they represent substructures, and the similarity of the substructures and the matrices are symmetrical, simplifying the formula.

Two variants of ϕ are described, the *rigid* and *elastic* similarity score.

The rigid form for ϕ is

$$\phi^R(D_A(i,k), D_B(j,l)) = \Theta^R - |D_A(i,k) - D_B(j,l)|,$$

where Θ^R is the zero level of similarity, defined to be 1.5 Å. This means that

- differences less than 1.5 Å give a positive contribution to the score,
- differences greater than 1.5 Å give a negative contribution to the score.

Example

Let us compare substructure 43–47 of 1rmd with substructure 26–30 of 1chc. That means comparing two submatrices around the main diagonal, hence two symmetrical submatrices. They are (to one decimal)

```
1rmd                                          1chc
        43    44    45    46    47                    26    27    28    29    30
43   0.0                                      26   0.0
44   3.8   0.0                                27   3.8   0.0
45   7.1   3.9   0.0                          28   6.3   3.8   0.0
46  10.6   7.1   3.8   0.0                    29  10.0   6.7   3.9   0.0
47  12.0   9.0   6.2   3.8   0.0              30  11.5   8.3   6.5   3.8   0.0
```

The score is

$$5(1.5 - 0) + 2[(1.5 - 0) + (1.5 - 0.8) + (1.5 - 0.1) + \cdots + (1.5 - 0)]$$
$$= 7.5 - 2[11.5] = 30.5.$$

Comparing the distance matrices $D_{1\text{rmd}}(35\ldots39, 72\ldots75)$ with $D_{1\text{chc}}(18\ldots22, 50\ldots53)$ means comparing the relation in 1rmd between the substructures of residue $(35\ldots39)$ and $(72\ldots75)$ with the relation in 1chc between substructures $(18\ldots22)$ and $(50\ldots53)$:

```
1rmd                                      1chc
        72     73     74     75                    50     51     52     53
35   15.3   11.5   10.7    8.5            18   15.6   16.0   13.1    9.3
36   13.1    9.4    8.6    6.5            19   12.2   12.4    9.5    5.7
37    9.3    5.7    5.2    4.7            20   10.1   10.5    8.2    5.3
38    8.1    5.6    5.2    5.8            21    6.3    7.4    6.0    4.6
39    6.1    5.3    6.8    8.9            22    5.9    5.7    5.2    5.8
```

$\triangle$

The rigid similarity score does not take the size of the corresponding distances into consideration. This means, for example, that the corresponding distances 3 and 5 count the same as the corresponding distances 16 and 18. The elastic form for ϕ also uses the size of the distances:

$$\phi^E(D_A(i,k), D_B(j,l)) = \left(\Theta^E - \frac{|D_A(i,k) - D_B(j,l)|}{D^*(i,k,j,l)}\right) w(D^*(i,k,j,l)), \tag{10.2}$$

where $D^*(i, k, j, l)$ is the average of $D_A(i, k)$ and $D_B(j, l)$. When comparing sub-structures, $D_A(i, i)$ and $D_B(j, j)$ are zero, so ϕ^E is set to Θ^E for the diagonal cells.

Θ^E is the elastic similarity threshold (chosen equal to 0.20, i.e. 20% deviation). The value of Θ^E is chosen from considering typical distances for SSE contacts. Adjacent strands in a β-sheet typically have distances of 4–5 Å, and should match to within 1 Å. Typical distances for strand–helix or helix–helix contacts are 8–15 Å, and should tolerate differences of 2–3 Å. For damping the contribution from the pairs in the long-distance range, which are abundant but less discriminative, the envelope function w is used, defined as $w(r) = \exp(-r^2/\alpha^2)$, where $\alpha = 20$ Å.

10.3 Exercises

1. This exercise illustrates geometric hashing. For simplicity we consider 'structures' in 2D space. Let A, B be two 'structures' of atoms with the following coordinates:

   A: a_1: (2, 11), a_2: (2, 7), a_3: (6, 7), a_4: (6, 11), a_5: (4, 1),
       a_6: (8, 5), a_7: (8, 1),

   B: b_1: (0.4, 5.2), b_2: (3.2, 10.8), b_3: (6, 13.6), b_4: (8.8, 10.8),
       b_5: (6, 8), b_6: (9, 6), b_7: (13, 6), b_8: (13, 2), b_9: (9, 2),
       b_{10}: (6, 5.2), b_{11}: (3.2, 2.4).

   (a) Plot the points in a diagram. (It might also be useful to draw lines between the points, to reveal squares and triangles.)

   (b) You are now to define reference frames, and the coordinates of the points in the reference frames are to be found. It is enough to find the coordinates of the points lying in a radius of 7 from the origin. You can use the procedure explained in Exercise 3 in Chapter 9. Define reference frames using the pairs of points: (a_2, a_3), (a_5, a_7) and $(b_1, b_{11})(b_2, b_5)(b_9, b_8)$. Define origin at the first point, and the x-axis through the second.

   (c) Define a two-dimensional hash table, and fill in data from both reference frames of A. An atom will be hashed to bucket (i, j) if its coordinates (x, y) satisfy $i - 0.5 < x \leqslant i + 0.5$ and $j - 0.5 < y \leqslant j + 0.5$. Fill only in those points laying inside a radius of seven from the origin.

   (d) For each of the reference frames of B, use the hash table to find similar 'substructures' of A and B consisting of at least three pairs of atoms.

2. In Section 10.1.1 on p. 214 (2D hashing) is written:

   > Using all reference frames may introduce redundancy. For example, let (a_i, a_k) be the basis for A, and (b_j, b_l) the basis for B, and let both (a_r, b_u) and (a_s, b_v) coincide in that (common) frame system. Then it is likely that we will get the same coincidence set if (a_r, a_s) and (b_u, b_v) are used as bases. Note, however, that similarity and

not exact equality is used in the calculations and comparisons; hence getting the same set cannot be guaranteed.

Show (graphically) that if the calculations and comparisons could be done exactly, the two cases would give the same coincidence set. Show that this does not need to be the case when the comparison is not exact.

3. Algorithm 10.2 assumes that the labels are included in the hashing function, and Algorithm 10.3 that the labels are saved in the hash table. Discuss the advantages and disadvantages for these two different ways of treating the labels.

4. Let A_1 and A_2 be two substructures of structure A, and B_1 and B_2 of B, represented as distance matrices. Let substructures A_1 and B_1 be equal (have equal submatrices) and the submatrix representing the relation between A_1 and A_2 be equal to the submatrix representing the relation between B_1 and B_2. Show that then the substructures A_2 and B_2 must have equal distance submatrices. Assume that there are at least four elements in A_1 and B_1.

5. Calculate the scoring between the substructures (43...45) of 1rmd and (26... 28) of 1chc, when the elastic form of the similarity measure ϕ is used.

6. At the end of Section 10.1.2 is written that for reducing the running time when comparing two structures one can claim that for a triplet (a_i, a_k, a_r) to be a reference frame there must be a triplet (b_j, b_l, b_s) in the query, such that the triangles that they form must be highly similar. Propose a way for doing such tests in an efficient way.

10.4   Bibliographic notes

An introduction to geometric hashing can be found in Wolfson (1997). Articles for use in structure comparison are Nussinov and Wolfson (1991), Fischer et al. (1994), Alesker et al. (1996) and Pennec and Ayache (1998). The method for SSE representation is from Holm and Sander (1995). A method for comparing flexible protein (around a 'hinge') is described in Shatsky et al. (2002a).

Use of distance matrices is described in Holm and Sander (1993b).

11

Clustering: Combining Local Similarities

In the preceding chapter we described some methods that are suitable for finding similar (local) substructures between two structures. The next step is then to try to combine two or several such pairs of similar substructures into larger similar substructures. This is done by clustering methods. Use of the word clustering here can be a bit misleading, since not all 'objects' have to be in at least one cluster. It is used, however, since it is widely used in the papers describing the relevant methods.

There exists a lot of different methods for structure comparison which are classified as clustering methods. In this chapter we describe the main common approach, and present some methods in more detail.

11.1 Compatibility and Consistency

The basic objects in the explanation of the clustering approach are pairs of *compatible elements*, one from each of the two structures. Two elements are compatible if they share some similarity, what kind of similarity depends on the details of the clustering method. Two residues may be defined to be compatible if they have the same amino acid type or have amino acid in the same amino acid group. Secondary structure elements may be compatible if they are of the same type, and for fragments similarity of internal distances might be required (distance matrices). This means, for example, that discovering compatible fragments can be done by finding similar submatrices around the main diagonals of the two structures, as explained in Chapter 10.

Compatible elements can then be grouped. For illustration, let A_i be compatible with B_j, and A_k with B_l, A_i and A_k being substructures of A, B_j and B_l being substructures of B. Two compatible pairs can be grouped if they are *consistent*. This means that (A_i, B_j) must be consistent with (A_k, B_l), which they are if the substructure consisting of (A_i, A_k) is compatible with the substructure (B_j, B_l), i.e. similar to a certain degree (the exact constraints being dependent on the method). Figure 11.1

Protein bioinformatics: an algorithmic approach to sequence and structure analysis
I. Eidhammer, I. Jonassen and W. R. Taylor © 2004 John Wiley & Sons, Ltd   ISBN: 0-470-84839-1

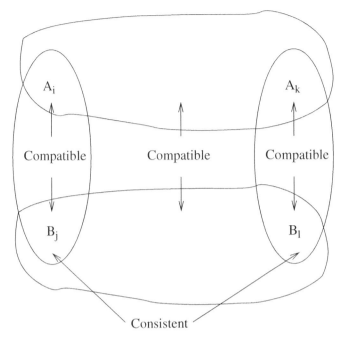

Figure 11.1 (A_i, B_j) are compatible, and so are (A_k, B_l). The pair (A_i, B_j) are consistent with (A_k, B_l), meaning that (A_i, A_k) is similar (compatible) with (B_j, B_l).

illustrates compatibility and consistency. Note that A_i, B_j, etc., may consist of one or several residues.

Note that *compatibility* is a binary relation between elements of *different* structures, and *consistency* is a binary relation between *pairs*, where a pair consists of one part (list of elements) from each of the two structures. Figure 11.2 shows an example of compatible pairs that are not consistent.

The methods using clustering first find *seed matches* (local similar substructures) between the structures, and then join these into larger clusters, a cluster representing one substructure from each structure. A seed match (also a cluster) consists of one or several pairs of compatible elements. (A_i, B_j) can, for example, be a seed match. A seed match on residue level can, for example, consist of the equivalence $[(a_1, b_2)(a_3, b_5)(a_5, b_3)(a_6, b_6)]$, found, for example, from using geometric hashing techniques.

Seed matches are joined into consistent clusters, and scores can be assigned to the clusters, depending on the level of similarity of the two substructures.

The clustering methods follow this scheme.

1. Find seed matches, each a pair of similar substructures, i.e. sets of compatible elements which are mutually consistent, $\{P_s\} = \{(A_{i_s}, B_{j_s})\}$. All elements in a seed match must be distinct, but an element can be in several seed matches.

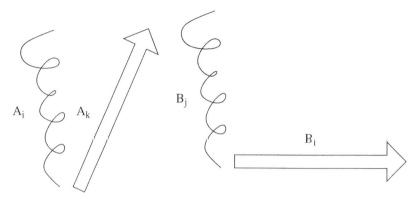

Figure 11.2 A_i is compatible with B_j, and A_k with B_l, but (A_i, B_j) is not consistent with (A_k, B_l) since (A_i, A_k) is not similar to (B_j, B_l).

Geometric hashing is one method for finding seed matches; use of distance matrices is another.

2. Iteratively group seed matches which are consistent into clusters, representing (*h*) substructure(s) with high score.

3. Optional refinement. This is often done using an iterative method alternating between superposition and alignment, as explained in Chapter 9.2.

Example

Assume we are given the structures $A = \{a_1, \ldots, a_{12}\}$ and $B = \{b_1, \ldots, b_{14}\}$, and that we find the following seed matches:

$$S_1 = \{(a_2, b_4), (a_5, b_7), (a_7, b_5)\},$$
$$S_2 = \{(a_1, b_3), (a_3, b_6), (a_5, b_4)\},$$
$$S_3 = \{(a_5, b_7), (a_6, b_3), (a_8, b_8), (a_9, b_{10})\},$$
$$S_4 = \{(a_4, b_1), (a_{10}, b_9), (a_{11}, b_{11})\}$$

(by, for example, geometric hashing). The clustering (grouping) now depends on the criteria for consistency. Suppose that we find that S_1 and S_3 are consistent, and group them into the cluster:

$$C_{1,3} = \{(a_2, b_4), (a_5, b_7), (a_7, b_5), (a_6, b_3), (a_8, b_8), (a_9, b_{10})\}.$$

Then we check whether $C_{1,3}$ is consistent with any of the other seed matches. This continues iteratively until no more clusters (or seed matches) can be grouped. △

The clustering methods vary in how they solve the following points.

1. How is compatibility defined, and how can the compatible elements and the seed matches be found?

2. How is consistency defined?

3. How is clustering performed (clustering algorithms)?

4. How are the clusters scored (the score is normally based on how similar and large the substructures are)?

5. How is refinement done?

11.2 Searching for Seed Matches

Different methods are used for finding the seed matches, e.g. straightforward search when each seed match consists of only one pair of elements. Geometric hashing can also be used. For example, the structures can be divided into small spheres, e.g. of 10–15 Å around each C_α, and seed matches then searched for in pairs of spheres, one from each structure. (This keeps the number of elements to be included in each hashing procedure low.) Geometric hashing has also been used in order to find seed matches when the elements are SSEs, as explained in Subsection 10.1.3.

When the elements are fragments, compatibility can, for example, be decided by requiring equal lengths, and using distance matrices. Similar submatrices around the main diagonal represent compatible elements.

11.3 Consistency

Two clusters C_1, C_2 (a cluster consisting of one or several seed matches) can be joined if they are consistent. Let the elements (substructure) from A in cluster 1 be C_1^A, the others denoted in the same way. Then the clusters can be joined if the substructure consisting of (C_1^A, C_2^A) is 'similar' to the substructure consisting of (C_1^B, C_2^B). Substructure similarity can be checked by seeing if the pair (C_1^A, C_1^B) is consistent with (C_2^A, C_2^B). To decide consistency, either *relations* between elements of the same structure are used or *transformations* between compatible elements from different structures. Figure 11.3 illustrates the concepts of relation and transformation, and Figure 11.4 illustrates with examples. By transformation is typically meant a transformation giving a good superposition, and in this case consistency exists if two transformations are similar enough.

Distance matrices are another way to represent relations, where similar submatrices off the main diagonal represents similar relations. Some comparison methods use both relations and translations for deciding consistency.

11.3.1 Test for consistency

Let the two clusters be $C_1 = \{P_r\}$ and $C_2 = \{Q_s\}$, where P_r and Q_s are pairs of consistent elements. Different methods check for consistency in different ways.

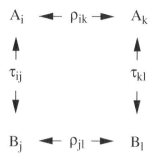

Figure 11.3   Illustration of relation (ρ) and transformation (τ). Relation is between elements or substructures of one structure, and transformation is between elements or substructures from two different structures.

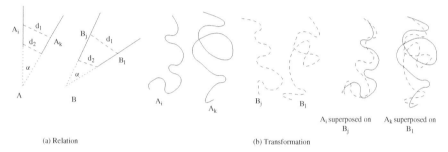

Figure 11.4   Examples of relation and transformation. A_i, A_k are two elements from a structure A, and B_j, B_l the same for structure B. Assume (A_i, B_j) and (A_k, B_l) each are compatible. (a) *Relation*, the elements represented as sticks. The relation is between elements from the same structure, here represented by two distances (d_1, d_2) and an angle (α, found by projection onto a plane). The pair (A_i, B_j) is consistent with (A_k, B_l) if the relation $(d_1, d_2, \alpha)_A$ is similar to the relation $(d_1, d_2, \alpha)_B$ (the substructure (A_i, A_k) is similar to substructure (B_j, B_l)). (b) *Transformation*, the elements represented by the 3D coordinate for each residue. The transformation is between elements from different structures, in the example it is the transformation for the superposition. If the transformation for the superposition of A_i onto B_j is similar to the one for A_k onto B_l, then (A_i, B_j) is consistent with (A_k, B_l). The transformations must refer to the same coordinate system. Reproduced from Eidhammer et al. (2000) by permission of Mary Ann Liebert.

1. C_1 and C_2 are looked upon as sets of pairs, and the test for consistency can be done on the pairs (known as *local clustering criteria*). Whether all pairs need to be tested depends on whether the criteria is transitive. Transitivity means that if an arbitrary pair P_r and Q_s is consistent, then all pairs are consistent, and one test is enough. Otherwise, all pairs must in principle be tested.

   Example. Let

   $$C_1 = \{(A_{i_1}, B_{j_1}), (A_{i_2}, B_{j_2}), \dots\} \quad \text{and} \quad C_2 = \{(A_{k_1}, B_{l_1}), (A_{k_2}, B_{l_2}), \dots\}.$$

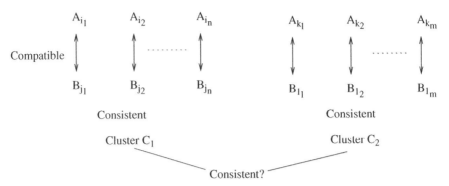

Figure 11.5 Cluster C_1 consists of n pairs, and cluster C_2 of m pairs. They can be joined if the substructure consisting of $\{A_{i_1}, A_{i_2}, \ldots, A_{i_n}, A_{k_1}, A_{k_2}, \ldots, A_{k_m}\}$ is similar (consistent) to the substructure consisting of $\{B_{j_1}, B_{j_2}, \ldots, B_{j_n}, B_{l_1}, B_{l_2}, \ldots, B_{l_m}\}$.

> If transitivity holds, it is enough to compare, for example, (A_{i_1}, B_{j_1}) with (A_{k_1}, B_{l_1}). For a simple illustration we use examples from sets of numbers. An example of a transitive relation is '='. An example which is not transitive is if the relation is difference. Let two sets be $\{7, 5, 4, 3, 8\}$ and $\{5, 8, 9, 4\}$, and the relation for consistency be that the difference must be less than 5. We see that 7 from the first set and 5 from the second satisfy this, but not 3 from the first and 9 from the second. Hence to decide if two clusters are consistent, it is not enough to test two pairs (how many depends on the context (see Exercise 2).△

2. C_1 and C_2 are looked upon as 'global units', and the test for consistency is done on these two units (known as *global clustering criteria*). With transformation, global criteria are often used. Substructures (one from each structure) are defined as consisting of all elements in a cluster, and for each cluster the transformation for the superposition of the substructures is found. Two clusters can then be joined if their transformations are similar.

Figure 11.5 illustrates clustering.

11.3.2   Overlapping clusters

In addition to similar transformations or relations, overlapping clusters must correspond for clusters to be joined. If two clusters contain some of the same elements, these elements should be paired in the same way in both clusters in order for the two clusters to be joined.

Example

Let $C_1 := \{(A_2, B_2), (A_4, B_1)\}$ and $C_2 := \{(A_2, B_3), (A_3, B_4)\}$. These two clusters cannot be joined, since A_2 is equivalenced with B_2 in C_1, but with B_3 in C_2.   △

11.4   Clustering Algorithms

The task is to group the seed matches such that the highest possible scoring pair of substructures is found. For some formulations of this problem, it can be shown to be equivalent to the well-known maximum clique problem, which is NP-hard. Heuristics are therefore used.

General clustering algorithms were described in Chapter 4; here we show how to use them for pairwise structure comparison. Figure 11.6 shows examples illustrating the different ways of clustering explained below.

11.4.1   Linear clustering

In linear clustering one has a current cluster, and in each cycle a comparison between the current cluster and a seed match is made. The procedure is shown in Algorithm 11.1. The result depends on the order in which the seed matches are presented. Since there are $m!$ orders for m seed matches, not all orders can be tested/used. One approach is to use each seed match to start a clustering, and in every cycle to join the seed match that best fits the current cluster, where a seed match is considered best if, for example, it results in the cluster with the highest score.

Algorithm 11.1. Linear clustering.

Ideally, the algorithm should be run for each seed match starting a new cluster
How to select the next seed match is not specified
var
$\mathcal{C}$ the current cluster
begin
   $\mathcal{C} := $ *one selected seed match*
   for *each other seed match sm* **do**
      if *sm is consistent with* $\mathcal{C}$ **then** *join sm to* $\mathcal{C}$
   end
end

Parallel linear clustering

Parallel linear clustering means that there are several current clusters. In each cycle, the seed match which 'best fits' (scores highest) to one of the clusters is joined to it. This means that in each cycle the remaining seed matches are tested against all current clusters. See Algorithm 11.2.

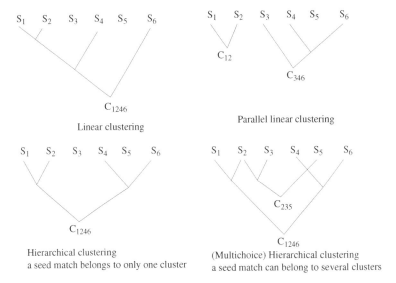

Figure 11.6   Examples of different clustering techniques.

Algorithm 11.2. Parallel linear clustering.

Each seed match is in this version in only one cluster
begin
   make a current cluster of each seed match, retaining the seed matches
  repeat
     find seed match (sm) which are consistent and 'best fits'
     to one of the clusters (C)
     if *there exist such a seed match sm* **then**
        join sm to C; remove sm from the set of seed matches
     end
  until *no more seed matches are joined to clusters*
end

11.4.2   Hierarchical clustering

In agglomerative hierarchical clustering each seed match initially forms a cluster and in each cycle the pair of clusters that are most similar (score highest) is joined. Algorithm 11.3 shows a procedure. A seed match belongs to only one cluster. However, we often want to find several of the best clusters. This can be achieved by not removing the clusters being joined, but recording the order in which the clusters were formed. In addition, some minor changes must be done in the algorithm. The result of the clustering can be drawn as a binary tree with the seed matches as leaves and each cluster as a subtree.

Algorithm 11.3. Hierarchical clustering.

var

$\mathcal{C}$	the set of current clusters
$\mathcal{H}$	the pairs of consistent clusters in $\mathcal{C}$ that increase score by grouping them
$S(C)$	the score of a cluster C
C_1, C_2	current highest-scoring pair of clusters in $\mathcal{H}$
$S(C_1, C_2)$	the score of grouping (C_1, C_2)

proc

$\quad$ incr$(C_i, C_j) := $ cons(C_i, C_j) **and** $[S(C_i, C_j) > \max(S(C_i), S(C_j))]$

$\qquad\qquad$ incr(C_i, C_j) is true if C_i and C_j are consistent,

$\qquad\qquad$ and the score increases by grouping them

begin

$\quad \mathcal{C} := $ *set of clusters, each seed match being a cluster*

$\quad \mathcal{H} := \{(C_i, C_j) \mid C_i \in \mathcal{C}$ **and** $C_j \in \mathcal{C}$ **and** incr$(C_i, C_j)\}$

$\quad$ *determine* C_1, C_2

$\qquad$ **while** $\mathcal{H} \neq \emptyset$ **do**

$\qquad\qquad C_B := group(C_1, C_2)$ *the highest-scoring pairs in* $\mathcal{H}$

$\qquad\qquad \mathcal{C} := (\mathcal{C} - \{C_1, C_2\}) \cup \{C_B\}$

$\qquad\qquad$ *update* $\mathcal{H}$ *and determine* C_1, C_2

$\qquad$ **end**

end $\qquad \mathcal{C}$ now contains the found clusters,

$\qquad\qquad$ and no pairs will increase the score

11.5 Clustering by Use of Transformations

Clustering by use of transformations can be used when the representation is both on the fine and coarse levels. One possibility is to use geometric hashing for finding the seed matches (see Figure 11.7).

When the elements are SSEs, a typical seed match consists of a (consistent) list of compatible SSEs, e.g. $\{(A_i, B_j)(A_k, B_l)\}$, typically found by using the relations between the SSEs. For each pair, the corresponding atoms (C_α) are determined (which gives the superposition). The transformation is then found.

The transformation referred to here is normally the one giving minimum RMSD, and can be described by two components: a rotation and a translation. Therefore, to examine if two clusters $C_1 = (S_1^A, S_1^B)$ and $C_2 = (S_2^A, S_2^B)$ can be grouped, we must compare the rotational and the translational components.

11.5.1 Comparing transformations

When comparing two transformations to decide if they are similar enough to justify consistency of their clusters, the rotational and the translational components must be compared. Quoting from Vriend and Sander (1991), 'if two proteins are perfectly superposable, then every pair of equivalent substructures is also perfectly superposable, with the same rotational component of the superposition transformation; devi-

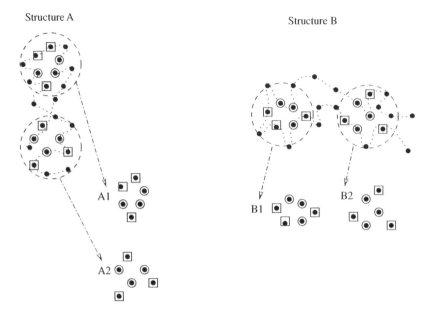

⬚ The residues used to define reference frames

◉ Residues which are found to have compatible residues in the other structure

Figure 11.7 The figure illustrates clustering from seed matches found by geometric hashing. The residues inside a sphere are hashed for each reference frame, and two reference frames are shown for each structure. Two seed matches are shown (A_1, B_1) and (A_2, B_2), where the residues used to define the reference frame are found to be part of the seed matches. The transformation for superposing (A_1, B_1) is found to be similar to the one for superposing (A_2, B_2). Therefore, those two seed matches can be grouped into one cluster, with the same transformation for superposing (A_1, A_2) onto (B_1, B_2).

ations from perfect superposability can be measured in terms of deviations in the rotational component. Because the same reasoning does not hold for the translational component of the transformation, we use the vector between the centres of mass...'. It is therefore common to compare the rotation and the translation separately.

Comparing rotations

One method for comparing the rotation of the transformation is presented by Vriend and Sander (1991). As stated in Chapter 9.1.1, the rotation can be described as a rotation around one axis. The difference between this angle of two rotations can then be used as a measure for the dissimilarity of two rotations. This difference can be found by multiplying one of the rotational matrices (R_1) with the inverse of the other (R_2), and quantifying the departure from the unit matrix (I) in terms of the resulting net rotation angle δ given by

$$\cos \delta = \tfrac{1}{2}[\mathrm{trace}(R_1 R_2^{-1}) - 1],$$

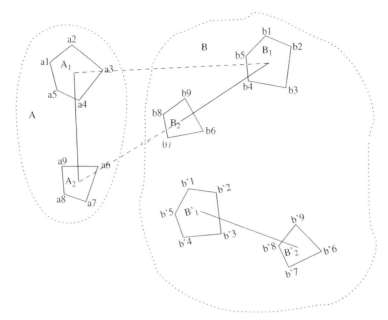

Figure 11.8   Seed matches examples, in two dimensions.
Nine of the atoms of A, and 18 of the atoms of B are shown.

where 'trace' means the sum of the diagonal elements. We see that $\delta = 0$ if R_1 and
R_2 represent identical rotations. Vriend and Sander set a value of 0.2–0.3 rad for
considering the two rotations as similar.

Alesker et al. (1996) describe a rotation by the angles around the three axes of the
coordinate system. For comparing two rotations, the sum of the absolute values of
the three-angle differences is compared with a threshold (the threshold used is 0.8).

Comparing translations

For two proteins being perfectly superposable, the translational component for every
pair of equivalent substructures does not need to be equal. To illustrate this, we visualize in two dimensions, where the rotation is around a point, but the main principles
will be the same as for three dimensions.

In Figure 11.8 is nine of the atoms of structure A, and 18 of the atoms of B, and
edges are drawn for clearness. We will find four seed matches:

$$\{(A_1, B_1)(A_1, B_1')(A_2, B_2)(A_2, B_2')\}.$$

Then we have to investigate whether we can cluster any of the seed matches, by
calculating the transformations for the seed matches: $T_{11} = \text{trans}(A_1, B_1)$, $T_{11}' = \text{trans}(A_1, B_1')$, $T_{22} = \text{trans}(A_2, B_2)$, $T_{22}' = \text{trans}(A_2, B_2')$, and compare these.

We first calculate the rotational components, and compare them. In two dimensions
the rotations can be described by one angle, for the example we let ϕ_{11} be the angle

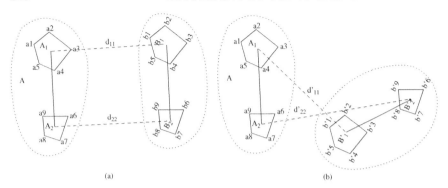

(a)   (b)

Figure 11.9 (a) (B_1, B_2) is rotated, and the vectors between the centres are compared.
(b) (B_1', B_2') is rotated, and the vectors between the centres are compared.

for the best superposition of (B_1, A_1), ϕ_{22} the angle for (B_2, A_2), and ϕ_{11}' and ϕ_{22}'
be analogously defined. We find that ϕ_{11} is similar to ϕ_{22}, and ϕ_{11}' is similar to ϕ_{22}'.
Therefore, it might be that (A_1, B_1) can be clustered to (A_2, B_2), and (A_1, B_1') to
(A_2, B_2'). We therefore need to compare the translations. The translations cannot be
compared directly, as shown in Figure 11.8. (A_1, B_1) is consistent with (A_2, B_2), but
the vectors between their centres (dashed lines) are different.

Therefore, we perform the rotations (found to be similar) on the combined substructures of B, and then compare the distances between the geometrical centres
of the seed matches. This is illustrated in Figure 11.9. The combined substructure
(B_1, B_2) is rotated ϕ_{11}° (a). Then the vector d_{11} is compared with the vector d_{22}, they
are found to be similar, and (A_1, B_1) can be clustered to (A_2, B_2) $((A_1, A_2)$ is similar
to $(B_1, B_2))$.

The same is done for (B_1', B_2') (see Figure 11.9(b)); it is rotated $\phi_{11}'^\circ$, and the vectors
d_{11}' and d_{22}' compared. They are not found to be similar, therefore (A_1, B_1') cannot be
clustered to (A_2, B_2') $((A_1, A_2)$ is not similar to $(B_1, B_2))$.

Ordinary measures for comparing vectors are used for comparing the translations
(e.g. the sum of the squares of the differences along each axis), and a threshold defined
for concluding similarity.

Alternative comparisons

Comparing the transformations by their parameters is very fast, but not accurate. Also,
one has to decide how to 'balance' the two thresholds against each other. Therefore,
alternative approaches are also used. Leibowitz et al. (2001) measure the distance
between transformations on the basis of the 'set theoretical' difference between the
sets of points that are superposed within some threshold. Define one of the structures
as the *source*. Each cluster C_i defines a transformation T_i for the source. Let $P\,P$ be the
union of all clusters. Each transformation is applied to the set of atoms from the source
appearing in $P\,P$. For each transformation all atom pairs (one atom from the source and
one from the other) in $P\,P$ which are sufficiently close after superposition are stored

in a list, L_i. The distance between T_j and T_k is then defined as $|L_j - L_k| + |L_k - L_j|$; this is the number of atom pairs which is found in one of the list, but not in both. This measure is clearly symmetric and rational, since similar transformations should find similar atom pairs. Two transformations are said to be similar if the distance between them is less than a given threshold.

Example

Assume two of the clusters are $C_j = \{(7, 3)(9, 5)(13, 7)(14, 10)(17, 13)\}$ and $C_k = \{(14, 11)(17, 13)(22, 17)(25, 21)\}$, where the number refers to atoms of the structures. Assume further that using T_j on all atoms of the source appearing in PP results in the list: $L_j = \{(7, 3)(9, 5)(13, 7)(14, 10)(17, 13)(21, 16)(22, 17)(25, 21)(26, 22)(29, 23)(30, 24)\}$. Let the corresponding list for T_k be $L_k = \{(9, 5)(11, 6)(17, 13)(21, 16)(22, 17)(25, 21)(26, 22)(33, 27)(35, 28)\}$. Then the difference between T_j and T_k is $|L_j - L_k| + |L_k - L_j| = 5 + 3 = 8$.   △

The clusters with the least difference are then commonly chosen for being clustered in the next step in a clustering procedure. Other more comprehensive methods can be used for deciding which to cluster next, for example, to give a score to the clusters resulting from joining each pair of clusters, and choosing the pairs with highest score.

Another alternative is presented in Dror et al. (2003), where the two transformations are used on the three farthest points. The RMSD value of the differences of the two results is calculated, and if it is under a given threshold (in the interval 1–3 Å), the two clusters are supposed to be consistent.

11.5.2   Calculating the new transformation

For each new cluster, a corresponding transformation has to be calculated, since a global clustering criteria is used. In order to obtain the most correct transformation, a new superposition should be done. If this is considered to be too time-consuming, one can, for example, use one of the original transformations. This is regarded by some people as more accurate than using a weighted average of the two original transformations. The reason for this is that a little change in rotational angle can result in a large change in the actual 3D position, especially when the centre of the coordinate system is far away from the mass centre. Taking the average might result in the mismatch of both the clusters, but by using one of the original transformations without changing it, at least one of the clusters is guaranteed to be matched.

11.5.3   Algorithm

Algorithm 11.4 describes the use of transformation and linear clustering.

Algorithm 11.4. Use of transformation and linear clustering.

var

C_{PQ}	denotes a cluster. P is the set of elements from structure A Q the corresponding from B, hence P and Q are ordered sets
$\mathcal{C}$	is an ordered set of 'current' clusters (each described by C_{PQ})
$\mathcal{T}$	is an ordered set of transformations T_{PQ} for the clusters in $\mathcal{C}$ Hence there is a one-to-one correspondence between $\mathcal{C}$ and $\mathcal{T}$ T_{PQ} is usually the transformation for superposition of the elements in P on the elements in Q
$\mathcal{H}$	is a data structure containing the pairs of consistent clusters in $\mathcal{C}$ and the scores if the pairs should be clustered. The score could be based on RMSD
S_{max}	the highest score of joining any two clusters in $\mathcal{C}$

procedure

join(C_1, C_2) join two clusters

begin

   $\mathcal{C} :=$ *all seed pairs (as singleton clusters)*
   $\mathcal{T} :=$ *the transformations for the clusters in $\mathcal{C}$*
   $\mathcal{H} :=$ *the pairs of consistent clusters in $\mathcal{C}$, and the scores if joining them*
   $(C_{PQ}, C_{RS}) :=$ *the consistent pair in $\mathcal{H}$ with highest score*
   $S_{max} :=$ *the score of* join(C_{PQ}, C_{RS})
   while $(C_{PQ}, C_{RS}) \neq \emptyset$ **and** score(join$(C_{PQ}, C_{RS})) \geqslant S_{max}$ **do**
      $S_{max} :=$ score(join(C_{PQ}, C_{RS}))
      $C_{uv} :=$ join(C_{PQ}, C_{RS}) the joined cluster
      $T_{uv} :=$ *the transformation for C_{uv}*
      $\mathcal{C} := \mathcal{C} - (C_{PQ} \cup C_{RS}) \cup C_{uv}$
      $\mathcal{T} := \mathcal{T} - (T_{PQ} \cup T_{RS}) \cup T_{uv}$
      find all $C \in \mathcal{C}$ which are consistent with C_{uv}
            and calculate the scores if joining them
      change $\mathcal{H}$ by removing all data for C_{PQ} and C_{RS} and adding the
            consistency data and scoring for C_{uv}
      $(C_{PQ}, C_{RS}) :=$ *the consistent pair in $\mathcal{H}$ with highest score*
   end

end

Example

In order to illustrate the algorithm, let the elements from A be $A_1, A_2, A_3, \ldots$ and from B be $B_1, B_2, B_3, \ldots$, and let the seed matches be $\mathcal{C} = \{C_{\{1\}\{2\}}, C_{\{2\}\{2\}}, C_{\{2\}\{4\}}, C_{\{3\}\{1\}}, C_{\{3\}\{3\}}, C_{\{3\}\{4\}}, \ldots\}$, where $C_{\{1\}\{2\}}$ means the pair (A_1, B_2).

   For illustration we show $\mathcal{H}$ as a table. '—' means noncorresponding overlapping elements, and '*' means supposed dissimilar transformations.

Initial $\mathcal{H}$	$C_{\{1\}\{2\}}$	$C_{\{2\}\{2\}}$	$C_{\{2\}\{4\}}$	$C_{\{3\}\{1\}}$	$C_{\{3\}\{3\}}$	$C_{\{3\}\{4\}}$	$\cdots$
$C_{\{1\}\{1\}}$	—	18.4	20.6	—	*	12.6	
$C_{\{1\}\{2\}}$		—	19.2	13.4	18.7	*	
$C_{\{2\}\{2\}}$			—	*	20.2	28.8	
$C_{\{2\}\{4\}}$				—	30.7	—	
$C_{\{3\}\{1\}}$					—	—	
$C_{\{3\}\{3\}}$						—	
$C_{\{3\}\{4\}}$							
.....							

Suppose the highest score in the whole table is for joining $C_{\{2\}\{4\}}$ and $C_{\{3\}\{3\}}$, resulting in the cluster $C_{\{2,3\}\{4,3\}}$. The transformation $T_{\{2,3\}\{4,3\}}$ is then found and compared with the other clusters (after $C_{\{2\}\{4\}}$ and $C_{\{3\}\{3\}}$ are removed). Then we might get the new table for $\mathcal{H}$.

$\mathcal{H}$ after one cycle	$C_{\{1\}\{2\}}$	$C_{\{2\}\{2\}}$	$C_{\{2,3\}\{4,3\}}$	$C_{\{3\}\{1\}}$	$C_{\{3\}\{4\}}$	..
$C_{\{1\}\{1\}}$	—	18.4	*	—	12.6	
$C_{\{1\}\{2\}}$		—	27.6	13.4	*	
$C_{\{2\}\{2\}}$			—	*	28.8	
$C_{\{2,3\}\{4,3\}}$				—	—	
$C_{\{3\}\{1\}}$					—	
$C_{\{3\}\{4\}}$						
.....						

$\triangle$

There might be a lot of current clusters. To reduce the running time, one can use techniques which will remove clusters which are likely to not be in the best alignments.

11.6   Clustering by Use of Relations

To decide whether two compatible pairs (A_i, B_j) and (A_k, B_l) are consistent, an internal (to each structure) relation ρ is used. The pairs are consistent if $\rho(A_i, A_k)$ and $\rho(B_j, B_l)$ are approximately equal.

ρ should satisfy the following: $\rho(A_i, A_k)$ and $\rho(B_j, B_l)$ should be approximately equal if and only if the substructures (A_i, A_k) and (B_j, B_l) are considered as similar. This is achieved if superposing substructures constructed from the set of consistent elements, results in small RMSD values. ρ should be invariant with respect to rotation and translation, and in most cases to sequential order. Local clustering criteria have to be used, and transitivity is generally not fulfilled (see Figure 11.10). Therefore, in general all pairs should be tested for consistency.

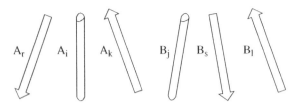

Figure 11.10 An example showing that relations are generally nontransitive. Assume a cluster $\{(A_i, B_j)(A_k, B_l)\}$. Using a suitable limit we find $\rho(A_i, A_r) \approx \rho(B_j, B_s)$ but $\rho(A_k, A_r) \not\approx \rho(B_l, B_s)$. This means that (A_r, B_s) is consistent with (A_i, B_j), but not with (A_k, B_l).

11.6.1 How many relations to compare?

Assume we have two clusters $C_1 = \{P_1, P_2, \ldots, P_m\}$ and $C_2 = \{Q_1, Q_2, \ldots, Q_n\}$, where each (P_i, Q_i) is a compatible pair. Local clustering criteria have to be used. If transitivity does not hold, then in the worst case mn relations have to be compared in order to decide whether C_1 and C_2 are consistent. However, the number of comparisons required in practice depends on the form of the relation formula ρ (see Exercise 2).

11.6.2 Geometric relation

By geometric relation we mean how two sticks representing substructures are related in space. Different methods use different numbers of parameters for defining ρ. The number depends on how stringent the constraint for being consistent is defined, i.e. given two sticks, how many parameters are needed to describe the 'geometric relation' between them? (This means changing the value of one of the parameters might change the geometry.)

We know that checking for consistency using transformation requires comparing six values, thus six values should also be enough when using relations. Usually, ρ is defined by distances and angles between the sticks, often one angle and two to four distances.

An example is taken from Alexandrov and Fischer (1996), where one angle and four distances are used. First, a medium line is defined by the middle point of the shortest line between axes of the SSEs. The angle δ between the axes (when projected on a parallel plane) and the minimum ($d^{\min}$) and maximum ($d^{\max}$) distances from each SSE to the medium line are used for the relation (see Figures 11.11(a)).

For two pairs of compatible SSEs to be consistent, they require

- $|\delta_{A_{ik}} - \delta_{B_{jl}}| < \delta$, usually $\delta = 50°$;

- $d_i^{\min} - d_j^{\max} < \epsilon$, where ϵ is, for example, 1.5 Å;

- $d_j^{\min} - d_i^{\max} < \epsilon$;

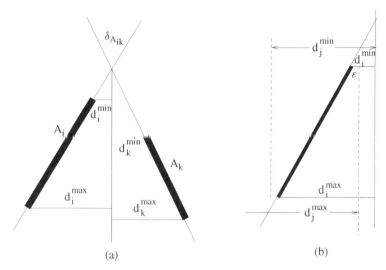

(a) (b)

Figure 11.11 (a) The parameters of ρ, visualized in 2D projection. A plane can be defined for each of the SSEs, such that the two planes are parallel (this is always possible). The SSEs are then projected on a third parallel plane, and the middle line drawn in this plane. The distances are then found. (b) Illustrating the distance constraints $d_j^{max} > d_i^{min} + \epsilon, d_j^{min} < d_i^{max} + \epsilon$. Reproduced from Alexandrov and Fischer (1996) by permission of Wiley-Liss.

- $d_k^{min} - d_l^{max} < \epsilon$;

- $d_l^{min} - d_k^{max} < \epsilon$.

Alexandrov and Fisher claim that these distance and angle restraints filter out differently arranged pairs of SSEs, leaving only superposable SSEs, and that the filtering procedure can detect difficult cases, when only relatively small segments of the SSEs are consistent, or when one of the SSEs is significantly larger than the other. The constraint is illustrated in Figure 11.11(b).

11.6.3 Distance relation

Compatible elements can be found by defining similar distance patterns along the main diagonals (intra-residue distances). To decide if two compatible pairs are consistent, we must compare the distance patterns relating the elements (inter-residue distances). These are the relations. In Figure 11.12, A_i is compatible with B_j, A_k with B_l and A_r with B_s. To see if (A_i, B_j) is consistent with (A_k, B_l), we must compare the distance pattern which relate A_i with A_k to the distance pattern which relate B_j with B_l. In the figure these are marked as similar.

To check whether (A_k, B_l) is consistent with (A_r, B_s), we must compare the distance patterns which relate A_k with A_r to the distance pattern which relates B_l with B_s. In the figure these are also marked as similar. Thus we have two clusters

$C_1 = \{(A_i, B_j)(A_k, B_l)\}$ and $C_2 = \{(A_k, B_l)(A_r, B_s)\}$. Can these be combined? We have to use local clustering, which means that each pair from the two clusters must be consistent. Here it is sufficient to check consistency between (A_i, B_j) and (A_r, B_s), and we need to compare the distance pattern relating A_i with A_r with the distance pattern relating B_j with B_s. In the figure these are drawn with empty patterns. If the two distance patterns are sufficiently similar, the two clusters can be combined. Figure 11.12(b) shows possible structures from the distance matrices in (a). Note that since the order of the compatible elements along the sequences are not the same, the structures have different topologies. (Figure 11.12 and the description above is based on Holm and Sander (1993b).)

Scoring

In order to decide which clusters to group, the clusters have to be scored. A function as defined in Equation (10.2) (Chapter 10.2.1) can be used. Note that the differences must be summed over the distances for both the elements and the relations. The score of the cluster $\{(A_i, B_j), (A_k, B_l)\}$ is therefore based on a sum over the differences in the intra-distances (A_i, A_i), (B_j, B_j) and (A_k, A_k), (B_l, B_l), and the differences in the inter-distances (A_i, A_k), (B_j, B_l).

DALI

DALI (Holm and Sander 1993b) divides the structures into (overlapping) backbone fragments of length $L = 6$ and for each pair of fragments constructs a distance matrix. Therefore, for a structure of length n there will be $(n - L + 1)^2$ overlapping distance matrices. Next DALI searches for similar distance matrices from the two structures, using the scoring in Equation (10.2). Since the search space is huge, some techniques are used for reducing it. Similar distance matrices are then stored in a list.

First, seeds are found with the help of the distance matrices. A seed is an equivalence of residues. Some technical operations are done, such that a seed might have a length of more than six residue pairs. The n (~ 100) best seeds are then chosen for one expansion (clustering cycle). After this step, the m ($m = 10$) best clusters are chosen for further expansions by a parallel linear clustering. The decision of which seed to add is done by a Monte Carlo search (simulated annealing). In this step the seed to be added to a cluster must overlap with the cluster. The overall approach is illustrated in Table 11.1.

11.6.4   Use of graph theory

Clustering by relation can be formulated with the help of graph theory in the following way: the structures are represented as graphs, where the nodes are the elements, and there is an edge between nodes (A_i, A_j) if the relation $\rho(A_i, A_k)$ satisfies some constraints. A constraint may be that they are not far from each other, for example, that some of the van der Waals volumes around the atoms of A_i overlaps with some of the van der Waals volumes around the atoms of A_k. The constraint might be empty,

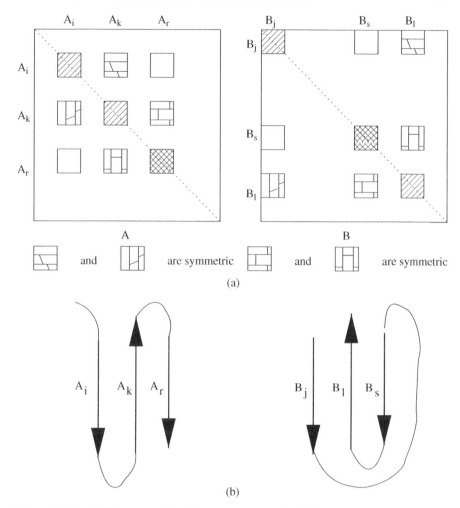

Figure 11.12 (a) Distance matrices for two structures with similar substructures. See text for explanation. (b) If the arrows represent SSEs, then the two structures have equal architecture, but different topologies (assuming equal handedness). Reproduced from Holm and Sander (1993b) by permission of Elsevier.

resulting in a complete graph. The nodes are labelled with element properties, and the edges with properties of the relations.

The problem is then, given two graphs, find a subgraph of each so that the subgraphs are similar (graph isomorphism) and have the highest (similarity) score which can be achieved between similar subgraphs. That means

- corresponding nodes must be compatible, and

- corresponding edges must be sufficiently similar.

Table 11.1 The overall approach of DALI. First, n seeds are found, and these are extended in one cycle (to clusters). Then the m highest-scoring clusters are extended further, and a final refinement is done for the highest scoring of these.

Make seeds	Expansion cycle	Choose the best	Expansion cycles	Refinement
s_1	se_1	\		
s_2	se_2	c_1	\	
.	.	c_2	C_1	CR_1
.	.	.	C_2	CR_2
.	.	.	.	.
.	.	.	.	.
.	.	.	C_k	CR_k
.	.	c_m	/	
s_n	se_n	/		

When the score is defined as the number of compatible pairs, the problem can be solved by finding the maximal cliques in a product graph (or connection graph), for which several algorithms exist. The (node) product graph is constructed by creating a node for each compatible node pair, and an edge from node (A_i, B_j) to node (A_k, B_l) if they are consistent, i.e. if the edges (A_i, A_k) and (B_j, B_l) have similar labels. See Figure 11.13 for an example.

In some methods (see, for example, Koch et al. 1996) an *edge* product graph is used. A node in an edge product graph corresponds to two edges, one from each structure graph.

It is also possible to assign a *continuous number* to the edges of the product graph (and to generalize from consistent/not consistent to giving a score of the consistency). In this way each cluster can be given a score $-\sum_{i \neq j} d((A_i, B_i), (A_j, B_j))$, where d is a measure of dissimilarity. Then search for the highest-scoring clusters is done, e.g. by hierarchical clustering.

11.7 Refinement

Depending on the clustering procedure, a final adjustment or refinement might be worthwhile. This is mainly done on residue level. The corresponding residues in the substructures are used, and an iteration of alternating superposition and alignment can be done, as explained in Chapter 9.2.

11.8 Exercises

1. Assume that A_i is similar to the mirror image of B_j, and that A_k is similar to B_l. By use of distance matrices we then find that A_i is compatible with B_j, and A_k

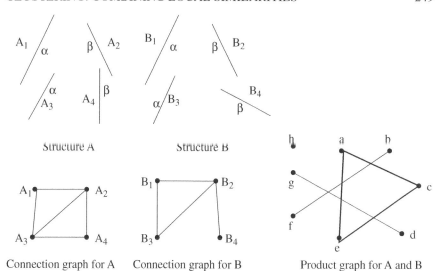

Structure A   Structure B   Connection graph for A   Connection graph for B   Product graph for A and B

Figure 11.13 Four SSEs are shown from each structure. We assume that the relations between $(A_1, A_2), (A_1, A_3), (A_2, A_3), (A_2, A_4), (A_3, A_4), (B_1, B_2), (B_1, B_3)(B_2, B_3), (B_2, B_4)$ satisfy the constraints for being edges in the connection graphs, as shown. Assume further that the constraint for compatibility is equal SSE type. We have the compatible pairs: $a : (A_1 B_1)$, $b : (A_1 B_3), c : (A_2 B_2), d : (A_2 B_4), e : (A_3 B_3), f : (A_3 B_1), g : (A_4 B_2), h : (A_4 B_4)$, where a is a node of the product graph, etc. Then there are eight nodes in the product graph. Assume that the relation $\rho(A_1, A_2)$ is similar to $\rho(B_1, B_2)$. This means that $(A_1 B_1)$ is consistent with $(A_2 B_2)$, therefore, there will be an edge from a to c. Assume further $\rho(A_1, A_3) \sim \rho(B_1, B_3)$, $\rho(A_1, A_3) \sim \rho(B_3, B_1), \rho(A_2, A_3) \sim \rho(B_2, B_3), \rho(A_2, A_4) \sim \rho(B_4, B_2)$. Therefore, there will be edges $a{-}e, b{-}f, c{-}e, d{-}g$ in the product graph. The maximal clique consists of a, c, e, hence the found similar substructures are $(A_1 A_2 A_3), (B_1 B_2 B_3)$.

with B_l. Then we use distance matrices to examine if the cluster (A_k, B_l) can be grouped with (A_i, B_j) (are consistent). Can the conclusion of the examination be wrong?

2. Suppose we have two clusters $C_1 = \{P_1, \ldots, P_m\}$ and $C_2 = \{Q_1, \ldots, Q_n\}$, and relations are used to test for consistency. If transitivity holds, it is only necessary with one test, if not, the worst case is mn. However, the real number of necessary comparisons depends on the form of the relation formula ρ. Suppose we have a theorem for a specific relation saying that if P_i is consistent with K (a constant number) of the pairs in C_2, then P_i is consistent with all pairs in C_2. How many tests are then needed in the worst case?

3. Change Algorithm 11.3 such that a seed match can belong to several clusters.

4. This exercise is to illustrate clustering by use of transformation. The seed matches can, for example, be found by geometric hashing, and we can use the 'similar' substructures found in Exercise 1 in Chapter 10.

(a) Each pair of seeds should now be examined to see if they are consistent. The first test could be to ignore those pairs which have overlapping atoms from one of the structures that are matched to different atoms in the other structure. For example, the pairs of seeds $[(b_1, a_2)(b_{10}, a_4)(b_{11}, a_3)]$ and $[(b_1, a_5)(b_2, a_2)(b_5, a_3)(b_{10}, a_6)(b_{11}, a_7)]$ cannot be grouped. Use this to find which pairs of seeds should be examined further.

(b) Now a transformation for each of the seeds should be found, and used to examine which of the seed pairs (from (a)) that are consistent. For the purposes of illustration, we will do it in a simpler way. We will say that two seeds (A_i, B_j) and (A_k, B_l) are consistent if the distance between the geometrical centres in A_i and A_k is approximately equal to the distance between the geometrical centres in B_j and B_l. We say that the distances are approximately equal if the difference is less than 0.1. Use this to find consistent seeds and write down the clusters.

(c) Compare the clusters with the diagram you drew in Exercise 1 of Chapter 10.

5. This exercise shall illustrate clustering by use of relations. We continue in 2D space, and consider SSEs. Let them have direction, and be represented as sticks. The 'structures' A and B are

A: $A_1(\alpha)$: $(4, 3) - (4, 8)$, $A_2(\beta)$: $(6, 3) - (10, 7)$, $A_3(\alpha)$: $(8, 1) - (11, 4)$, $A_4(\alpha)$: $(5.5, 4.5) - (7, 7.5)$, $A_5(\beta)$: $(8.5, 7.5) - (5, 5.5)$,

B: $B_1(\alpha)$: $(5.5, 4) - (5, 9)$, $B_2(\beta)$: $(7, 4.5) - (11, 8)$, $B_3(\beta)$: $(9, 2) - (12, 5)$, $B_4(\alpha)$: $(3.5, 4.5) - (1.5, 9)$, $B_5(\beta)$: $(0.5, 8) - (3, 7)$.

(a) Draw the SSEs in a diagram.

(b) Two SSEs are compatible if they are of the same type. Find the compatible pairs.

(c) As relation we shall use the distances between the end points, two values (d_1, d_2). (We ignore angles, to make it easy.) Find the relations between each pair of SSEs for each structure, to one decimal place. (If you wish, you can use a ruler.)

(d) We will say that two relations are 'similar' if the differences between corresponding distances are less than 0.6. Use this to find consistent pairs of the compatible pairs found in (b). You should find five consistent pairs.

(e) Examine if any pairs of the five consistent pairs found can be grouped into larger clusters.

(f) Write the largest (in number of SSEs) common substructures.

11.9 Bibliographic notes

Different methods using clustering are described by Vriend and Sander (1991), Alexandrov et al. (1992), Alexandrov and Fischer (1996), Fischer et al. (1995, 1994),

Alesker et al. (1996), Pennec and Ayache (1998), Verbitsky et al. (1999), Grindley et al. (1993), Madej et al. (1995), Rufino and Blundell (1994), Koch et al. (1996), Mizuguchi and Go (1995), Kleywegt and Jones (1997), Escalier et al. (1998) and Russel (1998). DALI is described in Holm and Sander (1993b). Other methods are cited in Chapter 13. Techniques for comparing transformations are presented in Vriend and Sander (1991), Leibowitz et al. (2001) and Dror et al. (2003).

Other methods

Several methods for structural comparison use general techniques for *searching* in a well-defined search space. One possibility is to define the search space as a set of transformations. A transformation T corresponds to one state, and the score given to the state is the score of the best alignment of the two structures under this transformation. The alignment is computed using dynamic programming where the score for a residue pair (a_i, b_k) is defined as a function of the distance $d(T a_i, b_k)$. Genetic algorithms can, for example, be used as the search method. Articles describing search methods are in Diederichs (1995), Falicov and Cohen (1996) and Holm and Sander (1996).

Statistical approaches are used when the proteins are characterized by some distributions of biological attributes, and compared by using a function (independent of protein length) that computes a measure of dissimilarity between the proteins. The methods using space-based descriptions are all statistical (Chapter 8.7.1). A common pattern can, for example, be found by comparing sets of corresponding cells (e.g. shells around a common centre).

The substructures can be described by the set of atoms within the cells, along with their 3D coordinates. In addition, user-defined properties can be included, such as the types of atoms, chemical groups, amino acids, secondary structures, charge, polarity, mobility and solvent accessibility. Another technique is using histograms for representing the distribution of the number of atoms in each cell, and the associated properties.

Examples of statistical methods are described by Bagley and Altman (1995) and Kastenmüller et al. (1998).

12

Significance and Assessment of Structure Comparisons

When comparing structures it is widely known from practical experience and from more systematic investigations that, beyond close similarity, there is no uniquely correct structural alignment of two proteins. Different alignments are achieved depending on which biological properties and relations are emphasized in the comparison. This adds a complicating element to the assessment of the result of a comparison. However, it is reasonable to assume that unique alignments exist for essential 'core' regions of homologous proteins.

The straightforward approach for assessing the significance of the result of comparing two structures is to do the same as for sequence comparison: to make a distribution over what scores can be expected only by chance. This could be done by constructing a set of random structures, and comparing those with one of the native structures. However, as for sequences, some basic nonrandom properties should remain. For sequences these include the length and amino acid distribution, for structures it would also be length and amino acid distribution, and in addition, overall shape (inertial axis), secondary structure content and packing density. As an extreme example, imagine that the proteins being compared contain only α-helices, and the unrelated random set of structures contains only β-strands. Relative to this background, the two α structures would appear more related than they should be. This demonstrates that the set of 'random' structures should have not only the 'nonrandom' properties intact but also include representatives of typical protein structures. Ideally, random structures (models) should be calculated for each comparison to match the nonrandom properties of the query structures.

12.1 Constructing Random Structure Models

When constructing models of random structures one has to ensure that the models can represent native structures; in steric terms, this implies a compact structure with

Protein bioinformatics: an algorithmic approach to sequence and structure analysis
I. Eidhammer, I. Jonassen and W. R. Taylor © 2004 John Wiley & Sons, Ltd ISBN: 0-470-84839-1

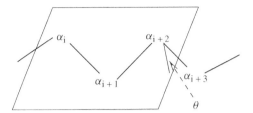

Figure 12.1   The torsion angle θ around the edge $\alpha_{i+1}, \alpha_{i+2}$ is the angle that atom α_{i+3} has relative to the plane defined by $(\alpha_i, \alpha_{i+1}, \alpha_{i+2})$.

nonoverlapping atoms. Several methods are used. The simplest procedure is a *self-avoiding random walk*. A chain is 'grown' by the addition of residues to the terminus so that steric clashing is avoided and residue packing is favoured (so that the structure remains compact). The length is also preserved.

12.1.1   Use of distance geometry

Dragon (Aszódi and Taylor 1994) is a program for constructing protein models in which only the C_α-atoms are used, and the structures are represented as distance matrices. The main constraints used in the procedure are the following.

- Each atom is modelled as a sphere.

- The bond length l between successive atoms is set to 3.8 Å.

- The minimal distance between two (not successive) atoms is $d_{bump} = 2r_{vdW}$ (van der Waals radius).

- The structure can then be described by two (virtual) angles for each residue: the bond angle β and the torsion angle θ. To define the torsion angle, one needs four points. The torsion angle θ around the edge $(\alpha_{i+1}, \alpha_{i+2})$ is the angle that atom α_{i+3} has relative to the plane $(\alpha_i, \alpha_{i+1}, \alpha_{i+2})$ (see Figure 12.1).

- The distance between atoms α_i and α_{i+2} is experimentally set to $d_2 = 6.0$ (± 0.4); hence $\beta = 2 \arcsin d_2/2l$.

- $(\pi/2) \leqslant \beta \leqslant \pi$.

- $-\pi \leqslant \theta \leqslant \pi$.

Most simply, this method can be used to 'randomize' a native structure by randomly changing some values in the distance matrix. Depending on the degree of 'noise' introduced, a range of structure variants (sometimes called 'decoys') can be generated at varying degrees of similarity to the starting structure. What is essential is that some of the main properties of the native structure can be retained during this changing, such as secondary structure elements and the presence of an hydrophobic core.

12.2   Use of Structure Databases

Random models are complex to generate and cannot be made for each individual comparison without extensive computation. Instead, it is easier to compare one of the structures (q) with a nonredundant structure database, and try to remove (or ignore) those scores coming from structures that are homologous to q.

12.2.1   Constructing nonredundant subsets

When using a database to generate a random background, one should work with a subset of structures which have limited similarity to each other. This is to avoid the results being biased, e.g. by the presence of many similar structures. Reasonable nonredundancy can be achieved by assuring that no two structures have sequence similarity over $T\%$. A common value for T in this context is 40% residue identity. Ideally, the threshold T should depend on the length of the structures. A proposed value for T is (Sander and Schneider 1991)

$$T(m, n) = 290.15(\tfrac{1}{2}(m + n))^{-0.562},$$

where m and n are the length of the structures, and sequence identity is computed from the structure alignment. T is about 25% identity for chains longer than 80 residues and is higher for shorter chains.

The nonredundant subsets are usually constructed by choosing the structures in PDB with the highest quality. The quality is assessed using the crystallographic resolution, secondary structure content, number of undefined atomic coordinates and degree of structure refinement. Web addresses for (constructing) nonredundant subsets are given in the bibliographic notes. For construction, some use local alignment and some use global alignment to allow for calculation of percentage identity.

12.2.2   Demarcation line for similarity

When RMSD value is used as a score, the number of residues over which the deviation is measured is critically important. When this measure is used, it is often plotted against the number of equivalent residues (N) and a curve drawn to separate 'true' (homologous) from 'random' (nonhomologous) similarities. Typically, the RMSD values of a large sample of pairs of nonhomologous structures (or fragments) are found, and the curve is calculated so that, for example, 99% of the values fall above the curve. The curve most often is of the form $a + b(N + c)^{1/x}$, where x is 1, 2 or 3. Figure 12.2 illustrates this.

12.3   Reversing the Protein Chain

When the comparison is based only on the α-carbon positions, it is possible to create a good random model by reversing the protein chain. The reversed model preserves

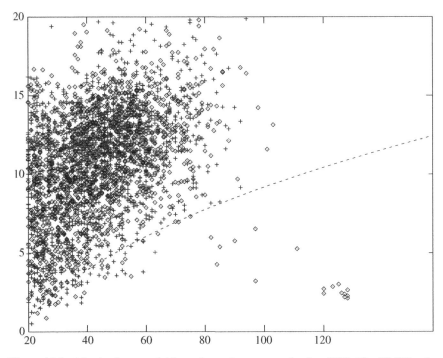

Figure 12.2 Match of a myoglobin probe against a nonredundant PDB. The RMSD value of each comparison (y) is plotted against the number of residues aligned (x) for comparisons using the native probe structure ($\diamond$) and for comparisons using the reversed probe as a control (+). The dashed line follows a function that has been scaled to include 99% of the reversed controls. The matches lying below this line include the probe matching itself (150 residues with an RMSD of 0), other globins (120–140 residues) and the phycocyanin family (80–115 residues), which has a similar fold.

the length and composition of the protein, including directionally symmetric correlations associated with secondary structure, while additionally in the reversed structural model, the bulk properties of the packing density and inertial axes are also preserved. (The latter are difficult to maintain in randomly generated structures.) What will be lost is any specific overall similarity to proteins that are homologous to the native probe. All scores from the reversed structure can be regarded as random scores. Figure 12.2 also illustrates the use of the reversed structure for constructing a random background of comparisons.

One might also think of reflecting the chain as a method for constructing a 'random' structure, but the reflected chain is clearly not suitable for generating random models as they contain both large- and small-scale chiral features which will change hand under reflection: including the hand of the α-helix (small scale) and the hand of connections between strands in a β-sheet (large scale).

12.4   Randomized Alignment Models

In general, the closer the random model is to preserving the properties of the native proteins, the more difficult it becomes to generate plausible random alternatives. This problem is particularly acute for the reversed-chain random model discussed above since, for any given protein, there is only one reversal. This problem can be partly circumvented at the stage of calculating the alignment. At this point the alignment with each random model can be expanded into a population of variants by introducing 'noise' into the score matrix and repeating the calculation of the alignment path from each noisy matrix. This generates a family of near-optimal subalignments and the spread of scores for this population can provide a measure of the stability or uniqueness of the answer. An advantage of this approach is that it can be applied not only to structures belonging to the set of randomized models but also to the native structure itself and the two resulting score distributions can be tested statistically to see if they are distinct.

If there is sufficient 'noise' and the population is large enough then almost all reasonable alignments for a pair of proteins can be sampled. Plotting these solutions by their number of aligned positions against RMSD has revealed a 'cloud' of points which is diffuse at high RMSD but have a sharp boundary on its lower edge (Taylor 1999a). This edge represents the limit, for a given number of aligned positions, below which a smaller RMSD cannot be found. As judged by the 'hard' edge to the distribution, this limit is not restricted by the method of comparing the proteins and so provides an absolute standard against which other methods can be compared.

An advantage of the use of large randomized alignment populations is that the alignments can be scored by any evaluation function and so long as the evaluation function does not radically contradict what would be considered a reasonable alignment (that is, one with a low, if not minimal RMSD) there is then a good chance that some members of the population will correspond closely with the minimum of the evaluation function. This can also be used to check if a method attains the minimum of its own scoring function (assuming that the low scores best).

12.5   Assessing Comparison and Scoring Methods

To test structural (or sequential) comparison and scoring methods, one must have sets of homologous and nonhomologous structures. To construct such sets, the SCOP database (see Chapter 14) has become the 'gold standard'. The SCOP database is a classification of the domains in the PDB into a hierarchical organization (done manually, based on sequence, structure and functional similarity).

It is first necessary to construct a nonredundant subset of the PDB structures (PDB*). Then each structure in PDB* can be used as query and compared with all the structures in PDB*, and the score of each comparison found. The scores can then be arranged in a list sorted after decreasing scores (requiring that the score must be consistent for each query), and the analysing methods explained in Chapter 3 can be used, for example, by sensitivity/specificity diagrams.

If the E value is calculated for each pair, one can use the E value in the assessment (Brenner et al. 1998). The pairs can then be arranged in a list sorted on increasing E value. Then work down the list and set a threshold T for assumed homology (positives) to be the point where the number of nonhomologous pairs (false positives) is, for example, 1% of the number of structures in PDB*. Thus, the number of false positives per query (called errors per query (EPQ)) is 0.01 at T. The E value at T is the expected number of false positives, hence if the scoring were perfect, the E value should be equal to EPQ, (0.01) at T. Thus for each method or scoring system we can calculate the sensitivity (the fraction of the homologs that have E value above the threshold) and the difference between the E value and the EPQ. One can of course do this for different values of EPQ.

Example

Assume a nonredundant database has 600 structures. We use two programs (and scoring systems) to perform all-by-all comparisons. Assume that the E value of the program P1 when six false positives are found is 0.022, and the sensitivity is 0.32. Assume further the corresponding values for the program P2 are 0.048 and 0.40. Then P1 has best specificity at this point, since its E value is closest to 0.01. $\triangle$

SCOP is also used as a standard for assessing sequence comparison methods.

12.6   Is RMSD Suitable for Scoring?

RMSD is an essential part of most scoring systems for structure comparison, though using RMSD alone has some serious pitfalls:

- best alignment does not always mean minimal RMSD;
- significance of RMSD depends on the size of the structures;
- significance of RMSD varies with the type of proteins;
- it is not a good measure when all equivalent parts of the proteins cannot be simultaneously superposed;
- all atoms are usually treated equally, though, for example, residues on the surface have a higher degree of freedom than those in the core (but weights might be used in the calculation).

Also, RMSD depends more on the worst-fitting atoms than on the best-fitting, and it does not (for alignments) penalize gaps. Therefore, alternative scoring schemes are used. The success of using RMSD in contrast to other scoring systems is investigated by use of the methods in the previous subsection. There are often different scoring systems for different comparison methods, and the investigations are therefore to compare a specific scoring system with RMSD when used on a specific structure comparison method. Experiments have shown that the other scoring systems usually are better, for example, by taking gaps into account.

12.7   Scoring and Biological Significance

When a structure is compared with every other structure (or with a representative selection), then scores will result which range from the clear relationships of homologous proteins to a large number of poor scores for obviously unrelated pairs. Between these extremes lies a twilight zone within which it is very difficult to assess the true significance of the score. This problem is exacerbated because many proteins contain similar substructures, such as secondary and supersecondary structures and the problem is to decide when a similarity is just a consequence of being protein-like and when it indicates a more specific relationship between the two proteins.

Because of its common currency, most considerations of this problem have focused on the significance of the RMSD measure based on comparison of proteins or protein fragments of equal length (see above). Others, such as the DALI method (Chapter 11), have adopted a similar approach based on the scores achieved over matches of protein fragments. Both these approaches require that the selected fragments are unrelated to the proteins being assessed; however, this raises the problem of what criterion can be used to make this distinction and, in principle, it should not be a weaker method than that used for the current comparison. However, as explained before, it is not acceptable to consider completely unrelated proteins.

12.8   Exercises

1. At http://www.fccc.edu/research/labs/dunbrack/pisces/ is a program for constructing nonredundant subsets of PDB. Use this program (or another) to construct such subsets for maximum percentage identity being 20, 30, 40, ... , 100. Then plot the number of nonredundant chains as a function of percentage identity.

   (a) Use this to estimate the number of sequence families in PDB.

   (b) Can you propose an explanation of the form of the curve? Can we draw any biological knowledge from it, or is the reason due to which proteins have been selected for structure determination, or is there any other reason?

2. Assume that we have a nonredundant database of 8000 structures. An all-by-all comparison is done, and the pairs of structures arranged in a list sorted on increasing score (assuming that the low scores best). The list is worked down until 16 nonhomologous pairs are found. What should the E value of the score at this entry of the list be?

12.9   Bibliographic notes

The uniqueness of structure alignments is discussed in Godzik (1996). Analysis of the use of RMSD is, for example, done by Alexandrov et al. (1992), Maiorov and Crippen (1994) and Betancourt and Skolnick (2001). Random structural models are discussed by Taylor (1997a). Dragon is described in Aszódi and Taylor (1994). Assessment methods and statistical frameworks are described in Brenner et al. (1998), Gerstein and Levitt (1998), Levitt and Gerstein (1998) and Stark et al. (2003). Comparing RMSD with other scoring schemes is done in Levitt and Gerstein (1998) and Taylor (1999a).

Some web addresses for (constructing) nonredundant subsets of PDB are
http://www.fccc.edu/research/labs/dunbrack/pisces/
http://www.ncbi.nlm.nih.gov/Structure/VAST/nrpdb.html#details)
http://www.cbrc.jp/papia/howtouse/howtouse_pdbreprdb_dynamic.html

13

Multiple Structure Comparison

The reason for doing multiple structure comparison is the same as for doing multiple sequence comparison: finding similarities in a set of (homologous) structures gives much more insight than pairwise similarities. We can approach the problem in at least three different ways.

1. Construct multiple alignments.

2. Discover common cores – an often larger part of the structure in the interior of the protein, e.g. the hydrophobic core.

3. Discover common local patterns (structure motifs) – a smaller part of the structure having a specific functional or structural role, e.g. the active site in enzymes or metal binding sites.

A fourth point could also be included: finding supersecondary motifs (constituted of several secondary structures).

These approaches are strongly related. When a multiple alignment is constructed, it is 'relatively easy' to iterate to a common core. Also, having found a common pattern, it can be used to initialize a procedure for constructing a multiple alignment.

When discussing (and developing) methods for multiple structure comparison, one can build on frameworks and methods for multiple sequence comparison and pairwise structure comparison. Methods for comparing pairs of structures can often be extended to the comparison of more than two structures, in the same way as pairwise methods for sequence comparisons are extended.

Superposition is an important part of structure comparison; therefore, we first look at the problem of multiple superposition, restricting ourselves to rigid-body superposition.

13.1   Multiple Superposition

Note that the input to a procedure for multiple superposition is a multiple equivalence. The simplest method for the simultaneous superposition of several structures is to

Protein bioinformatics: an algorithmic approach to sequence and structure analysis
I. Eidhammer, I. Jonassen and W. R. Taylor   © 2004 John Wiley & Sons, Ltd   ISBN: 0-470-84839-1

choose one structure as basis (pivot), and then superpose all the others onto it. An 'average' structure can then be calculated. Let $\{A^1, \ldots, A^m\}$ be the structures, and A^1 be (without loss of generality) the basis. First, the centres of mass of all the structures are superposed, so that the centres are located in the origin of the coordinate system. Then let A_i^j denote the ith coordinate set of structure A^j *after* the translation (defined by three values). Then for each structure, except the basis, the rotation matrix for the best superposition onto the basis is found by minimizing

$$\sum_{j=2}^{m} \sum_{i=1}^{r} w_{ij} (R_j A_i^j - A_i^1)^2$$

over all possible R_j, where r is the number of atoms in the equivalence. w_{ij} is a weight of coordinate set i of structure A^j. Note that w_{ij} can be defined as a combination of a structure weight (v_j) and a position weight (u_i).

An average structure A^* can then be defined by

$$A_i^* = \frac{\sum_{j=1}^{m} w_{ij} R_j^* A_i^j}{\sum_{j=1}^{m} w_{ij}},$$

where R_j^* is the optimal rotation matrix for structure A^j.

This procedure will, however, bias the average structure towards the structure of the basis. A way to reduce this bias is to let the basis change in an iteration, hopefully converging to an average structure. In each cycle, the average structure is taken as the basis for the next cycle, as shown in Algorithm 13.1.

Algorithm 13.1. Iterated multiple superposition.

var
p cycle number
A^{*p} the average structure in cycle p
R_j^{*p} the rotation matrix superposing A^j on A^{*p}
w_{ij} weight of residue i of structure A^j
const
K limit for convergence
begin
    $p := 0$; $A^{*p} =$ *one of the structures*
    repeat
        superpose all structures on A^{*p}, finding the rotation matrices $\{R_j^{*p}\}$
        $p := p + 1$
        calculate the new average by $A_i^{*p} = \dfrac{\sum_{j=1}^{m} w_{ij} R_j^{*p} A_i^j}{\sum_{j=1}^{m} w_{ij}}$
    until RMSD$(A^{*p}, A^{*p-1}) < K$ **or** *maximum number of cycles performed*
end

An alternative approach is to find the rotation matrices in a simultaneous optimization procedure, minimizing the sum of all pairwise RMSD values. First, all structures are translated so the centres of mass are at the origin, then find the rotation matrices $\{R_j\}$ that minimize

$$\sum_{j=1}^{m-1} \sum_{k=j+1}^{m} v_{jk} \sum_{i=1}^{r} u_i (R_j A_i^j - R_k A_i^k)^2,$$

where v_{jk} is a weight for the structure pair $A_j A_k$, and u_i is a weight for equivalence position i. References for the solution of this optimization problem are given in the bibliographic notes.

13.2   Progressive Structure Alignment

The easiest way to extend from pairwise to multiple structure alignment is to use one of the structures as a pivot, and then align all the others to this. The structure can be chosen arbitrarily or one can find a 'median' structure. A 'median' structure can be determined by first calculating all-by-all pairwise alignments, and selecting as a 'median' the structure which has the minimum sum of scores to all the others. The result is strongly biased towards the pivot structure. More advanced methods can be constructed by extending pairwise methods analogous to the way pairwise sequential methods are extended (see Chapter 4). The main problem when going from sequence to structure is to define the structural equivalent of a consensus sequence (or profile), for use when aligning (subset)alignments.

We will illustrate the approach, and discuss some of the challenges, by describing a fusion of the pairwise structure program SSAP (SAP can also be used) (Chapter 9), and the multiple sequence comparison program MULTAL (Taylor et al. 1994). First, each pair of structures is compared (using SAP or SSAP) to assess all pairwise similarities. The most similar structure pairs are aligned independently into consensus structures and then the pairwise similarities between all single and consensus structures are calculated, to progressively bring together the most similar of these at each stage (tree alignment). Note that the progression of the pairwise alignments is not determined by a pre-computed dendrogram, the similarities are re-computed at each stage and the most similar (subset)alignments aligned first. Algorithm 13.2 shows the procedure.

Algorithm 13.2. General progressive alignment of structures.

Progressive aligning of the structures $\{A^1, A^2, \ldots, A^m\}$
var
U current set of consensus structures
begin
    $C := \emptyset$
    for $i := 1$ **to** m **do** $U := U \cup \{A^i\}$ **end** one consensus of each sequence
    for *each pair* $(C^j, C^k) \in U$ **do** *align* (C^j, C^k) *to find the score* **end**
    repeat
        let(C^j, C^k) *be the pair in U with highest score*
        $C^l := consensus(C^j, C^k)$
        $U := (U - C^j - C^k) \cup C^l$
        for *each* $C \neq C^l \in U$ **do**
            find the score of aligning (C, C^l)
        end
    until $|U| = 1$
end

In SSAP the structures are represented as interatomic vectors between the β-carbons (distance matrix). In order to retain all information in an alignment, a bundle of vectors would be needed for each position. Aligning two alignments could therefore be computationally demanding. Therefore, the average of the component vectors is used as consensus. Let $\boldsymbol{v}_{ik}^{A^j}$ be the interatomic vector between atom i and k in structure A^j. Then the consensus vector for the aligned atoms becomes (m is the number of structures)

$$\boldsymbol{r}_{ik} = \frac{1}{m} \sum_{j=1}^{m} \boldsymbol{v}_{ik}^{A^j}.$$

Example

Assume four structures were aligned to a length of five positions. Let the second position be described by the following vectors (vectors from the second position to all the others; '—' means blank):

Structure	$\boldsymbol{v}_{2,1}$	$\boldsymbol{v}_{2,2}$	$\boldsymbol{v}_{2,3}$
A_2^1	$(-3.7, -0.2, 1.4)$	$(0, 0, 0)$	$(0.9, 3.1, -2.1)$
A_2^2	$(-3.3, -0.4, 1.6)$	$(0, 0, 0)$	$(0.7, 2.8, -2.3)$
A_2^3	$*$	—	$*$
A_2^4	—	$(0, 0, 0)$	$(1.1, 2.8, -2.2)$

Structure	$v_{2,4}$	$v_{2,5}$
A_2^1	$(1.6, 2.5, -6.1)$	$(5.1, 3.4, -5.8)$
A_2^2	—	$(5.3, 3.7, -5.6)$
A_2^3	*	*
A_2^4	$(1.6, 2.1, -5.7)$	$(4.9, 3.7, -5.4)$

Since the third structure has a blank in position 2, there is no vector to the other positions, illustrated by *. The blanks are not included in the calculation of the consensus, and the consensus becomes

Consensus	$r_{2,1}$	$r_{2,2}$	$r_{2,3}$
A_2^*	$(-3.5, -0.3, 1.5)$	$(0, 0, 0)$	$(0.9, 2.9, -2.2)$

Consensus	$r_{2,4}$	$r_{2,5}$
A_2^*	$(1.6, 2.3, -5.9)$	$(5.1, 3.6, -5.6)$

△

This representation raises one problem, it is not possible to distinguish a coherent bundle of vectors from a divergent bundle which, by chance, might have identical average vectors. Therefore, an error measure should be used, and a simple measure is the magnitude of the vector difference:

$$e_{ik} = \frac{1}{m} \sum_{j=1}^{m} (r_{ik} - v_{ik}^{A^j})^2.$$

This will then incorporate both the length and the direction of the component vectors.

13.2.1   Scoring

SSAP is used to calculate a similarity score between two alignments by using the average vectors (hence a pairwise scoring scheme is used). The scoring depends on the difference of the consensus vectors, and the associated errors. In order to calculate the score, the low-level scoring matrices of the SSAP algorithm must be computed.

For the low-level scoring matrix ^{ij}S (see Algorithm 9.2), the scoring between the atoms k and l in the two consensuses is defined as

$$^{ij}S_{kl} = \frac{a}{b + ^{ij}\delta_{kl} + we_{ij}e_{kl}},$$

where a and b are constants, $^{ij}\delta_{kl} = (r_{ik} - r_{jl})^2$ and w is an overall constant weight.

The log-normalized scores are used in the high-level scoring matrix (see Chapter 9).

Treating gaps

As shown in the example above, gaps are ignored in the calculation of the average vectors. However, gaps should be taken into account in the scoring. Gaps should decrease the score; one way of doing this is to add a constant to e_{ik} for each blank.

13.2.2   Construction of consensus

The result of the aligning process is the multiple alignment, and the average inter-atomic vectors between all the positions in the alignment. These vectors *are not necessarily mutually consistent*. Methods (which are outside the scope of this book) are then used for iterating to an appropriate consistent representation of the consensus structure.

Example

To illustrate the inconsistency problem, assume that we consider 'structures' in two dimensions, and have calculated some of the distances of the consensus to be (ignoring the direction, d is distance) $d(c_i, c_j) = 11$, $d(c_i, c_k) = 16$, $d(c_j, c_k) = 29$. This is impossible, but by small movements of the atoms we might achieve $d(c_i, c_j) = 12.5$, $d(c_i, c_k) = 15$, $d(c_j, c_k) = 27$, which is possible.   △

13.3   Finding a Common Core from a Multiple Alignment

When a multiple alignment is known, a core can be found by using an iterative procedure. First an average structure is created, using the columns, if any, without gaps in the alignment (for example, as described in Section 13.1). Then each structure is superposed onto the average core. The variability in the positions of the atoms are calculated for each position, and an initial core is defined by the positions where the variability is below a cut-off value. Then a new core is calculated as the average of the chosen positions, as shown in Algorithm 13.3.

Algorithm 13.3. Finding a common core, from a multiple alignment of m structures.

var

p cycle number

$\hat{M}$ the columns without gap in the multiple alignment

A_p^j the atoms in structure A^j used for the core in cycle p

G_p the columns in $\hat{M}$ used to define the core in cycle p

C_p the average core calculated in cycle p

const

V threshold for variability of the atom positions

begin

    $p := 0; G_p :=$ *the columns of* $\hat{M}$

    repeat

        $p := p + 1$

        $A_p^j :=$ *the atoms in structure* A^j *occurring in* G_p

        $C_p :=$ *the average structure of all* $\{A_p^j\}$

        for *each substructure* A_p^j **do**

            find the transformation T_p^j *superposing* (A_p^j, C_p)

            transform A^j *using* T_p^j

        end

        for *each column in* $\hat{M}$ **do**

            calculate the variability of the positions of the atoms

        end

        $G_p :=$ *the columns in* $\hat{M}$ *where the variability* $< V$

    until $G_p = G_{p-1}$ **or** *maximum number of cycles performed*

end

Note also that if a common core is discovered by use of other methods, it is possible to use this to construct a multiple alignment.

13.4   Discovering Common Cores

An intuitive way of extending the pairwise clustering methods to finding common cores is to first find multiple seed matches, and then cluster multiple seed matches to larger multiple clusters, using a general clustering procedure (see Chapter 11). In practice, it can involve a lot of work doing multiple clustering, especially the tests for consistency between clusters. Therefore, it might be worthwhile replacing the multiple clustering with a set of pairwise clusterings. This is used in the method we describe for finding common cores MUSTA (Leibowitz et al. 2001). The geometric hashing technique is used for finding multiple seed matches. One of the structures is chosen as a *reference*, the others are called *sources*. The procedure for finding cores consists of three major steps:

1. Find multiple seed matches. A multiple seed match is a multiple alignment of k columns. The m substructures (geometrical figures, one from each structure) defined by the atoms in the alignment are approximately similar (congruent).

2. For each source, extract from the multiple seed matches the pairwise seed matches between the source and the reference. A pairwise seed match for a source is the projection of a multiple seed match into the source and the reference. The result, for each (source,reference), is a set of pairwise seed matches. These seed matches are then used in a pairwise clustering procedure (using methods explained in Chapter 11) to find pairwise subalignments.

3. Make multiple subalignments (proposals for cores) from mutual consistent pairwise subalignments from all the sources.

13.4.1   Finding the multiple seed matches

In step 1 we need to choose a value for the parameter k (the length of the seed matches). This will be a compromise between the amount of work (increasing with increasing k), and the number of 'not useful' seed matches (decreasing with increasing k). Leibowitz et al. use $k = 5$ in their experiments. A substructure consisting of five atoms (one from each residue) can be completely specified (up to translation, rotation and mirror image) by 10 inner distances (see Chapter 8.3). However, by using nine inner distances, and classifying the type of symmetry (there are four—see Exercise 1), they avoid the mirror image problem. In the implementation of geometric hashing, four nine-dimensional hash tables can be used. The structures are represented by their C_α-atoms, and for each atom (a) is a neighbourhood consisting of a spherical shell around a defined. All possible k-tuples of atoms consisting of a and $k - 1$ atoms in its neighbourhood, are constructed, rejecting tuples consisting of only atoms consecutive in the protein's sequence. The atoms of the k-tuples are ordered internally in both decreasing and increasing order with respect to their positions in the sequence. (This ordering could be left out, and then constructing multiple equivalences which would not necessarily be alignments). All k-tuples of all structures are then hashed to the hash table for the appropriate symmetry type. Therefore, all k-tuples constituting similar substructures are hashed to the same bucket in the same hash table. Then the reference is used in an analogous way as a query. All its k-tuples are constructed and each is used to query the hash tables. Each 'queried' bucket containing a k-tuple from all the sources will then contain at least one multiple seed match.

Example

Assume we have four structures A^1, A^2, A^3, A^4, and let A^1 be the reference. Let a bucket in a hash table contain the following tuples [structure: tuples]:

$$[A^2 : (3, 5, 8, 9, 13); A^2 : (23, 26, 27, 31, 33); A^3 : (31, 35, 36, 39, 43);$$
$$A^4 : (7, 9, 13, 14, 17); A^4 : (37, 39, 43, 44, 48)].$$

Furthermore, let the k-tuple A^1 : (3, 5, 7, 10, 13) from the reference be hashed to this bucket. Then four multiple seed matches are found, where one of them is

$$
\begin{array}{cccccc}
A^1 & 3 & 5 & 7 & 10 & 13 \\
A^2 & 3 & 5 & 8 & 9 & 13 \\
A^3 & 31 & 35 & 36 & 39 & 43 \\
A^4 & 7 & 9 & 13 & 14 & 17
\end{array}
$$

∧

Now the multiple seed matches could be clustered to form larger equivalences. Alternatively, one could first do pairwise clustering for the reference and each of the sources, and then perform intersections of these in order to find 'good' multiple clusters (alignments).

13.4.2 Pairwise clustering

All multiple seed matches are projected onto each (of the $m - 1$) pairs of (reference, source) giving for each such pair a set of pairwise seed matches. For the example above, the pairwise seed matches for (A^1, A^4) from the considered bucket are (there are only two different projections for A^4 from the four multiple seed matches)

$$
\begin{array}{cccccc|cccccc}
A^1 & 3 & 5 & 7 & 10 & 13 & A^1 & 3 & 5 & 7 & 10 & 13 \\
A^4 & 7 & 9 & 13 & 14 & 17 & A^4 & 37 & 39 & 43 & 44 & 48
\end{array}
$$

Then for each (reference, source), clustering is performed in an iteration as explained in Chapter 11, using transformation as the test for consistency, for example, using Algorithm 11.4. Testing for consistency is done by one of the methods explained in Section 11.5.1.

Note that although clustering is performed pairwise, only atoms which occur in a multiple seed match take part in the clustering.

13.4.3 Determining common cores

The result of the second step is for each source a set of clusters (alignments) between the source and the reference. The third step is to group alignments from each source, producing the highest-scoring multiple alignment. In principle, all the clusters must be compared with each other, a procedure being of order $\prod_{j=2}^{m} t_j$, where t_j is the number of clusters found for source A^j.

However, by using the hash tables the time requirements can be reduced significantly. The clusters found in step two are being registered in the buckets from which they are being constructed. That means, if P_k is a seed match for source A^j, and a cluster C is constructed from (among others) P_k, then C is registered in the bucket where P_k is stored. This can be used to reduce the number of comparisons between clusters. A core (multiple alignment) should contain (part of, depending on how the

Table 13.1   Example buckets in hash tables, with seed matches and clusters.

	Bucket 1		Bucket 2		Bucket 3		Bucket 4	
	Seed	Cluster	Seed	Cluster	Seed	Cluster	Seed	Cluster
A^1								
A^2	P_1^2, P_2^2	C_1^2	P_3^2	C_2^2	P_4^2, P_5^2	C_1^2, C_3^2	P_6^2, P_7^2	C_2^2, C_3^2
A^3	P_1^3	—	P_2^3, P_3^3	C_1^3, C_2^3	P_4^3	C_1^3	P_5^3, P_6^3	C_1^3, C_3^3
A^4	P_1^4, P_2^4	C_1^4, C_2^4	P_3^4, P_4^4	$C_2^4,$	P_5^4, P_6^4	C_1^4, C_2^4	P_7^4	C_1^4

clustering is being done) a k-tuple from the reference, which has similar substructures in all the sources. That means one needs only to compare (and group) clusters which are registered in the same bucket.

First, clusters (from step two) which are of length less than a given threshold, are removed from the buckets. Next, only buckets containing clusters from all sources have to be investigated. These operations largely reduce the amount of work in this step.

Example

Suppose that we have three sources $\{A^j\}$, and that four buckets get multiple seed matches in step 1, denoted by P_i^j in Table 13.1. Furthermore, let step two result in the clusters C_k^j, after small clusters are removed.

Then only buckets 2, 3 and 4 need to be investigated, since bucket 1 does not have a cluster for A^3. The clusters compared for intersection are from bucket 2: (C_2^2, C_1^3, C_2^4) and (C_2^2, C_2^3, C_2^4) △

Doing the intersection

The intersection of $m-1$ clusters (one from each source) can now be done by using the atoms from the reference as a 'template'. If the cluster after the intersection is larger than a predefined threshold, the combination is stored as a candidate for a common core.

Example

Let the clusters be (from source A^2, A^3, A^4, respectively, the first numbers of the pairs refer to the reference)

$$C_2^2 = \{(7, 3)(9, 5)(13, 7)(14, 10)(17, 13)(21, 16)(22, 17)(25, 21)\},$$

$$C_1^3 = \{(7, 13)(9, 15)(13, 18)(15, 20)(21, 26)(22, 27)(25, 32)(27, 35)\},$$

$$C_2^4 = \{(7, 9)(9, 11)(13, 13)(14, 16)(17, 20)(22, 23)(25, 27)(27, 30)\}.$$

Then the common core is described by

From A^1	7	9	13	22	25
From A^2	3	5	7	17	21
From A^3	13	15	18	27	32
From A^4	9	11	13	23	27

13.4.4   Scoring clusters

For scoring clusters one can use the length of the clusters, some average of the pairwise RMSD values, and an evaluation of the similarity of the physio-chemical properties of aligned residues. If the method is able to detect cores shared by subsets (not all) of the structures, the number of structures in the subsets is also included.

13.5   Local Structure Patterns

One of the reasons for using structure patterns is that many of the most interesting functional and evolutionary relationships among proteins are so ancient that they cannot be reliably detected through sequence analysis and are apparent only through a comparison of the tertiary structures. These features often take the form of structure motifs, consisting of perhaps a few secondary structures or residues that adopt a common substructure, typically, in the structural core or functional centre of the proteins. As these relationships are ephemeral, confidence in their reality can be greatly boosted when they are found in more than a pair of proteins. Some protein families are difficult to describe adequately using sequence patterns.

- Sequence patterns do not perfectly discriminate between family members and nonmembers. The residues forming a site need not be close in sequence, and the sequences of the proteins sharing an active site are sometimes extremely divergent. For example, the bacterial and eukaryotic serine proteases share a very similar active site while having very different sequences. Especially when the residues involved in the site are nonlocal, sequence patterns (deterministic or probabilistic) cannot recognize such structures.

- Sequence patterns do not give complete descriptions of the features critical to the function/structure class associated with the family.

Here we concentrate on patterns consisting of a few residues close in space, but possibly with long distances in the sequence. Such patterns are called *(local) packing patterns*, and we describe a method, called SPratt, for discovering local packing motifs developed by Jonassen et al. (1999, 2000a). The methods explained up to now in this chapter use an progressive/clustering approach. SPratt use a 'simultaneous' (pattern-driven) approach.

13.5.1   Local packing patterns

A packing pattern is a tuple

$$P = p_1 \cdots p_n \tag{13.1}$$

for some $n \geqslant 1$ and each p_i (representing a residue) can be written in the form

$$p_i = (x_i, y_i, z_i, M_i, \alpha_i), \tag{13.2}$$

where x_i, y_i, z_i are real numbers (coordinates), M_i is an amino acid match set ($M_i \subseteq \mathcal{M}$) and α_i describes constraints on the secondary structure a, b, l, representing, respectively, α-helix, β-sheet and loop ($\alpha_i \subseteq \{a, b, l\}$). This definition can be extended to include other discrete properties (e.g. exposed/buried).

Matching a packing pattern

A packing pattern $P = (x_1, y_1, z_1, M_1, \alpha_1) \cdots (x_n, y_n, z_n, M_n, \alpha_n)$ is matched by a protein if there is an ordered list of residues $(a_1, \ldots, a_n)$ in the protein such that

1. the residues $a_1, \ldots, a_n$ are unique (each occurs only once);

2. the amino acid of a_i is in the match set M_i;

3. the secondary structure of a_i is in α_i;

4. the residues $a_1, \ldots, a_n$ are in order from the amino (N) to the carboxy (C) terminal;

5. the coordinates of $a_1, \ldots, a_n$ can be superposed onto the coordinates given in P within an RMSD of at most K (the coordinates of each a_i paired with (x_i, y_i, z_i) for $i = 1, \ldots, n$).

There are many ways to define a single set of coordinates (x, y, z) for a residue: for example, as the mean coordinate of the side-chain atoms of the residue (with the exception of glycine, for which the α-carbon is taken).

Example

Let a pattern be

$$P = [(-3.0, 32.0, 14.4, H, a)(-4.4, 5.4, 15.3, C, a)$$
$$(-6.5, 26.6, 15.4, [DE], l)(-9.9, 35.0, 11.2, S, l)],$$

where C_α atoms are used.

P is matched by the chain A of protein 1fon in PDB in residues 43, 44, 93, 187:

```
1fon

. . . . . .
CA   HIS A   43        45.780   18.860   22.414   helix
CA   CYS A   44        43.110   19.921   24.916   helix
```

```
. . . . . .
CA   ASP A   93        49.642  14.010  23.264  loop
. . . . . .
CA   SER A  187        38.379  13.995  22.874  loop
. . . .
```

Packing motifs

A *packing motif* is a packing pattern P that has multiple matches (occurrences) in a set of (or within one) protein structure(s). Packing patterns (or motifs) describe clusters of residues coming spatially close together but which are not necessarily local in the sequence.

13.5.2   Discovering packing patterns

Given a set of proteins with structures $\{A^1, \ldots, A^m\}$, the task is to find packing patterns occurring in at least k of the structures.

Each residue is represented by a point, representing the side chain centroid, and the distances between all pairs of residues inside a structure are calculated.

Neighbourhood

- A *neighbourhood* NH_a of a residue a is the set of all other residues in the structure with distances to a less than a pre-defined threshold (typically 8–12 Å).

- A *neighbourhood string* NS_a (or neighbour string for short) is the residue a and all residues in its neighbourhood, in the same order as they occur in the protein sequence. It is also possible to add an extra requirement. The neighbourhood residues must satisfy a direction constraint: a and the neighbourhood residue b must 'face' each other, that is, b must be inside a half sphere of a, and vice versa. Also, neighbourhood strings shorter than four are discarded. Figure 13.1 illustrates a neighbour string.

- a is the *anchor* of its neighbourhood string NS_a.

13.5.3   The approach

The task is to find a pattern matching at least k of the m structures. The approach is to incorporate a 'simultaneous view' of the structures, and the main idea is to start with one neighbourhood string, along with its coordinates, defined as a *probe*, p. A search is then performed to find generalizations of p which occur in at least k of the structures (is matched by at least k structures). The generalizations are packing patterns that satisfy the following constraints:

- the anchor must be in the pattern;

- the string of residues of the discovered packing patterns must be a subsequence of those in the probe;

- if packing patterns are allowed to have *amino acid match sets*, the amino acids in the neighbourhood string can be generalized to match sets.

Since the procedure starts with a probe, occurring in one of the structures, it is a pivot-based method. However, since the pattern must match at least k of the structures, it is enough to use $m - k + 1$ of the structures as pivot. Any pattern with minimum support will have an occurrence in at least one of any subset of $m - k + 1$ structures. The smallest (fewest residues) structures are used for reasons of efficiency. The approach is described in Algorithm 13.4

Algorithm 13.4. Discover a packing pattern matching at least k of m structures.

var

p	a probe
PP	patterns found for a probe
PP_set	contains at end all the discovered patterns
S	the set of structures used as pivot
A	a structure in S
a	a residue in A

procedure

$DF\_probesearch(p)$ a procedure using depth-first search to find all patterns with probe p

begin
    $PP\_set := \emptyset$
    make neighbour strings for all residues in all structures
    $S :=$ *the* $m - k + 1$ *smallest structures*
    for *each structure* $A \in S$ **do**
        for *each residue* $a \in A$ **do**
            $p :=$ *probe defined by* a *as anchor*
            $PP := \text{DF\_probesearch(p)}$
            $PP\_set := PP\_set \cup PP$
        end
    end
end

The depth-first search

Let us consider a neighbourhood string ACEWGGTGEA. It can be generalized to a large number of packing patterns, for example, G, GG, WG, and CWGT. The packing pattern derived from a neighbourhood string will inherit the neighbourhood string's coordinate sets.

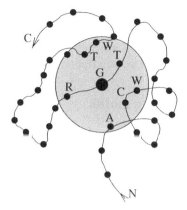

Figure 13.1   The chain around an amino acid G,
resulting in the neighbour string ACWT G RTW.

The search is a simple depth-first search used to find all generalizations of the probe that have occurrences in at least k structures, similar to the one used in the program Pratt, described in Chapter 7.

The simplest generalization of the probe only contains the probe's anchor and matches all neighbour strings whose anchor has the same amino acid type. This gives us a pattern P (equal to the anchor) with a list of matches M_P. M_P then contains all neighbour strings with anchor equal to the probe's anchor. A pattern P, can be extended by appending a residue a from the probe, forming Pa (note that a does not need to directly succeed the match of P in the neighbourhood). The matches to P (M_P) are analysed to see if they can be extended to matches of Pa, and it is checked whether Pa has sufficient support (occurs in enough structures). If it does, it is again extended in all possible ways by appending elements defined from residues to the right of a in the probe. At any point in this exploration, the patterns can be extended to the left, e.g. a pattern P extended to aP. To avoid analysing the same pattern twice, once a pattern has been extended to the left, all further extensions will be to the left.

Example

Let the neighbour sequence of the probe be ACEWG G TGEA. Let three other neighbour sequences be DELWG RTEA, DLEWR G TTEC, AERFW G LTREA. Then the sequence of a packing pattern can be EW G TE if the RMSD constraint is satisfied.   △

As the search proceeds all patterns satisfying the constraints given by the user are output and they are postprocessed by separate programs.

The search procedure used makes it likely to find the same pattern multiple times since several of the matching neighbour strings may be used as probes. Therefore, simple checksums are generated for each identified pattern and when new patterns are found, their checksum is compared with those of all previously discovered patterns before it is output.

In the search, the structural similarity of each match and the pattern is assessed by calculating the distance-based RMSD to save computations. When patterns are output, the structural similarity of each pair of matches is calculated using coordinate-based RMSD, and reported together with the description of the matches.

13.5.4   Scoring the packing motifs

To help the user interpret the output from the algorithm, a score is given for each reported pattern in which the potentially most interesting motifs are assigned the highest score: specifically, those that contain many residues and whose occurrences in the structures superimpose with low RMSD values. The score also depends on the amino acid constraints on each of the positions: the stricter the constraints, the higher score assigned.

A set of motif occurrences O, all matching a pattern P is scored by the formula

$$S(P, O) = \frac{I(P)}{\text{RMSD}(O)}, \tag{13.3}$$

where $I(P)$ is the information content of the sequence pattern (Chapter 6) and $\text{RMSD}(O)$ is the maximum RMSD between any pair of occurrences in O.

13.6   Exercises

1. In Section 13.4.1 it is asserted that using nine inner distances and a symmetry variable with four possible values is enough to describe a structure consisting of five atoms. Explain this. Hint: a 3-tuple, viewed as a triangle, is determined by its three distances; a 4-tuple can be seen as a 3-tuple which forms a triangular base, with an additional point that forms a pyramid with the base. Three distances and a 'position' (above or below the base) uniquely define the 4-tuple. A 5-tuple can be considered as a triangular base with two additional points.

2. This exercise illustrates the expansion of methods for pairwise comparison of structures. In this chapter we have seen an example of double dynamic programming to align several structures (call it MSAP). In this exercise we shall consider how a method for aligning of pairs of structures using alternating superposition and aligning can be expanded to aligning several structures. Let us call the pairwise alternating method SupSam, and the expanded, multiple one MSupSam. In MSAP, every structure is represented by a vector matrix V, where $V(i, j)$ is the vector for the residue pair (i, j). The result of an alignment is a consensus structure, and this is represented by a vector matrix, which for each (i, j) contains the average of the corresponding structure vectors. In addition, an error matrix is used in the representation.

    (a) Explain how you will represent structures and consensus structures in SupSam and in MSupSam?

(b) Explain when and how you will generate the consensus structure in MSup-Sam.

(c) Discuss how SupSam can be expanded to MSupSam in a way similar to CLUSTAL or MSAP.

3. The results from a method for finding local structural motifs in a set of structures can be used to guide a pairwise progressive method. The results from the geometric hashing step of the method in Section 13.4 (k-tuples common for all the structures) is, for example, used for finding larger pairwise alignments, which with higher probabilities can be 'intersected' to a multiple alignment. In a similar way results from SPratt can be used to guide SAP and MSAP to produce alignments consistent with the found patterns.

(a) Propose a way to use SPratt patterns in SAP and MSAP.

(b) How can (e.g. SPratt) patterns be used to give SupSam and MSupSam (Exercise 2) a 'flying start'?

4. This exercise illustrates the SPratt algorithm.

(a) The patterns used in SPratt are all generalizations of probes, and each pattern inherits the coordinates of the probe. How does this make it possible to lose patterns which are matched by the structures to the given RMSD limits?

(b) When a pattern P is expanded to Pa (or aP), every match to P is investigated to see if it can be expanded to match Pa (or aP). In this analysis, the first a is always used (aligned to the a in the pattern). Explain why we can lose 'good' patterns in this way. Why is it a larger problem if we (in the positions in the patterns) use amino acid groups instead of single amino acids? Explain how the procedure should be changed to avoid the problem.

(c) We have the probe ACD*E*GIJKL (the anchor is marked), and the neighbour strings:

$$
\begin{array}{l}
\text{AWDF}E\text{GGKL} \\
\text{KKLSAWC}E\text{GJHHIJSK} \\
\text{WGD}E\text{CDEGKWG} \\
\text{AGESDW}E\text{GSLIJK} \\
\text{AGEGCD}E\text{GISWLEK}
\end{array}
$$

Show the depth-first search for finding all patterns matched by at least four of the neighbour strings.

13.7   Bibliographic notes

MNYFIT is a program by Sutcliffe et al. (1987) that uses pairwise superposition to iterate to a common average structure. Kearsley (1990), Shapiro et al. (1992) and Diamond (1992), describe methods for simultaneously superposing all pairs of structures.

The fusion of SSAP and MULTAL is described in Taylor et al. (1994). Gerstein and Levitt (1998) use a similar but simpler method. The multiple alignment of Sali and Blundell (1990), Russel and Barton (1992), Ding et al. (1994) and May and Johnson (1995) all follow the same general algorithmic scheme for progressive alignment.

Two pairwise clustering methods using relations are extended to the multiple case, both using the pivot-based extension (Escalier et al. 1998; Koch et al. 1996).

Methods using search in a space of multiple alignments have also been developed. Godzik and Skolnick (1994) use a multidimensional lattice chain and a Monte Carlo (simulated annealing) approach. Holliday and Willet (1997) has developed a method using the genetic algorithm.

A method for finding local motifs is in Kasuya and Thornton (1999). Wallace and Thornton have constructed the PROCAT database containing manually defined patterns describing active sites:

http://www.biochem.ucl.ac.uk/bsm/PROCAT/PROCAT.html

(Wallace et al. 1997, 1996).

Similar patterns discarding the sequence order of the aligned residues can be discovered by the methods using geometric hashing (Fischer et al. 1995, 1994; Nussinov and Wolfson 1991). Yet another similar description formalism called 'fuzzy functional forms' (FFFs) is proposed by Fetrow and Skolnick (1998).

The described program for finding common cores is Leibowitz et al. (2001). Two programs for finding common cores in subsets of the given structures are MultiProt (Shatsky et al. 2002b) and MASS (Dror et al. 2003), the latter based on SSEs. Both use the pivot approach.

SPratt is described in Jonassen et al. (1999, 2000a). Finding patterns from multiple alignment can be done as described in Gerstein and Altman (1995) and de Rinaldis et al. (1998).

Other methods using simultaneous search (in addition to SPratt) are in Wako and Yamato (1998). Su et al. (1999) and Cook et al. (1996) represent the structures as linear graphs (hence sequential), one version on residue level and one on SSE level. Gilbert et al. (1999b) have developed a program for discovering supersecondary motifs in TOPS cartoons (Flores et al. 1994).

14

Protein Structure Classification

The number of proteins with known 3D structure has grown to several thousands (19 691 in the PDB of January 2003), and to be accessible this large number of structures needs to be organized and classified. Like any Natural history collection a classification or taxonomy of the objects is especially helpful for the understanding of their evolution (as structure is more conserved than sequence). In addition, when a taxonomy exists, new protein structures can be placed into this, helping to understand the function of the protein. If it is discovered that there exists only a fixed number of structure classes, structure prediction will be easier (finite search).

Since the evolution of protein structure is not fully understood, there is no definitive taxonomy that can be used to derive a classification and, as a result, several systems have been developed. The most widely used classification systems are CATH, SCOP, Dali-FSSP and Dali-DD. They are all hierarchical, and most use the *protein domain* as classification unit. Their databases are all accessible via the World Wide Web.

14.1   Protein Domains

A domain is part of a polypeptide chain of a protein or the whole chain. They are compact, local and semi-independent units, but there is no general agreement as to the exact definition of what a domain is. One definition is that it is part of the chain that can independently fold into a stable structure, and that it is also a unit of function (different domains of a protein are often associated with different functions). This implies that a domain should contain an hydrophobic core and should therefore be larger than, roughly, 40 residues (the rule is that a domain consists of one hydrophobic core, but in rare cases it might consist of two). The number of domains in proteins can be from one up to several dozens, and also a domain does not need to comprise a sequential region of the polypeptide chain.

Example

The PDB entry 3grs contains three domains. In CATH, domain 1 is constituted of the residues 18–160 and 290–365, domain 2 of 161–289, and domain 3 of 366–478. The

Protein bioinformatics: an algorithmic approach to sequence and structure analysis
I. Eidhammer, I. Jonassen and W. R. Taylor   © 2004 John Wiley & Sons, Ltd   ISBN: 0-470-84839-1

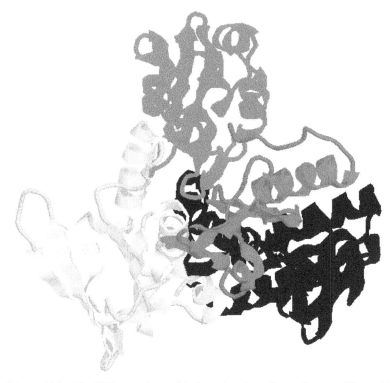

Figure 14.1   The PDB entry 3grs with three domains. (Drawn by use of Rasmol.)

same three domains are classified in SCOP, with slightly different residues (domain 1 is 18–165 and 291–363). See Figure 14.1.   △

Since different classification methods generally use different methods for domain identification, the domains will not always correspond. Several approaches have been used when developing methods for domain identification. Some of the concepts used are the following.

- Local compactness, which means that a domain will make more intra-domain contacts than contacts to the residues in the remainder of the structure. Almost all methods use this, but can in addition use some of the other points.

- Domains must contain at least one hydrophobic core.

- Minimizing the number of chain-breaks needed to separate domains while also measuring the degree of association (number of contacts) between the separating units. This implies a trade-off in trying to both minimize the number of chain-breaks and the number of contacts.

- Solvent area calculation. Let D_1 and D_2 be two potential domains. If the solvent area calculated when the potential domains are split is almost equal to the one calculated when not split, then it indicates two domains.

- Secondary structures (including β-sheets) should rarely cross between different domains.

14.2 An Ising Model for Domain Identification

Taylor (1999b) presents a bottom-up approach for domain identification, also using the preferable higher interdomain contact, though indirectly. A general model, referred to as an Ising model is used. (The model takes its name from the physicist who developed it.) An Ising model consists of nodes, which can be in one of several states. At the start, each node is initialized. The whole idea is then to change states in an iteration, during which nodes can influence the state of their neighbours to adopt their state and grow into groups with the same state.

In our application, the nodes represent the residues, a group is a domain and the states are represented by numeric values. During the iteration, it is hoped that residues belonging to the same domain should get the same state value. One (simple) way of doing this is from each residue to look at its (spatially) neighbouring residues, which are likely to be in the same domain. If the average state value of the neighbours is higher than the value of the residue, then the state value of the residue is increased, and decreased if the average is lower, and unchanged if equal. Then it remains to decide

- the initial state values (the values of the states at start);

- how to define the neighbourhood;

- how much should the state values increase/decrease.

These can be chosen in different ways. One possibility for choosing the start values is to use the sequential residue number itself. This naturally embodies the desired property that sequentially adjacent residues will be predisposed to belong to the same domain. The neighbourhood can most easily be defined by a radius around the residue.

Let the states of a protein of length n be represented as $S = \{s_1, s_2, \ldots, s_n\}$, where s_i is the state of residue i. Then a (simultaneously) changing iteration can be stated as

$$s_i^{t+1} = s_i^t + \mathcal{U}\left(\sum_{j=1}^{n} \mathcal{J}(s_i^t, s_j^t)\right), \quad \forall i, \ i = 1, \ldots, n, \qquad (14.1)$$

where $\mathcal{U}$ is a function returning $+1$ if its argument is greater than 0, -1 if its argument is less than 0, and 0 if the argument is 0.

The function $\mathcal{J}$ is called the coupling function, and calculates the contribution from each neighbouring residue as

$$
\mathcal{J}(s_i, s_j) \equiv
\begin{cases}
p_{ij} & \text{if } s_j > s_i \wedge d_{ij} < r, \\
-p_{ij} & \text{if } s_j < s_i \wedge d_{ij} < r, \\
0 & \text{otherwise,}
\end{cases}
\tag{14.2}
$$

where d_{ij} is the interatomic distance between the α-carbons of residues i and j, p_{ij} is the inverted distance r/d_{ij}, and r the neighbourhood radius.

Example

We have the following distance matrix extracted from a structure:

```
        1    2    3    4    5    6    7    8    9   10   11   12   13
     ----------------------------------------------------------------
 1 |  0.0
 2 |  3.9  0.0
 3 |  6.0  3.8  0.0
 4 |  7.0  6.7  3.8  0.0
 5 |  9.4 10.0  7.2  3.8  0.0
 6 | 11.2 12.3 10.0  6.2  3.8  0.0
 7 | 13.5 15.3 13.1  9.5  6.3  3.8  0.0
 8 | 16.9 18.3 16.0 12.2  9.2  6.1  3.8  0.0
 9 | 18.2 20.7 18.5 14.7 12.0  8.6  6.5  2.9  0.0
10 | 16.3 18.8 17.2 13.5 11.3  7.7  5.9  4.4  3.9  0.0
11 | 14.1 15.5 13.8 10.1  8.3  4.5  4.8  4.6  5.7  3.8  0.0
12 | 15.2 13.4 12.4  9.1  7.9  5.0  5.5  7.3  8.7  5.6  3.8  0.0
13 |  8.1  9.8  8.6  5.4  5.3  4.0  6.5  9.2 11.2  9.0  6.0  3.8  0.0
14 |  6.0  8.6  7.8  5.5  5.2  6.1  7.5 11.1 13.5 11.3  9.0  6.3  3.8
15 |  4.5  6.0  4.9  4.4  5.5  8.3 10.3 13.9 16.5 14.9 12.1 10.0  6.7
16 |  8.1  8.2  5.3  4.5  4.3  8.1 10.1 13.3 16.2 15.4 12.6 11.5  8.3
17 | 10.4 11.5  9.0  7.3  5.0  8.0  8.6 11.9 14.8 14.4 12.3 11.4  9.2
```

Assume the following state values:

```
Residue: 1  2  3  4  5  6  7  8  9 10 11 12 13 14 15 16 17
State:   4  4  2  2  5  5  5  8  8  8  5  5  5  2  2  5  5
```

Let $r = 10$. Then the neighbours of residue 1 are residues 2, 3, 4, 5, 13, 14, 15, 16, and the new state for residue 1 becomes

$$
s_1^{t+1} = 4 + \mathcal{U}(0 - 1.7 - 1.4 + 1.1 + 1.2 - 1.7 - 2.2 + 1.2) = 4 + \mathcal{U}(-3.5) = 4 - 1 = 3.
$$

The new states of the other residues are found in the same way, and then the states are changed. $\triangle$

Terminating the iteration

In all iteration algorithms there must be a stop condition; often it is when there are now further changes. In this program for domain identification, it possible for the domain boundaries to 'flicker' between two domains. Therefore, an average of the labels over two cycles are kept, and the squared deviations of this between successive cycles are summed. Any simple oscillation is then averaged out. The iteration stops when the mean squared deviation of the average is less then 10^{-6}, or if the number of cycles reaches half the number of residues in the chain. This gives sufficient opportunity for both the amino and carboxy termini to evolve to a common label if they lie in the same domain.

Example

Let us look at some consecutive residues, having the following values in three successive cycles. $\{\ldots,3, 3, 5, 5,\ldots\}$, $\{\ldots,3, 5, 5, 5,\ldots\}$, $\{\ldots,3, 3, 5, 5,\ldots\}$. Then the averages between two cycles become $\{\ldots,3, 4, 5, 5,\ldots\}$ and $\{\ldots,3, 4, 5, 5,\ldots\}$, and the flicking is 'averaged out'. $\triangle$

The Ising model described, though simple, produces good results when the parameters are tuned and some other modifications made.

14.3 Domain Classes

The core of the proteins is made by packing of the secondary structure elements. Since there are only two types of SSE taking part in the packing, there are only three types of pairwise combinations:

1. α with α;

2. β with β;

3. α with β.

All these combinations might exist in a domain. In nature, however, it is observed that many domains contain almost exclusively one of the combinations, and these lead to the definition of three (main) classes of domains: mainly-α, mainly-β and α–β. Mainly-α is used interchangeably with all-α, and analogously for β. The α–β class is sometimes divided into two classes (see below). The border between the classes is not sharp, so doing classification is not always straightforward. Figure 14.2 shows examples of the main classes.

14.3.1 Mainly-α domains

A typical α-domain structure consists of a bundle of α helices. Loop regions on the surface of the domain connect the α helices, and they are packed to each other pairwise. One side of each helix faces the hydrophobic core, where the packing connections take place. The other side faces the solution. The connections are usually antiparallel, in that adjacent α helices are connected by a loop. The mainly-α domains are generally smaller than the mainly-β and α–β domains. Some of them also contain a minor part of β-strands (e.g. allowing up to 5% β-strands).

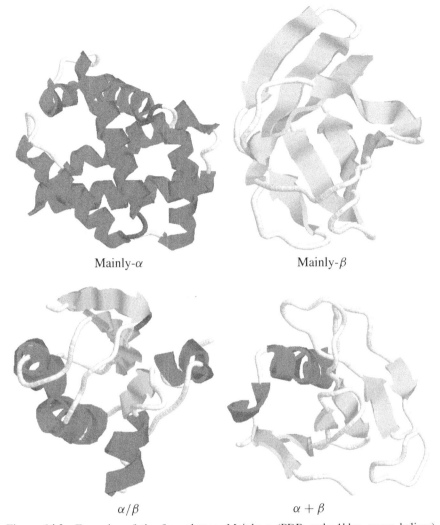

Mainly-α Mainly-β

α/β $\alpha + \beta$

Figure 14.2 Examples of the four classes. Mainly-α (PDB code 1hbg, seven helices), mainly-β (1who, eight strands in two sheets), α/β (1auz) and $\alpha + \beta$ (1c54). (Drawn by use of Rasmol.)

14.3.2 Mainly-β domains

The mainly-β domains consist of one or more β-sheets, and possibly a small number of α-helices (e.g. less than 5%). They are often classified by the number of sheets and the number and direction of the β-strands in the sheets. This makes it easier to further classify the β domains than the α domains.

14.3.3 α–β domains

In some classifications the α–β domains are divided into two: α/β-domains have a mainly alternating arrangement of α-helices and β-strands along the sequence. The β-sheets are mainly parallel. $\alpha + \beta$ domains are more segregated into two (or more) parts, and the β-sheets are mainly antiparallel.

14.4 Folds

The way the secondary structure elements are packed (arranged in an architecture), together with how the chain 'runs' through the secondary structures, is referred to as the fold of the chain.

It is assumed that there is a relatively small number of different folds (that it really is finite comes from the fact that there is a finite number of different proteins). That only some of the possible packing and topological arrangements are observed probably comes from the physical and chemical constraints on the chain. Several people have tried to predict the number of different folds, with results ranging from some hundreds to a few thousands. Most groups seems to agree that among the currently known structures (in 2002) there are around 800 different folds.

Proteins having the same fold are assumed to have a greater probability of having a common ancestor (being homologous), but they might also have the same fold due to convergence to the same fold from different ancestors (being analogous).

14.5 Automatic Approaches to Classification

The different systems for classification differ in how much automation is used: from fully automatically to almost fully manually.

The techniques used for automatic classification of structures depends on at which level the classification is to be done. At class level, some direct methods are used, for example, by summing up the numbers of SSEs satisfying some properties, and looking at the relations between SSEs. At lower levels, e.g. for folds, pairwise sequence and structure comparisons are used. The classification then inherit all the problems associated with structure comparison, as described in previous chapters.

The criteria used for classification will cluster together evolutionarily related structures but also structures which have adopted a similar fold because it represents a favourable arrangement of secondary structures without being evolutionarily related (convergent evolution). For automatic classification, it depends on the cut-offs used for placing structures in the same group. These cut-offs have been determined empirically and are probably not optimal.

14.6   Databases for Structure Classification

The three most popular databases for structure classifications are (all accessible via the web) the following.

FSSP-DaliDD.   FSSP is a fold classification based on structure–structure alignment of proteins (or protein chains). See http://www2.ebi.ac.uk/dali/fssp/. FSSP classification is done fully automatically, by use of the pairwise structure alignment program, DALI. The pairwise alignments of a representative subset of PDB are scored by the Z values, and a hierarchical classification is done based on the Z values. A Z value of 2 is used to divide into different folds (two structures obtaining an alignment $Z \geqslant 2$ are assumed to belong to the same fold type).

Dali Domain Dictionary (see http:/www2.ebi.ac.uk/dali/domain) classifies domains fully automatically. It has five levels: class, fold, functional family, sequence family and PDB entry of representative domain.

CATH: Class, Architecture, Topology, Homologous superfamily.   See

http://www.biochem.ucl.ac.uk/bsm/cath_new/.

CATH classification is done by using both automated and manual approaches. It has six levels: class, architecture, topology, homologous superfamily, sequence family, PDB entry.

SCOP: a Structural Classification of Proteins Database.   See

http://scop.mrc-lmb.cam.ac.uk/scop/.

SCOP classification is essentially done manually, and has seven levels: class, fold, superfamily, family, protein domain, species and PDB entry. It has become the gold standard for assessing sequence and structure comparison methods.

Since the methods used for classification are different, the resulting classifications are different. A systematic comparison of the results of these three classifications has been made which fortunately shows a high degree of agreement. Most of the discrepancies arise from different domain definitions.

Tables 14.1 and 14.2 show some comparisons of the classifications.

14.7   FSSP-Dali Domain Dictionary

The FSSP database includes all protein chains from the Protein Data Bank which are longer than 30 residues. First a sequence-unique representative subset of the PDB is generated. No pair of proteins within this set is more than 25% identical in sequence, and all removed structures are more than 25% identical to a representative. The percentage is calculated as the number of residue identities in the aligned region, divided by the average length of the two proteins. The representative set is then fully automatically classified by performing an all-against-all comparison of the structures

Table 14.1 A comparison between CATH, SCOP, FSSP and DaliDD of 3 October 2002. The first five levels of CATH are noted, and the corresponding levels in the other classifications. The numbers are compared on the fold-level.

CATH		SCOP		FSSP		DaliDD	
Class	4	Class	11			Class	7
Architecture	38						
Topology	775	Fold	830	($Z = 2$) 696		Fold	1083
H. superfamily		Superfamily				Funct. family	
Seq. family		Family				Seq. family	

Table 14.2 A comparison between CATH, SCOP and DaliDD at class level of 3 October 2002. The numbers of different folds (topologies) inside different classes are shown. The numbers of different architectures are also shown for CATH.

CATH			SCOP		DaliDD	
Mainly α	(5)	219	All α	151	All α	257
Mainly β	(119)	133	All β	110	All β	69
$\alpha\,\beta$	(13)	245	α and β (a/b)	113	α/β	97
Few SSs	(1)	77	α and β (a+b)	208	α-β meander	19
			Multi-dom. (α and β)	34	Antipar. β	28
			Membr. and cell surf.	12	Ambiguous	208
			Small	50	Unclassified	405
			Coiled coil	4		
			Low resolution	8		
			Peptides	63		
			Designed proteins	17		

with the structure comparison program Dali (based purely on the 3D coordinates of the proteins). Alignments are constructed for each pair and scored by the Z value (the number of standard deviations above the database average). A hierarchical clustering (average linkage) is performed based on the Z values. Family indices are constructed by cutting the tree at Z values of 2, 3, 4, 5, 10, 15. That means, for example, that the Z value of the alignment of the proteins with index 1.1.1.2.1.1 and 1.1.1.3.1.1 satisfies $4 \leqslant Z < 5$. Cutting the tree at the $Z = 2$ defines different folds. That means two structures scoring $Z < 2$ belong to different folds.

While Dali/FSSP is about proteins, or protein chains, the Dali Domain Dictionary is about domains. The domains are delineated automatically using the criteria of recurrence and compactness. Four levels are defined.

Class. Each structure is positioned relative to the others in an abstract, high-dimensional shape space (by a method related to principal component analysis). Five densely populated regions are identified, which they call attractors, namely, α/β, all-β, all-α, antiparallel β-barrels and α–β-meander. These define five classes. Domains which are not clearly closer to one attractor than the others are assigned to the mixed class six. Domains which are not connected to any attractor are assigned to class seven.

Folds. Fold types are defined as clusters of structural neighbours with average pairwise Z values (the average over all pairs in the cluster) above 2.

Functional families. Domains belonging to one functional family have strong structural similarities accompanied by functional or sequence similarities. A neural network is used, which predicts the likelihood of being homologous. The neural network use local sequence similarities, clusters of identically conserved functional residues and SwissProt keywords. It is trained using the fold-to-superfamily relation in SCOP.

Sequence families. Domains having at least 25% sequence identity are assigned to the same sequence family.

14.8   CATH

In CATH, classification of protein structures is currently generated on the basis of scores and overlap ratios returned by global structure comparison algorithms. The emphasis is on clustering proteins which have a similar fold. All the classification is performed on individual protein domains.

14.8.1   Domains

Multidomain proteins are subdivided into their constituent domains using a consensus procedure, based on three independent algorithms for domain recognition. This currently allows over 50% of the proteins (i.e. those for which these algorithms agree) to be defined as single or multidomain proteins automatically. The remaining structures are assigned domain definitions manually. The multidomain proteins are then split into their separate domains.

14.8.2   Class

CATH has four classes, those three based on alpha–beta combinations as described above, and a fourth small one. Class is automatically assigned (using the method by Michie et al. (1996)) according to the secondary structure composition and packing within the structure. A small number (under 10%) has to be manually classified. The secondary structures are treated in the following way.

Representation. The secondary structure elements are represented by sticks (vectors). For finding the best stick representation one estimates the curvature of an

SSE at any point. Those SSEs that deviate too much from the ideal straight helix or strand can then be split at the point of maximum curvature, resulting in a pair of straighter structures.

Composition. The secondary structure composition is represented by the content, the relative number of residues falling in each of the three secondary structures: helices, sheets or coils.

Contact. In addition to the number of each secondary structure type one in many cases also should use the pairwise contacts between the SSEs, the number and the types (helix–helix, helix–strand and strand–strand). This will reflect where the elements are, in the core or on the periphery. The number of contacts between two SSEs is measured as the number of C_α–C^α distances less than a given cut-off, where the cut-off is dependent upon the types of the interacting SSEs.

The classification

The compositions and contacts are then used to determine the class. For example, to be a mainly-α structure a structure should have at least a 60% alpha composition, and less than 5% β composition. In addition, these proteins must have more than 50% α–α and less than 5% β–β contacts. Similar constraints exists for the other two main classes. The α–β structures can be further subdivided into α and β and α/β categories using a secondary structure alternation score and a cut-off for the percentage of parallel β-strands.

14.8.3 Architecture

The architecture describes the arrangement of secondary structures as determined by the orientations, independent of topology. At the moment (October 2002) there are 38 different architectures. These are currently assigned manually using a simple description of the secondary structure arrangement, e.g. barrel or three-layer sandwich. Reference is made to the literature for well-known architectures (e.g the beta-propeller or alpha four helix bundle).

14.8.4 Topology (fold family)

Structures are grouped into fold families at this level depending on both the overall shape and connectivity of the secondary structures. This is done using the structure comparison algorithm SSAP (see Chapter 9.3). Parameters for clustering domains into the same fold family have been determined by empirical trials throughout the databank. Structures which have an SSAP score of 70 and where at least 60% of the larger protein matches the smaller protein are assigned to the same fold family.

Some fold families are very highly populated, particularly within the mainly-β two-layer sandwich architectures and the α–β three-layer sandwich architectures.

In order to appreciate the structural relationships within these families more easily, they are currently subdivided using a higher cut-off on the SSAP score (75 for some mainly-β and α–β families, 80 for some mainly-α families, together with a higher overlap requirement (70%)).

14.8.5   Homologous superfamily

This level groups together protein domains which are thought to share a common ancestor and can therefore be described as homologous. Similarities are identified first by sequence comparisons and subsequently by structure comparison using SSAP. Structures are clustered into the same superfamily if they satisfy one of the following criteria.

- Sequence identity of at least 35% and 60% of larger structure equivalent to smaller.

- An SSAP score of at least 80.0 and sequence identity of at least 20% and 60% of larger structure equivalent to smaller.

- An SSAP score of at least 80.0, 60% of larger structure equivalent to smaller, and domains which have related functions.

14.8.6   Sequence families

Structures within each superfamily are further clustered on sequence identity. Domains clustered in the same sequence families have sequence identities greater than 35% (with at least 60% of the larger domain equivalent to the smaller), indicating highly similar structures and functions.

14.8.7   The CATH classification procedure

1. Group the chains from PDB into sequence families by using 35% sequence identity.

2. Choose one representative from each chain sequence family, and find the domains of the chain.

3. Find the domains of each chain. This is for a chain done by calculating the alignment (using SSAP) between the chain and the representative of its (sequence) family, and let the chain inherit the domain definition from the representative.

4. Split the chains by domains, given the domain boundaries found above.

5. Group the domains into sequence families by using 35% sequence identity.

6. Assign a class to the domains (90% automatic).

7. Cluster into superfamilies and fold by using SSAP.

8. Assign architecture, manually.

Note that in point (1) chains are grouped, and in point (5) domains are grouped.

14.9   Classification Based on Sticks

When secondary structures are reduced to line segments (sticks), it is clear that many proteins can be represented as layers of secondary structure (with only one type of SSE in each layer). At the architectural level, proteins can be classified by how many secondary structures they have in each layer. For example, in the alternating $\beta\alpha$ class, a β-sheet with helices above and below constitutes three layers (BAB) and if there are two helices 'above' the sheet, five strands in the sheet and three helices 'below', then the architecture can be summarized as 2–5–3. (As there is no real distinction between 'above' and 'below', the smaller number of helices is given first.)

With different combinations of layers, plus a method of representing circular sheets (barrels), this system can capture almost all the architectures found in globular proteins. The 'filling' of these layers with secondary structures has been compared with the filling of atomic orbitals with electrons, giving rise to a 'periodic table' of proteins (Taylor 2002).

The path of the chain through these secondary structure layers can be recorded in the form of a string. This provides an alternative to the graphical TOPS representation. While a graphical representation is good for visualization, a string representation is better for automatic processing and can be used to analyse the large numbers of folds generated by some structure prediction methods (Petersen and Taylor 2003).

Assuming an organization of secondary structure elements in layers, an idealized template library of possible arrangements of secondary structure elements has been compiled. When a structure is scanned over such a library, the best matching template is used to define the topology together with the connectivity information of the original structure. Using this approach, both known and model proteins can be analysed and their fold distribution analysed.

14.10   Exercises

1. In this exercise we shall look at the PDB structure 1b8t. Find the entry for the structure in SCOP and CATH, and see how the structure is divided into domains. Then download the structure description (it consists of 37 models, determined by NMR), and use, for example, Rasmol to visualize the structure. The special form of the structure might be due to uncertainties in the structure determination. Find where the domains are, defined by SCOP and by CATH. Can you give an explanation of why the domains are defined so differently?

2. Consider the example in Section 14.2. Calculate the new state of residue 2.

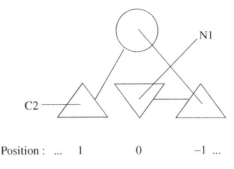

Position :   ...   1   0   −1 ...

Topology string :   +B0.−B−1.A0.−B1.

Figure 14.3 *Ring finger topology string.* (Upper) The topology of a small protein is shown as a TOPS diagram orientated in the reference frame used to extract a string description (Taylor 2002). (Lower) The derived topology string in which the first strand represented by +B0 defines the reference frame. The following element in the chain is an anti-parallel strand (−B) in position −1. Next, the first helix is entered (A0) and followed by a strand anti-parallel to +B0 in position 1. As seen, both direction and positions of β-strands are made with reference to the first beta strand in the sheet and the orientation is chosen so that the first helix is on top of the sheet.

3. Find the entry in CATH for the PDB code 1fon. You will find that it consists of two chains, and each are divided into two domains.

   (a) Find where the domains are on chain A.

   (b) Perform an alignment between chain A of 1fon and chain B of 1slu, by using SSAP (you will find a button for that in the CATH window). How much sequence identity do you get?

   (c) In the CATH classification procedure, in the first step the chains in PDB are grouped into families by using 35% sequence identity. Therefore, 1fonA and 1sluB will be in the same group. Suppose that 1fonA is chosen as the representative for the group. Use the constructed alignment to find the domains of 1sluB.

   (d) Compare the domain definitions with those given in CATH, and comment the differences.

14.11   Bibliographic notes

Properties for domain definition are discussed in Taylor (1999b), and methods are described in Jones et al. (1998), Swindells (1995), Holm and Sander (1993a), Holm and Sander (1998a), Islam et al. (1995) and Siddiqui and Barton (1995). Holm and Sander (1998a) describes a method which does not determine the domains of a structure independently, but use a recurrent method for getting consistent domain definitions between structures that are evolutionary related members of a protein family.

SCOP is described in Murzin et al. (1995), CATH in Orengo et al. (1993), Orengo et al. (1997) and Pearl et al. (2000), and FSSP/DaliDD in Holm and Sander (1996), Holm and Sander (1997), Holm and Sander (1998b) and Dietmann and Holm (2001).

The different classifications are compared in Hadley and Jones (1999) and Dietmann and Holm (2001). Elofsson and Sonnhammer (1999) compares SCOP and Pfam.

Classification based on sticks is described in Taylor (2002).

Part III

SEQUENCE-STRUCTURE ANALYSIS

.

15

Structure Prediction: Threading

In order to understand a protein under study one would like to obtain information on its structure. Since determination of the structure experimentally (by X-ray crystallography or NMR) is expensive and time-consuming, there has been much work in trying to develop good methods for predicting the structure. The simplest and most reliable is to use homolog sequences with known structures. Sequence alignment is performed, and the known structure is adopted to the new sequence, in accordance with the alignment. One should also search sequence-derived databases such as Inter-Pro containing patterns or profiles associated with protein (or domain) families.

If a homolog sequence is not found, one can try to compare the new sequence directly with protein structures. Since the number of naturally occurring protein folds is limited, there should be a good chance that the new protein has a fold which is already known, and this could be revealed by comparing the sequence against a database of structures. If a significant match is found, then the result provides a good idea of what 3D structure the novel sequence might adopt based on the structure of the identified target protein. For the more distant relationships, this allows the structure of the sequence to be predicted but even among the more certain relationships, an alignment based on structure (rather than a sequence/sequence alignment) is potentially more biologically meaningful and more accurate when compared with the equivalent structure/structure alignment obtained in test cases where both structures are known.

The third prediction method is the *ab initio* structure prediction method. This is method which uses only the information in the novel sequence itself, and general information relating sequence to structure.

In this chapter we mainly focus on the sequence/structure comparison methods, commonly called *threading*. Some of these use the secondary structure of the new sequence. Therefore, it is appropriate that we first describe secondary structure predictions. Such methods predict the location of secondary structure elements along the protein's backbone and the degree of residue burial.

Protein bioinformatics: an algorithmic approach to sequence and structure analysis
I. Eidhammer, I. Jonassen and W. R. Taylor   © 2004 John Wiley & Sons, Ltd   ISBN: 0-470-84839-1

15.1   Protein Secondary Structure Prediction

There are a lot of methods for secondary structure prediction—we will describe one of them, using machine learning. The problem of secondary structure prediction can be formulated as a pattern classification problem and methods from statistics or machine learning are suitable. For this problem, *supervised* learning has proved to be best. Such methods are used to develop classifiers taking as input a description of an instance and producing as output a prediction of which class that instance belongs to. For secondary structure prediction it is common to let each instance be a residue and its class be the secondary structure type to which it belongs. The classes are simply alpha, beta, or loop/turn (or $\alpha/\beta/l$). The input to the classifier is a set of features describing or related to the residue. A challenge in pattern classification problems is to find a good set of features that allows prediction of the instances' class labels. It has been found that the most important determinant of the secondary structure of one residue is local sequence, that is, the amino acid types of the residues 'in the neighbourhood' of the residue.

15.1.1   Artificial neural networks

Artificial neural networks (ANNs) are frequently used for classification problems, and are also used for secondary structure prediction. We will discuss this by regarding a special sort of ANN called *feedforward ANN*. It consists of a number of *nodes* (also called *neurons*) organized in a number of *layers*. There is one input layer, one output layer and zero or more hidden layers, as shown in Figure 15.1, where there is one hidden layer, consisting of four nodes.

Each node in the input layer gets a value for a feature, in Figure 15.1 it is a value describing an amino acid type. Each node in a layer is connected to all nodes in the preceding layer. In this case the layers are *fully* connected (as opposed to *partly* connected). A *weight* is associated with each connecting line. For each node an *activation value* is calculated based on the node's input value. Denote a node by j, and let the nodes on its predecessor layer be $i, i = 1, \ldots, K$. Furthermore, let the weight between nodes i and j be w_{ji} and the activation value for node i be x_i. Then a value $S_j = \sum_{i=0}^{K} w_{ji} a_i$ is calculated ($i = 0$ corresponds to an extra node with $a_0 = 1$, used as a threshold). The activation level at node j is then calculated as a 'soft threshold function' $f(S_j)$, and this value is sent on all connection lines out from j. A common function is $f(S_j) = 1/(1 + e^{-S_j})$.

Training the network

Training the network means finding values for the weights associated with each line in the ANN so that the network gives correct predictions as often as possible. For this purpose a supervised learning approach is used. In a supervised approach a training set is constructed as a set of labelled instances, i.e. a set of residues (encoded by their feature vectors) along with their correct labels (here each label being α, β or l).

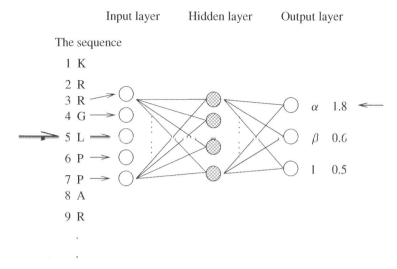

Figure 15.1 A simple artificial neural network for prediction of the secondary structure. Residues $i - 2, i - 1, i, i + 1, i + 2$ are used for the SSE prediction of residue i. The output is, for each type of SSE, a number describing the strength of predicting that the residue belongs to this class. Residue five is predicted to be a helix.

During the training phase of the ANN, the weights associated with each of the lines in the network are adjusted.

Each instance y in the training set is presented to the ANN, the ANN produces an output using its current weights and the output is compared with the correct label of y, and an error is calculated for each of the nodes in the output layer. Now the weights of the lines into the output layer can be adjusted to reduce the error for this one instance. Then the so-called *back-propagation* algorithm is used to propagate the errors to the nodes in the layers 'before' (providing inputs to the output layer) and to adjust the weights of the lines into this layer. The back-propagation algorithm can then be used recursively to propagate the errors to all layers and adjust the weights until all weights in the ANN have been adjusted. All instances in the training set are presented to the ANN many times, and the order in which the instances are presented is randomized. Different criteria can be used to decide when to end the training phase.

Evaluating performance

A danger of training an ANN is that the weights become very well adjusted to the training set—producing a very low error on this set—but fail to generalize to new examples. To test the performance of a prediction system, a test set must be defined that is independent of the training set (does not contain any of the sequences in the training set or any sequence similar to a training set sequence).

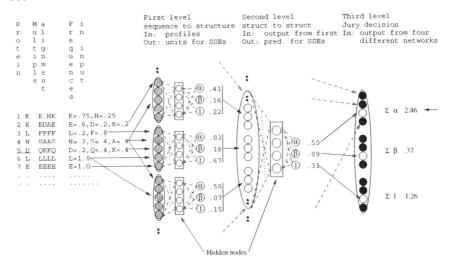

Figure 15.2 An illustration of the PHD program. Stippled arrows are output from networks not shown in the figure. See the text for explanation. Residue five is predicted to be α. Based on a figure in Rost and Sander (1993).

15.1.2   The PHD program

One of the best programs for secondary structure prediction is the PHD (Profile network from HeiDelberg) developed by Rost and Sander (1993).

Instead of the single sequence, family (alignment) information is used as input. First, a set of sequences homologous to the sequence under study is searched for, and a multiple alignment is constructed. One assumes that in most cases all residues in each alignment column belong to the same kind of secondary structure, and this is utilized, as shown to the left in Figure 15.2. It then becomes possible for the neural network to 'discover', for example, columns with predominantly hydrophobic amino acids and patterns of such columns. The PHD program consists of different ANNs on two *levels*, and a third level for a decision based on results from several independently trained networks on the first two levels.

On the first level a prediction of the SSE type of residue i is done based on a window around residue i of length w (= 13), as shown in the figure where $w = 5$. The window slides down the sequence, making a prediction for each residue. The input to the network is the frequencies of the amino acids occurring in each position of the window of the multiple alignment (including the sequence under consideration). Such inputs can be represented as real values or as bit vectors.

The second level takes into consideration that the SSE type of consecutive residues are correlated, in that, for instance, a helix consists of at least three consecutive residues. The input is the output from v networks on the first level; thus for the prediction of the SSE type of residue i at the second level, $w + v - 1$ columns of the multiple alignment are used.

The third level is calculating an average over the prediction values from several independently trained networks; in the figure there are four such networks (called the 'jury').

15.1.3   Accuracy in secondary structure prediction

A common measure used to determine the accuracy of a method is simply to use the percentage of the residues that are correctly classified, taken as average over a set of test cases. This means that for some sequences it might be higher, and for some lower, and this measure does not penalize for over-predictions (false positives) or under-predictions (false negatives). For typical proteins, one will, for example, in many cases achieve an accuracy of only 50% by predicting all residues to be loop residues. A better measure is to calculate a value for each of the three types, by using the correlation coefficient as defined in Chapter 3.5.1. A final value can then be found by taking, for example, the geometrical mean of those three values.

None of the measures above considers more physical properties: that the lengths of the SSEs are reasonable or that there is the correct number of SSEs. This can be better monitored, for example, by using a measure called SOV score (segment overlap score), presented in Rost and Sander (1993). It is however, shown that for accuracy levels over 70% the SOV score and the percentage of correctly predicted residues are highly correlated.

15.2   Threading

With the pressing need to have a structural interpretation of the large number of sequences being determined in the many genome projects, a variety of structure/ sequence comparison methods (threading methods) have been developed in recent years. These range from those that are basically sequence alignment methods augmented with structural data (called *3D–1D* methods), through those that consider pairwise (3D spatial) interactions between residues (true *threading* methods), to those that construct detailed atomic models and evaluate an overall energy-like scoring function for the whole model (*modelling* methods). This range thus forms a bridge from the area of sequence comparison to the area of molecular modelling. Some of these methods will be considered, beginning with the simpler methods through to the more advanced. This also roughly corresponds with their historical development.

15.3   Methods Based on Sequence Alignment

The simplest way of performing threading is to encode the structure as a string and then compare it as if it were a protein sequence. To reduce the three-dimensional structure to a one-dimensional string clearly loses a lot of information and all that can be retained is the local structural properties of each amino residue. This gives rise to a string of distinct structural states that can then be compared with a true protein sequence using sequence/sequence alignment methods. The components that need to be defined are the set of structural states and a log-odds-like scoring matrix relating these to the 20 amino acid residue types.

The most basic properties of a residue in a globular protein structure are its secondary structure state: α-helix, β-sheet or neither, and its degree of burial in the protein core (away from solvent)—a property sometimes quantified as its solvent accessible surface area (SASA or SAS). As each of the three secondary structure states can be either buried or exposed, the number of distinct substates is the product of the number of independent properties. In this simple example, we have only six (3×2) distinct states in which a residue can exist but it can be seen that the number of distinct states will grow rapidly as further properties are introduced. This becomes a serious problem for some of the more advanced methods considered below.

15.3.1   The 3D–1D matching method

In this method of Bowie et al. (1990, 1991), 18 different states (6 burial $\times$ 3 secondary structure states) were used. Let the probability of finding an amino acid (a) in an environment (j) be represented as $P_{a:j}$, while the probability of finding amino acid (a) anywhere be P_a. They then calculate the information value score:

$$s_{aj} = \log\left(\frac{P_{a:j}}{P_a}\right). \tag{15.1}$$

These values were calculated from multiple sequence alignments based on a small database of 16 proteins (sometimes called the *training set*). The boundaries (on SASA values) between the six buried states were adjusted so that the sum of s-scores over the training set was maximized.

When applied to proteins that were not part of the training set (the *test set*), the positions in a structure were assigned to one of the structural states. By this assignment, the position then takes the propensities for each amino acid type as defined by the score s (in Equation (15.1)). The resulting table of propensities for each position in the structure was referred to as a *3D structural profile*, being similar to earlier work on multiple sequence profiles (Chapter 7).

A protein sequence can then be aligned against this profile and the maximum alignment score taken as a measure of how well the sequence 'fits' the structure. As is common, the alignment scores were calibrated against a background of randomized scores and transformed into a Z-score. As an example, the method was applied to the globin family using the structure of a myoglobin as a probe. This test clearly identified myoglobins from nonglobins but the other globins was less well separated from the nonglobin background.

In later developments of this approach (Rice and Eisenberg 1997), the comparison of predicted secondary structure states with observed structure was included.

15.3.2   The FUGUE method

This approach use a greater variety of residue states, each of which has an 'environment' specific substitution table (Overington et al. 1992; Shi et al. 2001). The structural state of a residue is determined by its hydrogen-bonding pattern, its secondary structure state and its solvent exposure. A fourth SSE state was introduced to represent residues with an unusual positive ϕ main-chain torsion angle. Side-chain hydrogen bonds were classed as side-chain/side-chain, side-chain/main-chain, with the latter being further distinguished by donors (to a C=O group) and acceptors (to an N–H group). Solvent exposure, however, was only divided into two states. Combining these has resulted in a set of 64 tables. However, as not all amino acids can form hydrogen bonds, some tables are restricted.

As each new property at least doubles the number of tables, and statistics must be gathered for each of the residue states over a limited structural databank, the inclusion of too many states can reduce the residue counts for the less common states to a level that is too low to give an accurate representation. This explains the restricted set of solvent exposure states used. Statistical methods can be used to partly overcome this problem and they will be considered below in the context of pairwise interactions, where the problem becomes more serious.

15.4   Methods Using 3D Interactions

The inherent limitation of matching structure strings with protein sequences is that, in principle, it is possible for the same structure strings to adopt three-dimensional structures with completely different folds. This has led to the development of methods that explicitly consider 3D interactions. The simplest interactions that can be incorporated into a threading method are those involving only two residues (pairwise interactions). For each candidate alignment of a sequence onto a structure, the resulting pairwise interactions can be evaluated and, typically, summed to give an overall match score. In this approach, two problems must be solved: the nature of the interactions and how the candidate alignments are generated prior to evaluation.

In considering the nature of the interactions, it must be remembered that most amino acid side-chains are flexible, so two residues, say an arginine (R) and an aspartate (D), can interact in a variety of ways involving both different separations and orientations of the backbone peptide chain. Statistics gathered on this type of interaction will therefore have a spread of properties. This means that if we intend to use these data to predict the type of residues that occupy a pair of positions in a protein structure, there will be many equally likely contending pairs, giving considerable uncertainty in the resulting choice and hence alignment. It can be hoped, however, that the final overall summed score of all pairwise interactions will average out some of the individual pairwise noise and give a significant final match score.

The simplest way to incorporate 3D interactions is to define a measure of contact between residues (say, two atoms in the pair must lie within 5 Å) and then count the frequency of contact between all residues for each of the 400 (20×20) possible types of amino acid pairs observed in the protein structure databank (or a reduced subset of structures). After suitable normalization for the frequency of amino acid occurrence, the normalized values can be used to evaluate the contribution of each pairwise interaction when it is observed in a sequence/structure alignment. In this form, these contact propensities could be used to evaluate any alignment in which one of the sequences has a known structure. A rough threading method could be based on calculating a series of sequence alignments, using different scoring matrices and gap penalties. The alignment with the best 3D score is then kept, as illustrated in Algorithm 15.1.

Algorithm 15.1. Threading by simple 3D interactions.

Threading of the sequence q on structure A

const

C A scoring matrix constructed from observed interactions

$\{\mathcal{R}\}$ a set of amino acid scoring matrices

R a scoring matrix from $\mathcal{R}$

var

A a structure

s_A the sequence of A

q a sequence

S the score of a sequence/structure alignment

S_H the highest sequence/structure score found

$\mathcal{A}$ an alignment

a, b amino acids from q

begin

    $S_H := 0$; *choose a scoring matrix $R \in \mathcal{R}$*

    repeat

        $\mathcal{A} :=$ sequence alignment of (q, s_A) using R

                           calculate the score of the alignment:

        $S := 0$

        for each residue $(a_j, a_k) \in A$; $j \neq k$ **do**

            let a and b be the amino acids from q aligned with a_j and a_k

            if (A_j, A_k) interacts **then** $S := S + C(a, b)$ **end**

        end

        if $S > S_H$ **then** $S_H := S$ **end**

        choose a new scoring matrix $R \in \mathcal{R}$

    until *satisfied alignment found*

end

In our simple threading method, it is possible that the sequence alignment might never generate the optimal scoring 3D model and it would be better to have a method in which the 3D interactions are used more directly to determine the alignment in the first place.

15.4.1   Potentials of mean force

As the definition of a 'contact' is somewhat arbitrary, the contact approach can be generalized to provide statistics on the likelihood of a pair of residues (of given types) occurring at any distance separation and not just for those in contact. This approach was developed by Sippl (1990) to derive an empirical function for the propensity of each of the 400 pairs of residues to be any given distance. (Symmetric pairs are not equivalent because of the direction of the polypeptide chain.) In practice, the values are stored as a table for a set of discrete distance intervals. Although the long-range 'interactions' do not reflect any direct physical interaction of the residues in a structure, their combined effect is to ensure there is a reasonable distribution of residue types in the structure.

Sippl used techniques drawn from statistical mechanics to represent the occurrence frequency of different residue pairs as a pseudo-energy called a 'potential of mean force' (PoMF). For example, consider two small hydrophobic residues, alanine (A) and valine (V), both of which 'favour' packing in the core of the protein. The PoMF for A and V, plotted against increasing distance, will have a 'dip' around 5 Å, corresponding to the frequently observed separation when A and V pack together. Conversely, an aspartate (D) and a valine will have a peak around 5 Å as these residues are not frequently packed (see Equation (15.2)). It should be noted that there will be a slight drop in the level of the long-range interactions for D and V, indicating that separations that avoid placing these two residues together are favoured. Specifically, consider two amino acid types, a and b, which have a frequency distribution $f^{ab}(s)$ (over the distance interval s). These frequencies are transformed to PoMF as

$$\Delta E^{ab} = -k \log \left(\frac{f^{ab}(s)}{f(s)} \right), \tag{15.2}$$

where $f(s)$ is the frequency of all pairs in the interval s. In this simplified representation of Sippl's original formulation it can be seen that there is a correspondence with Equation (15.1) used by Bowie and co-workers.

This simple idea of residue pair packing preferences has been elaborated into a large variety of related measures. Sippl compiled separate tables for all of the above distance intervals for different separations of residue pairs in the sequence. (The separation is the difference in residue number along the sequence; thus residues eight and two have a separation of six.) These were roughly divided into some specific short-range separation (to further distinguish secondary structure types) and a medium and a long range class. As with the simple methods, each of these distinctions multiplies the number of tables and hence spreads the observed data ever more thinly, with the danger that few observations will be available for the less common states (such as Met–Trp).

A general approach to the problem of sparse data was used by Sippl in which the table entries for structure states with low counts defaulted towards the less specific values of the background count for all residue pairs. If the 'true' frequency distribution for acids a and b that we want to use in calculating the potentials is $f^{ab}(s)$ and the raw counts that have been obtained from the structure databank is $g^{ab}(s)$, the $f^{ab}(s)$ frequency can be estimated as

$$f^{ab}(s) \approx \frac{1}{1 + m\sigma} f(s) + \frac{m\sigma}{1 + m\sigma} g^{ab}(s), \tag{15.3}$$

where $f(s)$ is the frequency (for distance interval s) for all amino acid pairs and m is the number of observation. The factor σ acts as a weight that balances the relative contributions of the general background and the specific pair (a, b) observations. When the number of observations is large (big m), then the first term of the sum vanishes and the factor on the second approaches 1, giving $f^{ab}(s) = g^{ab}(s)$, while, at the other extreme, if there are no observations ($m = 0$), then $f^{ab}(s) = g^{ab}(s)$. (The factor σ also plays a role in the transform to PoMF.)

Separate tables were compiled for the separation of most combinations of main-chain atom types including (C_α–C_α, C_β–C_β, N–O, N–O, N–N and O–O. (Other pairings such as C_α–C_β were also suggested.) These sets of distances provide a feature that reflects the relative orientation of a pair of residues: residues 'facing' each other (for example, with side-chains packed) will have a shorter C_β–C_β distance separation to their C_α–C_α separation. The principal advantage of the N and O distances is to provide an indirect measure of the secondary structure state. Residue pairs with close N–O distances will be hydrogen bonded. It should be noted that this elaboration of the method to different atom types does not affect the problem of data coverage of the states as each residue pair contributes equally to each distinct atom-pair table.

The large number of values in the tables also has the more worrying aspect that the tables can become so detailed that there is the possibility that a sequence/structure relationship might be recognized through specific rather than generic interactions. For example, consider a structure that contains some quite rare interactions: say, an R_α/W_α, W_β/C_α and a D_c/V_β (where the subscripts specify their secondary structure states[1]). If this structure is used to compile the tables, then these interactions will be stored in the tables. Because of their rarity, there may be few contributions from other proteins with these interactions and none with this particular combination. When a sequence that is related to this structure (or the sequence of the structure itself) is scanned against the database for structures, the rare residue combinations will 'click' into place with a high score. This will produce the expected result that a sequence will 'recognize' its own structure with a high score and also that related sequences will attain a high score. When the sequence is novel (say from a recent genome) there is no harm in this effect—it will correctly identify the structure. However, during the testing of the method, the effect will give the false impression that the sequence is being recognized by its fit to a structure, not a fit to sequence features 'hidden' in the structure tables. During the testing of a threading method it is therefore important that none of the test sequences are related to the structures that were used to compile the tables (the training set). Most workers go to elaborate lengths to avoid this 'memory'

---

[1] In the method of Sippl, the secondary structure state is indirectly encoded in the different atom tables, but, for simplicity, is represented here as a single conformational state.

effect in the testing but it is so difficult to be certain that there is no residual effect that some would argue that the only proper test is to use novel sequences that have not been seen before. This has led to the CASP exercise and the automated version CAFASP, which will be considered along with other evaluation procedures below.

15.4.2 Towards modelling methods

There are a number of other methods that can be used to extract and encode the structural properties of proteins, some of which also incorporate modelling features (constructing atomic models and evaluating an overall energy-like scoring function for the whole model). In all these approaches, there remains the problem of keeping a balance between the size of the tables (equivalent to the number of parameters in some of the methods) and the size of the database from which they were derived— along with the ever present danger of over-fitting (or memorization). Examples of such methods are briefly explained in the bibliographic notes.

15.5 Alignment Methods

The simple methods considered above, in which the 3D structure is encoded as a string, can be implemented using any of the existing single or multiple sequence alignment methods. Similarly, pattern-matching methods can also be used. For alignment, all that is necessary is the use of a sequence/structure-based matrix (instead of a Dayhoff-like matrix) and the standard dynamic programming algorithm. These methods are particularly suited to the use of different gap penalties with the different structure states.

When pairwise interactions are considered, the situation is not so simple: indeed the first principle of dynamic programming—that the score obtained for a match should not depend on other positions—is immediately broken by the incorporation of pairwise interactions. To avoid this problem, a variety of heuristic and iterative methods has been developed.

15.5.1 Frozen approximation

The simplest of these methods is called the 'frozen approximation' (Flockner et al. 1995). In this method, a substituted residue is assessed using the interactions with those already found in the structure. Clearly, this is not ideal as many of the interactions considered will not be found in the true structure of the sequence that is being modelled. It can be argued, however, that if the sequence does find a match with the structure, then many of these false interactions will probably be the same or similar to the true interactions anyway.

To encode this method into a dynamic programming algorithm is straightforward as each residue match in the score matrix can be evaluated independently of what

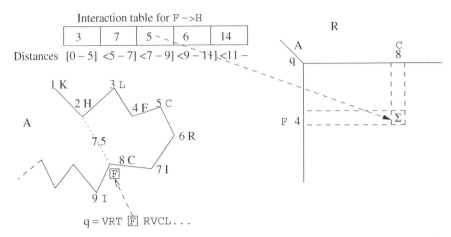

Figure 15.3 Illustration of the calculation of the scoring of R for $i = 4$ and $j = 8$. q_4 (F) is substituted for the position 8 of the structure. Then the pairwise interactions are summed up. The figure shows the contribution in this sum from position 2 (H) in the structure. The interaction table for F → H is used. The distance is 7.5, hence the contribution is found to be 5. ($< r - s$] means the interval greater than r and less or equal to s.)

residue is eventually assigned to the other positions. The usual matrix is constructed with the sequence q (of unknown structure) along one dimension and the sequence of known structure A along the other. The score for each matrix element R_{ij} is then calculated by summing the pairwise interactions obtained when q_i is substituted at position j in A. The pairwise interactions are found from a set of interaction tables, of which the values are calculated from a database of known structures. There is one table for each (ordered) amino acid type pair, and each table has an entry for a distance interval, the normalized number of observed interactions at the specified distance for the actual pair. The method is illustrated in Figure 15.3. Having filled the matrix, the highest-scoring path (or lowest energy for PoMF measures) is extracted by ordinary dynamic programming.

The frozen approximation algorithm (FAA) lends itself well to iteration and this was used by Skolnick and Kihara (2001). The first cycle can proceed as described above but in the second cycle, the positions assigned in the first cycle can be used to assess the pairwise interactions, as shown in Figure 15.4. This helps to solve the problem of how false interactions contribute to the score. False interactions are introduced in the first iteration, and further iterations might then refine these assignments.

When the novel sequence and the target structure are closely related (homologous), the FAA will work well as the initial cycle will provide a good starting point for further refinement. Problems are more likely to be encountered for distant relationships where the sequence of the known structure is a poor imitation of the true interactions. Unfortunately, it is in the latter situation where the added power of threading methods is needed most. In the easy problem, a simple sequence alignment or a simpler threading algorithm might equally provide the correct answer.

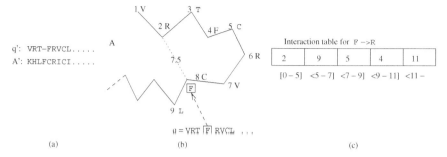

Figure 15.4   Illustration of the iteration: (a) is the alignment found in the first cycle; (b) shows how the residues from q are replacing that of A. (c) Illustration that this implies use of other interaction tables. (No substitution was made at position 5, which retains its original residue.)

As with these simpler methods, the robustness of the solution obtained by this method can be tested by introducing an element of randomization into the scoring and analysing repeated runs for common solutions.

15.5.2   Double Dynamic Programming

An alternative approach to calculating a threading alignment with 3D interactions can be derived from the methods used for structure comparisons. Sequence/structure comparison faces the same problem as structure/structure comparison in that the calculation of an element in the score matrix is not independent of the other positions. While heuristic methods, such as simulated annealing, can always be applied to such a situation, we will concentrate below on a methods based on the double dynamic programming (DDP) algorithm (Chapter 9).

This approach was initially applied to threading by Jones et al. (1992b) using the potentials of mean force (PoMF) combined with an explicit residue exposure term. As in structure comparison, the basis for calculating the low-level scoring matrices of the DDP algorithm is to make the assumption that a sequence residue q_i does indeed belong on a position A_j in the structure. With this working assumption, the scoring matrix cell $^{ij}S_{rs}$ can be assigned with the value of the propensity (or PoMF) for residue type q_r to be found at a distance $A_j - A_s$ from residue type q_i. Remember that only the cells in the submatrices $r < i, s < j$ and $r > i, s > j$ need be evaluated. The best path through ^{ij}S gives a score which can be interpreted as follows: 'If sequence residue i lies on structure position j, what is the best score that can be attained?' If the i, j assumption happened to be true (meaning one that would be found were the structures of both proteins known), then it can be hoped (if the score function is at all realistic) that true assignments will attain a high (good) score while spurious (false) assignments will only get a low (poor) score. The low-level alignments are summed up in the high-level scoring matrix (as described in Chapter 9). Unlike the FAA, at no point was the alignment calculated using the sequence of the protein of known structure.

Iterated DDP

As for comparing structures, the DDP algorithm can be iterated by taking the results of the current high-level alignment to bias the alignments from the low-level matrices in the next cycle. Depending on the strength with which the previous solution is imposed, the algorithm can be guaranteed to converge.

This approach is of additional interest as it allows the low-level pairings to be made selectively. In the initial phase, pairs can be selected on the basis of local structure. In structure comparison, this was made on the basis of observed secondary structure state and observed burial (in both proteins); however, in threading, the local structure states for the protein of unknown structure must be predicted. This allows an initially sparse selection to be made and the alignments made from these can be summed into the high-level matrix as described above. The highest-scoring cells in the matrix can then be taken as the selected pairings for the next cycle.

The method is very flexible as the selected pairs can be augmented by external criteria—such as residues known to be equivalent for some functional reason (perhaps identified by motif matching).

15.6   Multiple Sequence/Structure Threading

With the increasing number of related sequences and structures, it is desirable to have a threading method that can make use of multiple sequence and structure alignments. In sequence alignment, various heuristic methods were described for multiple sequence alignment, the most popular of which follow a progressive strategy in which the most similar pairs are firstly aligned and represented as a consensus sequence that can be further aligned with other single sequences to other consensus sequences. As mentioned above, for simple threading methods that use only string representations of a structure, this approach can be followed directly.

For methods using 3D (pairwise) interactions, the problem is more like that found in structure comparison. In this application, the progressive approach can still be followed but the resulting consensus structure can prove difficult to represent without losing information. In threading, we can begin by considering the more limited problem of aligning a sequence alignment with a single structure. Clearly, this can use the methods described above and the only change needed is to change the evaluation function to deal with a set of residues on each structure position.

Following multiple sequence alignment methods, the score for aligning two sets of residues can be calculated as the sum of their individual pairwise interactions (sum-of-pairs or SP score). In threading, however, a sum-of-pair(wise) may not be ideal, as can be illustrated by the following example. Consider two positions in a sequence alignment, with the residues GW and WG. If we assume that GG has an ideal separation of 3 Å, WW 9 Å and GW 6 Å, then if these two positions are placed on two positions in a structure which is 6 Å apart, an SP score will include two unfavourable interactions (GG and WW), whereas the interaction within each individual sequence is ideal. To avoid this, interactions can be summed only within a sequence but this misses an important aspect of the sequence alignment which is the degree of conservation at each position (which is implicitly captured in an SP score).

15.6.1   Simple multiple sequence threading

To avoid the problem sketched above and also the problems associated with very large sets of tables, Taylor (1997b) developed a simple multiple sequence threading method (MST) based on the single pairwise property that conserved hydrophobic positions in a multiple alignment should pack in the core. This was augmented by an explicit matching of the local properties of predicted versus observed secondary structure state and predicted versus observed residue burial. Weights on each of these components were optimized on a training set to obtain an ideal match.

The secondary structure matching combined with the exposure matching by themselves produced very good results with apparently only a minor contribution from the pairwise term. However, the method was tested mainly on small globular proteins and when applied to larger (multi-domain) proteins it was found that the pairwise packing term was essential to maintain a compact match within one domain.

15.7   Combined Sequence/Threading Methods

The threading methods described above have been aimed at finding distant relationships between sequence and structure. This has been driven by a need to extend the limits of recognition beyond what can be achieved using sequence alignment (or multiple sequence alignment). As most threading methods have been optimized for distant relationships, some can behave surprisingly poorly when there is a reasonably clear sequence similarity. It has therefore been necessary to make a decision when to use sequence alignment and when to use a threading method. When the methods need to be applied automatically to a large volume of sequences, as in a genome analysis or in an automatic web server, this need for manual decision-making is not ideal.

To circumvent this clumsy situation, Jones (1999b) trained a neural net to recognize the degree of sequence similarity in the current comparison problem. For difficult problems the parameters/methods shifted towards the threading side while for simple comparisons a sequence alignment was used. The resulting program called GenTHREADER can thus be run automatically over genomes and provides an easy-to-use web server.

15.8   Assessment of Threading Methods

A wide range of methods for threading have been described above, ranging from those that are effectively 'only' sequence alignment with a specialized amino acid substitution matrix to those that construct realistic atomic models. It can be argued that the more complex methods are only justified if they can be shown to perform better than the simpler methods, and in turn the simpler methods are only justified if they too can perform better than basic sequence alignment. This is an oversimplification because distantly related proteins often retain a conserved motif that might be missed by a threading method that is over-focused on general properties such as hydrophobicity or secondary structure propensity. It is also possible that sequence alignment using a profile of homologous sequences might perform better than a threading method using a single sequence.

Some simple measures have been reassessed recently by McGuffin et al. (2001), including pure sequence alignment and the alignment of simple secondary structure strings. However, this analysis only uses known secondary structure assignments and an added degree of uncertainty will be involved when predicted secondary structures are used. In this limited situation it was shown that secondary structure alignment at least has the potential to be better than sequence alignment.

15.8.1   Fold recognition

As threading methods are generally aimed at finding distant relationships, one of the main ways of assessing their power is by measuring the number of relationships that can be identified among proteins with the same fold. There are difficulties in this approach as the limit to which two proteins can be said to have the same fold is not well defined for distant relationships. Nevertheless, a criterion based on the CATH or SCOP databases is usually selected. The power of fold recognition can then be monitored by the number of correct matches (true positives) plotted against the number of 'misses' (false positives) or functions of these quantities, such as sensitivity and selectivity (see Chapter 3). Other comparisons are cited in the bibliographic notes.

15.8.2   Alignment accuracy

Despite proven power in recognizing the relationships between distantly related proteins, threading methods are not equally good at producing accurate alignments between sequences. This may be due to different reasons but it is likely that the more general properties that are used in threading methods have few specific features (such as motifs) that can act as 'place-holders' in an alignment. Without any such markers, the whole alignment can drift, and providing this happens in a structurally meaningfully way, then a good score will still be returned by the threading methods. Examples of such drift range from the common ± 4 (or ± 8) shifts in α-helices or ± 2 (or ± 4) shifts in β-strands to shifts in which a number of strands or helices are displaced to the position of the adjacent secondary structure element. Clearly, the latter will have a disastrous effect on the alignment accuracy but in terms of structure, all that might be lost is a helix (or strand) from the edge of a domain.

15.8.3 CASP and CAFASP

Given the range of methods involved, and the inherent difficulties in assessing them, there are few comparative studies that have not been compiled from an unbiased point of view covering a wide range of methods. In those that have been attempted, comments are often made that one of the methods is not up-to-date, or was trained on a different database or was not run using the best parameters. Similarly, when threading methods are assessed by the same workers who developed the method there can sometimes be disputes over the independence of the training and testing datasets and whether the latter might have been selected to reveal the method in its best light.

To circumvent these doubts and difficulties, it was suggested that the only way to properly assess these different and complex methods is on new structures that have never been seen before and cannot therefore be part of anyone's training set. This exercise, called the *critical assessment of structure prediction* (CASP), was begun in 1994 and involves the collection of protein structures prior to their publication. The sequences are then broadcast (via the internet) and anyone with any sort of method (even guesswork) can submit their predictions.

The organizers then provide an extensive analysis of the results based on fold recognition, modelling accuracy and alignment accuracy. The results have tended to confirm that no one method or even general approach is obviously superior. However, in the region of difficult threading problems, the number of proteins (referred to as targets) is often limited and so a large variation can be expected based on just a small number of results. If there is one approach that has consistently performed well it is the application of a broad knowledge of protein structure embodied in a person and not a computer program.

While CASP attempts to circumvent the problems associated the 'self-assessment' approach, it has raised its own complexities as many predictors use a variety of measures including not only their own insight but also the results of other methods that can be found as servers on the internet. This has led to a complex mix and the CASP exercise has become more of an assessment in powers of data integration rather than any individual prediction method. To avoid some of this complexity, a parallel exercise has been set up (called CAFASP) that is open only to fully automated servers. However, this does not exclude automatic methods that harvest the results of other automatic servers (called meta-servers) and, indeed, in the last exercise it was a meta-server that produced the best results.

15.9 Bibliographic notes

PHD is described in Rost and Sander (1993). Other methods for secondary structure prediction using neural networks are described in Qian and Sejnowski (1988) and Jones (1999a).

The threading method of Bowie et al. is described in Bowie et al. (1990, 1991) and Rice and Eisenberg (1997), and the method of Blundell et al. (FUGUE) in Overington et al. (1992) and Shi et al. (2001). Potential of mean force is described in Sippl (1990). A method including site-specific gap penalties and the HMModel is in Karplus et al. (1998).

Frozen approximation is described in Flockner et al. (1995) and Skolnick and Kihara (2001), and DDP in Taylor and Orengo (1989) and Jones et al. (1992b). Use of multiple sequences is in Taylor and Orengo (1997), combined sequence/threading in Jones (1999b), and comparisons of programs for fold recognition are found in Shi et al. (2001) and Rice and Eisenberg (1997).

Modelling methods

Contact-based methods   A threading method similar to the Sippl method, based on contact frequencies, has been developed by Godzik et al. (1992) but without the proliferation of different tables. Skolnick and Kihara (2001); Skolnick et al. (2001) have developed an alternative method.

Associative memory Hamiltonians   A related threading-like approach was also developed by Friedrichs and Wolynes (1989); Friedrichs et al. (1991) in which the 'tables' of propensities take the form of values in an associative memory Hamiltonian (AMH). This is a matrix based on a Hopfield neural network in which the average pairwise residue interactions from the training set are stored. A protein model is then adjusted until its current set of internal interactions attain a maximum when scored through the Hamiltonian.

Artificial neural nets   More recently, the propensities for residue contact have been captured as the weights in an artificial neural net (ANN)—both as a 3D–1D method (Lin et al. 2002a) and incorporating 3D information (Lin et al. 2002b). The ANN approach can be used for sequence alignment or for assessing how well a sequence fits a structure. It is related to the AMH, and, as with most of these approaches, care must be taken to ensure there are enough data for training the network and that the training and test sets are independent.

Appendix A

Basics in Mathematics, Probability and Algorithms

Some sections in the book assume a basic knowledge of mathematics, statistics and algorithms. In this chapter we give a brief overview of the most basic of these topics.

A.1 Mathematical Formulae and Notation

$\displaystyle\sum_{i=1}^{n} x_i$ means the sum of the n numerical variables $x_1, x_2, \ldots, x_n$.

Example: $\displaystyle\sum_{i=1}^{n} i = \frac{1}{2}n(n+1)$ is the sum of n variables when $x_i = i$.

$\prod_{i=1}^{n} x_i$ means the product of the n variables $x_1, x_2, \ldots, x_n$.

$n!$ is faculty: $n! = 1 \cdot 2 \cdot 3 \cdots n = n(n-1)!$. Note that $0!$ is defined as 1.

$\binom{n}{x}$ is the binomial coefficient, defined as $\dfrac{n!}{x!(n-x)!}$ (n choose x)

Example: $\binom{6}{4} = \dfrac{6!}{4!(6-4)!} = 15$.

$n^{(r)}$ is defined as $\dfrac{n!}{(n-r)!}$.

Example: $6^{(2)} = \dfrac{6!}{4!} = 30$.

Protein bioinformatics: an algorithmic approach to sequence and structure analysis
I. Eidhammer, I. Jonassen and W. R. Taylor © 2004 John Wiley & Sons, Ltd ISBN: 0-470-84839-1

Table A.1   Definition of the basic logical operators.

P	Q	P *and* Q	P *or* Q	*not* P
false	*false*	*false*	*false*	*true*
false	*true*	*false*	*true*	*true*
true	*false*	*false*	*true*	*false*
true	*true*	*true*	*true*	*false*

A.2   Boolean Algebra

Boolean algebra makes use of logical variables and operators. Thus, a logical variable may take on the value *true* or *false*. The basic logical operators are *and*, *or* and *not*. The operations are defined in Table A.1.

Example

Let three logical variables have the values P = *false*, Q = *true*, R = *true*. Then the expression (P *and* Q) *or* ((*not* P) *and* R) = (*false and true*) *or* (*true and true*) = *false or true = true*   △

A.3   Set Theory

A *set* is a set of *elements*; unless otherwise stated, all elements are unique. The elements can be of any type: number, letters, (sub)sequences, (sub)structures, compound data, etc. All elements in a set are of the same type. The elements are enclosed by {}.

   Example: $\{2, 7, 4, 5\}$, $\{a, e, f, c, h\}$, $\{s^1, s^4, s^8\}$, $\{(A_i, B_j), (A_k, B_l)\}$.

   Different operators are defined for sets.

$x \in U$ is *true* if x is an element of the set U.

$x \notin U$ is *true* if x is not an element of U.

$U \cup V$ is the *union*, the set of all elements which are in either U or V, or in both.

$U \cap V$ is the *intersection*, the set of all elements which are in both U and V.

$U - V$ is the *difference*, the set of all elements in U which are not in V. Note that generally $U - V \neq V - U$.

The priority of the operators in a formula are specified by parentheses.

Example

Let the elements be the alphabet of amino acids, and the following two sets are given: $U = \{C, K, L, R, S, Y\}$, $V = \{A, C, R, T, W\}$.

Table A.2 The number of different samples when
r elements are drawn from n different elements.

	Replaced	Not replaced
Ordered	n^r	$n^{(r)}$
Not ordered	$\binom{n+r-1}{r}$	$\binom{n}{r}$

$L \in U$ is *true*.

$L \in V$ is *false*.

$A \notin U$ is *true*.

$U \cup V = \{A, C, K, L, R, S, T, W, Y\}$.

$U \cap V = \{C, R\}$.

$U - V = \{K, L, S, Y\}$.

$V - U = \{A, T, W\}$.

$U \cup \{R, W\} = \{C, K, L, R, S, Y, W\}$.

$U \cup (V - \{A, R, T\}) = \{C, K, L, R, S, W, Y\}$.

$\triangle$

Note that if $\mathcal{M}$ denotes the set of one-letter amino acid symbols, then $\mathcal{M} - \{C, D, E, H, K, R, Y\} = \{A, F, G, I, L, M, N, P, Q, S, Y, V, W\}$.

A.4   Probability

A.4.1   Permutation and combination

Random sampling is basic for statistical theory. Assume that we have a set of n different elements. The number of different samples which can be formed when r elements are drawn depends on whether the elements are replaced and whether the samples are ordered. The number for the different cases is shown in Table A.2.

Example

Assume $n = 4, r = 2$ and the elements are A, C, G, T. Then the different samples are as follows.

Ordered and replaced. 16 samples: AA, AC, AG, AT, CA, CC, CG, CT, GA, GC, GG, GT, TA, TC, TG, TT.

Ordered and not replaced. 12 samples: AC, AG, AT, CA, CG, CT, GA,
GC, GT, TA, TC, TG.

Not ordered and replaced. 10 samples: AA, AC, AG, AT, CC, CG, CT,
GG, GT, TT.

Not ordered and not replaced. 6 samples: AC, AG, AT, CG, CT, GT.

△

A.4.2   Probability distributions

Consider a *discrete random variable* X with the value set $x_1, x_2, \ldots, x_n$. The probability distribution (or simply the distribution) of X is the list $x_1, x_2, \ldots, x_n$ together
with the probabilities $p_1, p_2, \ldots, p_n$, where $p_i = P[X = x_i]$, the probability that X
has the value x_i. Note that $0 \leqslant p_i \leqslant 1$ and that $\sum_{i=1}^{n} p_i = 1$.

Example

Let X be the result of throwing a (true) die. Then the value space is $\{1, 2, 3, 4, 5, 6\}$,
and the probability distribution is $\{1/6, 1/6, 1/6, 1/6, 1/6, 1/6\}$.   △

A.4.3   Expected value

The expected value of a random variable is the weighted (by the probability) average
over all possible values of the variable, and is denoted by μ. Then $\mu = \sum_{i=1}^{n} p_i x_i$.
For example, the expected value of throwing a true die $\sum_{i=1}^{6} \frac{1}{6} i = 3.5$. Note that the
expected value need not be among the possible values.

Example

When comparing protein sequences we score pairs of amino acids. Let the score
between amino acids a and b be R_{ab}. Let us randomly pick two amino acids, where
the probability for a to be picked is p_a. Then the expected score between the two
picked amino acids is

$$E = \sum_{a,b \in \mathcal{M}} p_a p_b R_{ab}.$$

△

The expected value is often approximated by the *mean* of some observed values of
the variable in question.

A.5   Tables, Vectors and Matrices

A *table* is a structure containing elements. It may have dimension one (a *vector*), or two or higher (a *matrix*). The elements of a vector are specified by one index, T_8 denotes the eighth element of the vector T.

A two-dimensional matrix consists of *rows* and *columns*. $M_{3,8}$ denotes the element in row three and column eight.

A.6   Algorithmic Language

An algorithm is a stepwise description of a method or procedure. It is written as sentences in a simple language with keywords. We describe the most basic structures/elements in the notation we use in this book.

Constant. const is used for specifying constants, and = is used for specifying the value. *Example:* **const** $C = 4$ means that the constant C has value 4, and this never changes.

Variable. var is used for specifying variables. *Example:* **var** x.

Assignment. := is used for variable assignment. *Example:* $x := 4$ assigns the value 4 to the variable x.

Delimiters. ; is used as a delimiter between sentences on the same line. A line break is also a delimiter. *Example:* $x := 4$; $y := 3$.

A.6.1   Alternatives

Alternative 1. if logical expression **then** statements **end**
   If the logical expression is true, then statements are executed.
   Example: **if** $x = 4$, **then** $y := 2$ **end**
   If the variable x has the value 4, then 2 is assigned to the variable y.

Alternative 2. if logical expression **then** statements-1 **else** statements-2 **end**
   If the logical expression is true, then statements-1 are executed, otherwise statements-2 are executed.
   Example: **if** $x = 4$ **then** $y := 2$ **else** $z := 4$; $y := 3$ **end**
   If the variable x has the value 4, then 2 is assigned to y, otherwise 4 is assigned to z and three to y.

Alternative 3. if logical expression-1 **then** *A generalization of Alternative 2*
      statements-1
   elseif logical expression-2 **then**
      statements-2

   . . .

elseif logical expression-n **then**
    statements-n
else
    statements-$n + 1$
end

A.6.2 Loops

Loop 1. while logical expression-1 **do** statements **end**
    statements are executed as long as the logical expression is true.
    Example: $i := 1$; sum $:= 0$; **while** $i < 5$ **do** sum $:=$ sum $+ x_i$; $i := i + 1$ **end**
    sum will get the value $\sum_{i=1}^{4} x_i$.

Loop 2. repeat statements **until** logical expression
    statements is executed until the logical expression becomes true.
    Example: $i := 1$; sum $:= 0$; **repeat** sum $:=$ sum $+ x_i$; $i := i + 1$ **until** $i = 5$
    sum will get the value $\sum_{i=1}^{4} x_i$.

Loop 3. for the values of a variable **do** statements **end**
    statements are executed for each specified value.
    Examples:
    sum $:= 0$; **for** $i := 1$ **to** 5 **do** sum $:=$ sum $+ x_i$ **end**
    sum $:= 0$; **for** $i \in \{2, 7, 3, 5\}$ **do** sum $:=$ sum $+ x_i$ **end**
    for each element $x \in U$ **do**
        if $x \notin V$ **then** $V := V \cup \{x\}$ **end**
    end
    In the last **for** sentence V becomes the union of V and U.

A.7 Complexity

Though we get faster and faster computers, there still exist a lot of problems which take an unsatisfactorily long amount of time to solve. It is often possible to solve the problem for small datasets, but it might be impossible for larger ones. The *time complexity* of an algorithm describes how the processing time grows with increasing data size (input size). For simplicity we assume that the size of the dataset can be described by the value of a variable n, and the time for solving the problem can be given as a function of n. We can also analyse the *space complexity*, the space needed for executing an algorithm.

Example

Let T be the time it takes to add two numbers. The time it takes to add n numbers is then $(n - 1)T$. The time increases *linearly* with n, and we say that the time complexity is $O(n)$ (of order n).

Table A.3   The time for different polynomial, exponential and factorial time complexity functions, when the time for running the program for $n = 1$ is 0.00001 s.

Complexity function	$n = 10$		Size n of input $n = 20$		$n = 30$		$n = 50$	
n	0.0001	sec	0.0002	sec	0.0003	sec	0.0005	sec
n^2	0.001	sec	0.004	sec	0.009	sec	0.025	sec
n^4	0.1	sec	1.0	sec	8.1	sec	22.5	sec
n^5	1	sec	32	sec	4.1	min	52.1	min
2^n	0.005	sec	52	sec	89.4	min	178	years
3^n	0.3	sec	290	min	32.6	years	$11 \cdot 10^8$	cent
$n!$	36	sec	$7.6 \cdot 10^3$	cent				

A method (dynamic programming) for comparing two sequences (protein or DNA) of length n, involves filling in values in the cells of a two-dimensional table with n rows and n columns. Filling in one cell takes a constant time (C), so the time for filling in all cells takes time Cn^2. The time for the comparison grows as n^2, hence the time complexity is $O(n^2)$. The time complexity for comparing k sequences each of length n, using a trivial extension of this algorithm, is $O(n^k)$.   △

If the complexity is $O(n^k)$ (for a constant k), we say that it is polynomial, and if it is $O(k^n)$, it is exponential (for $k > 1$). Suppose that it takes 0.00001 s to run a program for data size $n = 1$. The times it takes for other values of n, for different time complexity functions, are shown in Table A.3.

We see that the time grows rapidly for performing algorithms with exponential time, hence problems which cannot be solved by polynomial time algorithms can be impossible to solve when the size of the input grows.

A special class of problems is the class of *NP-complete* problems. The problems in this class are all equivalent in a strong sense: *if there exists an algorithm which can solve one of the problems in polynomial time, all the problems in the class can be solved in polynomial time*. It is generally believed that none of the NP-complete problems can be solved in polynomial time, hence all the problems are believed to require exponential time. NP-hard problems in a sense at least as hard as NP-complete problems; NP-complete problems are decision problems while NP-hard problems may be optimization problems.

Appendix B

Introduction to Molecular Biology

Biology is the study of life and molecular biology is then the study of life at a molecular level of resolution. The area has undergone a revolution during the last few decades. This is largely due to the development of new technology making it possible to study biological mechanisms at the molecular level in a way never possible before. In this chapter we give an overview of the basic principles of molecular biology that should be accessible to readers with only a minimal training in biology. This means that we have avoided many details, some of which even constitute exceptions to the simple picture we have portrayed.

B.1 The Cell and the Molecules of Life: DNA–RNA Proteins

All living organisms are made of *cells*, where (in higher-order organisms) groups of cells are linked to perform specialized functions. A cell is bounded by a lipid membrane, and filled with an aqueous solution of molecules. The *eukaryotic* type of cells (cells of higher organisms) contain a *nucleus*, surrounded by the cytoplasm. The fluid matrix of the cytoplasm is called the cytosol. The rest of the cell is constituted by structural and functional units called organelles; see Figure B.1 for a schematic figure of a eukaryotic cell. In lower organisms (consisting of one cell only) such as bacteria, the cells have no nucleus, and such cells are called *prokaryotic*.

Living organisms encode information about their form and function in a molecule called DNA (deoxyribonucleic acid). The DNA molecule is so large that each cell has just one (in prokaryotes) or several (in eukaryotes) and the same DNA molecules are found in all cells of the organism. The DNA molecules contain the genes of the organisms and are located in the nucleus in eukaryotic cells. The term gene was introduced by Gregor Mendel in 1866 as the unit responsible for expression and passing on of a single characteristic to the next generation. In molecular terms, the

Protein bioinformatics: an algorithmic approach to sequence and structure analysis
I. Eidhammer, I. Jonassen and W. R. Taylor © 2004 John Wiley & Sons, Ltd ISBN: 0-470-84839-1

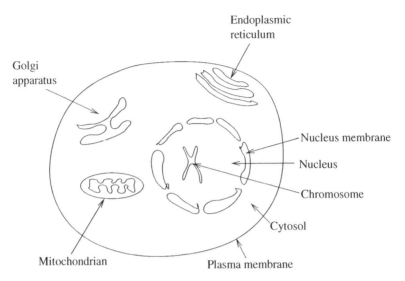

Figure B.1   A typical eukaryotic cell with some of the most important organelles.

DNA is copied into duplicate molecules at cell division with one copy going into each daughter cell. When these cells are sperms or eggs, then the genes get passed to a new individual.

DNA is constructed by four different basic units called *nucleotides*, by which information is stored. Each nucleotide contains phosphate, sugar, and one of the four bases: adenine, guanine, cytosine and thymine (denoted by A, G, C and T). The structure of DNA is a double helix (which Watson and Crick discovered in 1953). The double helix is formed by two intertwined chains (each forming a helix) with each strand being a polymer consisting of nucleotides chained together by phosphodiester bounds. The two strands are held together by *hydrogen bonds*. These bonds are formed by pairs of bases, with each base pair consisting of one purine base (A or G) and one pyrimidine base (C or T) paired according to the following rule: A pairs with T and G pairs with C. This means that if one is given the sequence of bases along one of the helices in a double helix, it is straightforward to deduce the sequence of the other. See Figure B.2 for a schematic of DNA.

This pairing is used by the cell when it needs to make a copy of its DNA, for example, when dividing into two. The copying 'machine' separates the strands of the double helix and 'reads' the bases of one of them to assemble a complementary chain. For example, the sequence AGTTCCT assembles complementary bases which pair up with TCAAGGA. This is called *DNA replication*.

The DNA is then responsible for storing and passing on to the next generation the information encoded in the sequence of bases along the organism's DNA molecules. Collectively, the DNA molecules of an organism (or a cell in a multicellular organism) is called its *genome*. However, the building of the structures of a living organism, supervision and catalysis of chemical processes, etc., is mostly performed by another

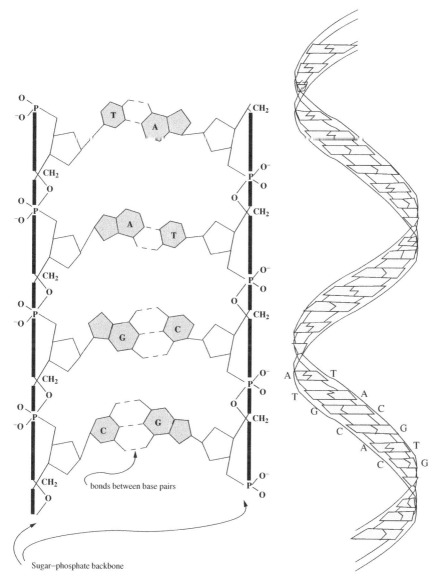

Figure B.2 Illustration of DNA with the different nucleotides and the hydrogen bonds to form the double helix.

type of molecule, called *protein*. A protein is built from one or more *polypeptides*. A polypeptide is a molecule consisting of a chain of *amino acids* (see Figure B.3(c)). There are 20 different types of amino acids, all with an amino group, a carboxyl group, a central carbon atom, and a side chain, which are different for each amino acid, see Figure B.3(a) and (b). The amino acids along the polypeptide chain are often referred

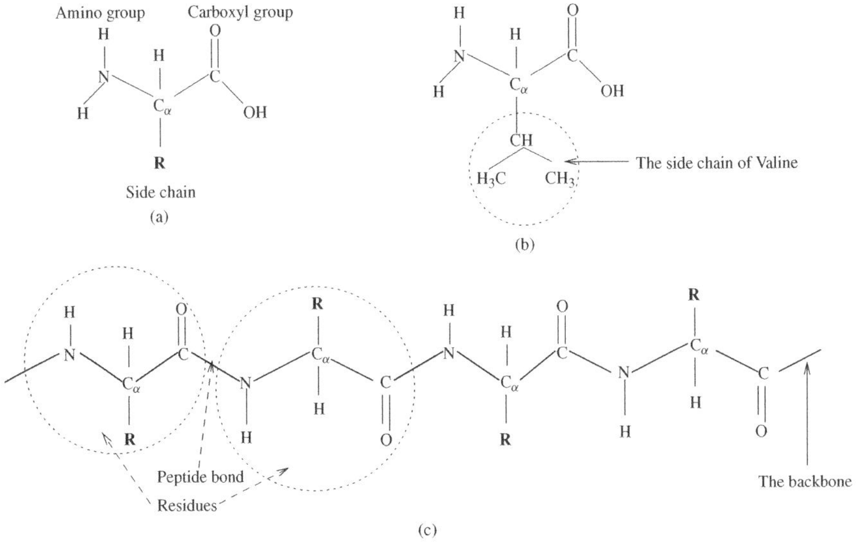

Figure B.3 (a) All 20 amino acids have the same form, but different side chains. (b) The amino acid valine with its side chain. (c) A polypeptide is formed by chaining amino acids. The bonds between consecutive amino acids are called peptide bonds. When two amino acids bond, a water molecule (H_2O) is released.

to as *residues*. The bond between two consecutive residues is called the *peptide bond*, and the chain along the atoms N–C_α–C–N$\cdots$ is called the *backbone*. The chain is said to run from the amino (or N) terminus (where there is a free NH_3 group) to the carboxy (or C) terminus (where there is a carboxylic acid group).

A third type of molecule present in all living cells is RNA, ribonucleic acid. Like DNA, RNA is also formed by purine and pyrimidine nucleotides but uracil takes the place of thymine, so the bases in RNA are denoted A, G, C and U. While DNA normally forms a double helix, RNA normally exists in single-stranded structures. RNA molecules are much shorter and much more numerous than DNA and perform a variety of different functions in the cell. One of the most important is mRNA (messenger RNA), which carries genetic information from the DNA to the machinery where it is translated into protein.

B.2   Chromosomes and Genes

The nucleus of the eukaryotic cells contain *chromosomes*. A chromosome is a long DNA molecule packed together with proteins (mainly histones and together the complex is called chromatin). Every somatic (body) cell usually includes two copies of each chromosome (excluding the sex X, Y chromosomes). The number of chromosomes vary among species. Humans have 22 pairs of chromosomes plus the sex chromosomes.

A gene is a specific sequence of nucleotide bases along a chromosome carrying information for constructing a protein (or an RNA, collectively called a gene product). Genes are parts of the chromosomes. The proportion of the genome constituting genes varies a lot from organism to organism. In yeast (a primitive eukaryotic organism), genes occupy about 70–80% of the DNA, in some bacteria over 90%. In humans, genes constitute only about 5–10% of the DNA. The rest of the DNA has sometimes been called 'junk DNA' since it is unclear what its role is.

In the simplest case the gene consists of a contiguous stretch of bases along the DNA. This is the case in prokaryote organisms. In eukaryotes the gene is often made up of smaller pieces (exons) interspersed by so-called noncoding regions (introns). When the gene is expressed, the whole DNA region containing all the exons and introns of the gene is transcribed (e.g. 'copied to mRNA'). Then the introns are spliced out and the remaining (mature) mRNA is translated to a protein (see Figure B.5). The cell's machinery 'knows' how to do this, but it is not easy to figure out the rules used: given the sequence of bases of all the DNA of a eukaryotic organism, finding all the genes and their structure in terms of exons and introns is an extremely difficult task. To complicate things further, the splicing may be done in alternative ways (including different subsets of the exons in a mature mRNA) giving rise to a set of different proteins from the same gene. An example of a description of a DNA fragment containing a gene is shown in Figure B.4.

B.3   The Central Dogma of Molecular Biology

The central dogma states the relationships between DNA, RNA and protein. DNA is read to produce an mRNA molecule (messenger RNA) containing the same information. This process is called *transcription*. Transcription produces mRNA molecules that are transported to the *translation* machinery (ribosome) which takes as input the information encoded in the sequence of bases along the mRNA and produces a polypeptide which is going to be (part of) a protein. See Figure B.5 for illustration. There are a number of other processes including post-transcriptional and post-translational modifications making it possible for one gene to give rise to a high number of different products (proteins).

B.4   The Genetic Code

Most genes code for proteins. Since there are only four different bases and 20 different amino acids, each base cannot determine one amino acid on its own. Three consecutive bases in the gene determines one amino acid to be in the protein. The *reading frame* of the gene determines which three bases from a triple go together. For example, in the DNA sequence *atggcag...*, the three alternatives would be (1) *atg, gca,...*, (2) *tgg, cag,...* and (3) *ggc,....* Both strands can also be used as coding, resulting in six different reading frames.

```
ID   HSINSU      standard; DNA; HUM; 4992 BP.
.........
DT   30-MAR-1995 (Rel. 43, Last updated, Version 6)
DE   Human gene for preproinsulin, from chromosome 11. Includes a highly
DE   polymorphic region upstream from the insulin gene containing tandemly
DE   repeated sequences.
KW   germ line; insulin; repetitive sequence; signal peptide; tandem repeat.
............
............

FT   conflict        2101..2101  /note="G is AGG in [2]"
FT   misc_feature    2156..2161  /note="TATAAA (Hogness) box"
FT   precursor_RNA   2186..3615  /note="preproinsulin primary transcript"
FT   mRNA            2186..2227  /note="preproinsulin mRNA (part 1)"
FT   intron          2228..2406  /note="intron"
FT   variation       2401..2401  /note="T can be A (see [2])"
FT   mRNA            2407..2610/note="preproinsulin mRNA (part 2)"
FT   CDS             join(2424..2610,3397..3542) /db_xref="SWISS-PROT:P01308"
FT                   /product="preproinsulin" /protein_id="CAA23828.1"
FT                   /translation="MALWMRLLPLLALLALWGPDPAAAFVNQHLCGSHLVEALYLVCGERGF
FT                   FYTPKTRREAEDLQVGQVELGGGPGAGSLQPLALEGSLQKRGIVEQCCTSICSLYQLENYCN"
FT   sig_peptide     2424..2495  /note="peptide pre (signal peptide)"
FT   mat_peptide     2496..2585  /note="proinsulin peptide B"
FT   mat_peptide     join(2586..2610,3397..3476)  /note="proinsulin peptide C(part 1)
FT                   (2610 is 1st base in codon)"
FT   intron          2611..3396  /note="intron"
FT   conflict        3068..3068  /note="G is GG in [2]"
FT   variation       3229..3229  /note="G can be C (see [2])"
FT   mRNA            3397..3615  /note="preproinsulin mRNA (part 3)"
FT   mat_peptide     3477..3539  /note="proinsulin peptide A"
FT   variation       3551..3551  /note="T can be C (see [2])"
FT   variation       3564..3564  /note="A can be C (see [2])"
.............
SQ   Sequence 4992 BP; 849 A; 1553 C; 1755 G; 835 T; 0 other;
     ctcgaggggc ctagacattg ccctccagag agagcaccca acaccctcca ggcttgaccg        60
     ................
     agaccgccg ggaggcagag gacctgcagg gtgagccaac cgcccattgc tgcccctggc       2640
     cgcccccagc caccccctgc tcctggcgct cccacccagc atgggcagaa gggggcagga       2700
     ggctgccacc cagcaggggg tcaggtgcac ttttttaaaa agaagttctc ttggtcacgt       2760
     cctaaaagtg accagctccc tgtggcccag tcagaatctc agcctgagga cggtgttggc       2820
     ttcggcagcc ccgagataca tcagagggtg ggcacgctcc tccctccact cgcccctcaa       2880
     acaaatgccc cgcagcccat ttctccaccc tcatttgatg accgcagatt caagtgtttt       2940
     gttaagtaaa gtcctgggtg acctggggtc acagggtgcc ccacgctgcc tgcctctggg       3000
     cgaacacccc atcacgcccg gaggaggcg tggctgcctg cctgagtggg ccagacccct        3060
     gtcgccagcc tcacggcagc tccatagtca ggagatgggg aagatgctgg ggacaggccc       3120
     tggggagaag tactgggatc acctgttcag gctcccactg tgacgctgcc ccggggcggg       3180
     ggaaggaggt gggacatgtg ggcgttgggg cctgtaggtc cacacccagt gtgggtgacc       3240
     ctccctctaa cctgggtcca gcccggctgg agatgggtgg gagtgcgacc tagggctggc       3300
     gggcaggcgg gcactgtgtc tccctgactg tgtcctcctg tgtccctctg cctcgccgct       3360
     gttccggaac ctgctctgcg cggcacgtcc tggcagtggg gcaggtggag ctgggcgggg       3420
     gccctggtgc aggcagcctg cagcccttgg ccctggaggg gtccctgcag aagcgtggca       3480
     ttgtggaaca atgcgtgtacc agcatctgct ccctctacca gctggagaac tactgcaact       3540
     agacgcagcc tgcaggcagc cccacacccg ccgcctcctg caccgagaga gatggaataa       3600
     agcccttgaa ccagccctgc tgtgccgtct gtgtgtcttg ggggccctgg gccaagcccc       3660
     ..........
     agagacaggc gc                                                          4992
//
```

Figure B.4 Example (extract) of a DNA region as described in the EMBL database of the human gene for preproinsulin, from chromosome 11. Note where the introns and mRNAs are, and from where the protein sequence (marked /translation) comes. Note also that this sequence has entry P01308 in the SWISS-PROT protein database.

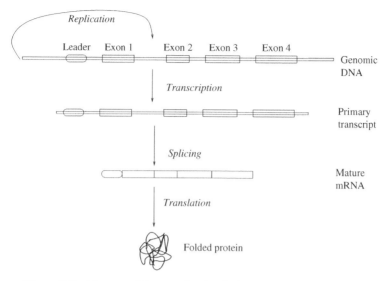

Figure B.5   The central dogma of molecular biology with transcription, splicing and the resulting protein for an eukaryotic gene.

There are 64 ($= 4^3$) different triples of bases, so some triples code for the same amino acid (the code is said to be redundant). The mapping from triples of bases to single amino acids is determined by a genetic code. More or less the same code is used by most organisms and it is therefore called the *universal genetic code* (see Table B.1). A database description of a protein sequence resulting from an eukaryotic (human) gene is shown in Figure B.4.

B.5   Protein Function

The 3D structure of a protein is determined by the sequence of amino acids along the polypeptide, and the amino acids can be characterized by their physio-chemical properties. For example, some amino acids are polar and 'like' water (hydrophilic amino acids), others are most 'happy' when not in direct contact with water (hydrophobic amino acids). Hydrophobic amino acids are more frequently found buried within the protein's 3D structure while hydrophilic ones are more frequently found on the exposed surface of proteins. The amino acids also differ in size, charge, polarity, and a number of other properties. The similarity of two amino acid's physio-chemical properties has been found to be related to how often one finds one replaced with the other in related protein structures.

Proteins have different tasks in different components (compartments) of the cell. They can be classified according to their biological function, the largest class being the enzymes. Others have structural or storage roles or serve as transport proteins while again others are 'messengers', of which insulin is an important example. The

Table B.1 The one- and three-letter codes and names for
the amino acids and the universal genetic code.

A	Ala	Alanine
C	Cys	Cysteine
D	Asp	Aspartic acid
E	Glu	Glutamic acid
F	Phe	Phenylalanine
G	Gly	Glycine
H	His	Histidine
I	Ile	Isoleucine
K	Lys	Lysine
L	Leu	Leucine
M	Met	Methionine
N	Asn	Asparagine
P	Pro	Proline
Q	Gln	Glutamine
R	Arg	Arginine
S	Ser	Serine
T	Thr	Threonine
V	Val	Valine
W	Trp	Tryptophan
Y	Tyr	Tyrosine

First position	Second position				Third position
	U	C	A	G	
U	Phe	Ser	Tyr	Cys	U
	Phe	Ser	Tyr	Cys	C
	Leu	Ser	Stop	Stop	A
	Leu	Ser	Stop	Trp	G
C	Leu	Pro	His	Arg	U
	Leu	Pro	His	Arg	C
	Leu	Pro	Gln	Arg	A
	Leu	Pro	Gln	Arg	G
A	Ile	Thr	Asn	Ser	U
	Ile	Thr	Asn	Ser	C
	Ile	Thr	Lys	Arg	A
	Met	Thr	Lys	Arg	G
G	Val	Ala	Asp	Gly	U
	Val	Ala	Asp	Gly	C
	Val	Ala	Glu	Gly	A
	Val	Ala	Glu	Gly	G

proteins are created outside the nucleus, and then transported to the compartment in
which they are to function.

B.5.1   The gene ontology

During the last few years there have been several initiatives attempting to standardize
and formalize descriptions of protein function. This is important to allow for cross-
species (and cross-discipline) analysis and for automated analysis, e.g. data mining.
One such initiative is that from the gene ontology (GO) consortium. They propose to
characterize three aspects of function for each protein: biological process, molecular
function and cellular localization. For each category they give a number of defined
terms and their relationships (e.g. is-a, is-part-of relationships). Using GO a protein's
function can be described by attaching to it at least one term (as precise as possible)
from each of the three aspects in GO.

B.6   Protein Structure

For a protein to perform its structural and/or functional role it must be in its *native*
state. Proteins fold into a three-dimensional structure which is more or less stable in
the cellular environment. To a first approximation, it is assumed that the 'native' state
of a folded protein corresponds with the minimum energy of the structure.

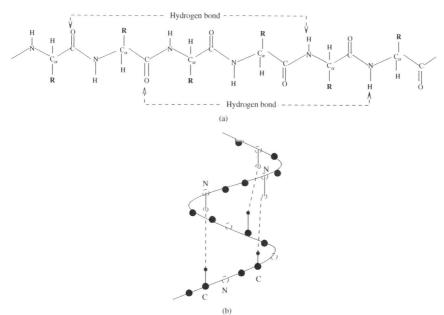

Figure B.6   (a) Schematic of the hydrogen bonding forming an α-helix. (b) For the hydrogen bonding to take place, the sequence must be formed as a helix in the space.

Since the structure is determined by the sequence of amino acids (and the environment), it means that Nature folds the protein into its three-dimensional structure simply given the sequence of amino acids along the chain. This 'folding' of the protein is not very well understood. However, it is possible (but difficult and expensive) to determine the three-dimensional native structure of (at least some) proteins by using X-ray crystallography or NMR spectroscopy.

For the protein to obtain minimal energy, bonds between different parts of the protein molecule are established (making a bond releases energy). The bonds can be between residues close in the chain (sequence)—local interactions—or between residues far away from each other in the chain, but close in the three-dimensional structure—global interactions. The local interactions that are most favourable fall into two main patterns, α-*helices* and β-*strands*, and these are called the *secondary structures* of proteins. Most proteins are made up of a number of these secondary structure elements packed against each other. The α-helices are formed by consecutive hydrogen bonds between the carboxyl group of residue i and the amino group of residue $i + 4$, as shown in Figure B.6.

The β-strands are formed by hydrogen bonds between regions on the peptide which might be far away from each other. The regions are usually from five to ten residues long, and several strands constitute a β-*sheet*. Two adjacent strands in a sheet might be parallel (going in the same direction) or antiparallel, as shown in Figure B.7.

One level of describing the structure of a protein is to specify its secondary struc-

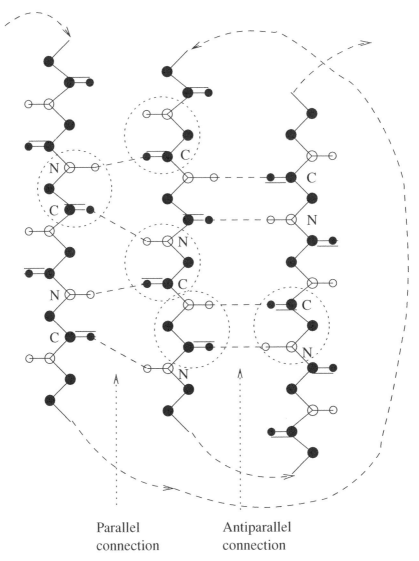

Parallel   Antiparallel
connection   connection

Figure B.7   A β-sheet formed of three β-strands, with one parallel and one antiparallel
set of H-bonds. Note that strands near in space do not need to be near in sequence.

tures, which can be a mix of helices and strands. Figure B.8 shows the secondary
structure of a small protein (insulin).

The parts of a structure not being helix or strand are called *loops*, *turns* or *coils*.
There is no universal agreement of the understanding of these concepts, and they are
used interchangeably. However, turn is often used for small regions, for example,
between two consecutive strands, forming an antiparallel connection.

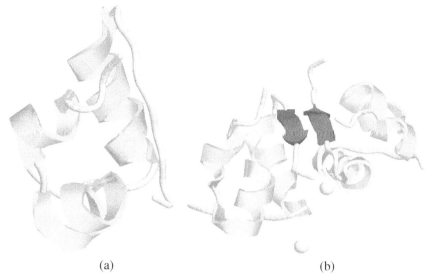

(a)   (b)

Figure B.8   The structure of insulin consists of two chains, with two alpha-helices in one of the chains and one in the other, as shown in (a). (b) They often exist in dimeric forms. Then the chains with one helix contain strands, which together make a β-sheet. Note also that there are two zinc atoms, used to keep the structure intact along with some disulphide cross-links (not shown). (Illustrated by Rasmol.) For a fuller description of the insulin structure, see http://www.rcsb.org/pdb/molecules/pdb14_1.html.

The structure of a protein can then be described at different levels.

Primary.   This is simply the order (sequence) of amino acids along the polypeptide chain.

Secondary.   Specification of the secondary structure elements along the polypeptide chain, which can be specified by giving type, the start (residue number) and the end of each secondary structure element.

Tertiary.   Description of the three-dimensional appearance of the protein (one poly-peptide), which can be given by (x, y, z) coordinates for each atom in the molecule. Each chain in Figure B.8 represents a tertiary structure.

Quaternary.   If the protein consists of more than one chain, the quaternary structure describes how these come together in space, which can be given by the (x, y, z) coordinates for each atom in the molecules.

The structures in Figure B.8 represent quaternary structures.

Many proteins bind to other molecules, and this is done at the *active site* or the *binding site*. The former is found in enzymes and contains the residues that perform the catalysis on a *substrate* (the substance upon which the enzyme acts). Binding sites cover other types of interaction. Both types of site are constituted by a set of residues, near in space, but not necessarily near in sequence. For example, the enzyme trypsin is a *protease* (cuts up other proteins) and has an active site comprising three residues (His, Ser, Asp), all brought together from different parts of the sequence.

B.7   Evolution

As explained above, the hereditary material is DNA which is transmitted from one generation to the next. However, the fidelity of this transmission is not perfect and in sexually reproducing species, the DNA from two individuals (father and mother) is further recombined (or shuffled) to form a hybrid DNA copy in the offspring. In nonsexually reproducing species, the DNA is simply an (imperfect) copy of the 'parent' DNA. Changes in the genes of DNA will sometimes make changes in the proteins they code for. This change will sometimes make the individual perform slightly better in its environment so giving them an advantage over their 'competitors' and hence a higher chance of passing on their version of the genome to the next generation (survival of the fittest).

So, through evolution, changes are made to the DNA in the form of *mutations*: alterations in the arrangement, or amount, of the DNA. The most important changes are *substitution*, *insertion* and *deletion*. Substitution means that one or several bases in the DNA are changed, for instance, a T is changed to G. An example of an insertion is that AT is inserted at a position in the gene. *Inversion* may also happen, for example, an occurrence of ATG might change to GTA. In sexual organisms, *crossovers* occur in meiosis (cell division to produce cells, egg and sperm cells, with single copies of chromosomes) and effectively allow exchange of genetic material (stretches of DNA) between (corresponding) chromosomes.

The species we see today have evolved from more primitive organisms over billions of years. This has happened through substitutions, insertions/deletions, crossing-overs, and other events, producing individuals with variant DNA, and through selection driving the evolution towards species that are well adapted to their environments. The same is true for genes and hence for proteins. Genes are randomly changed by mutations and combined by crossovers. The variants producing proteins good at their job give rise to individuals with high fitness, and the new gene variant is more likely to succeed.

Proteins (and genes) can be grouped into *families*, which consist of proteins (or genes) that have evolved from the same protein (gene) in an ancestor species. The structure and function of proteins/genes in the same family is often the same (conserved through evolution) depending on how long ago their most recent common ancestor lived (their evolutionary distance). The distance can be estimated by comparing the sequences: high sequence similarity means that their most recent common ancestor is relatively recent.

It is not possible to trace the evolutionary events giving rise to a family of genes since the genetic material of the ancestors of today's species normally is not known. However, based on models of evolution one can search for the series of changes (represented as a tree) most likely to give rise to today's species and genes/proteins given the model. Trees explaining the evolutionary history of a set of species can be built using (nongenetic) traits, the sequences of well-conserved genes (e.g. ribosomal genes), or, for example, using gene contents and gene order.

B.8   Insulin Example

Throughout the following text, most methods and algorithms are illustrated with idealized or simplified examples (a 'toy' example). However, where possible, a real biological example is included and for consistency in this we have focused on the small protein, insulin. It is desirable to have a small protein (if only so that the sequence can be printed easily) but of the many small proteins, insulin exhibits probably the greatest variety of features, besides being a molecule of considerable medical importance. To avoid the background information on insulin appearing in scattered pieces throughout the text, an overview of the structure and function of the protein is provided here.

Physiologically, insulin, with its partner molecule *glucagon*, controls the level of sugar in the blood. When there is too much glucose in the blood (say after a meal), then insulin is secreted from the pancreas and is dispersed through the blood. From the blood, it gains access to cells generally, but acts especially on the liver, skeletal muscle and fat tissue, stimulating them to absorb glucose. Such 'messenger' molecules are called *hormones* and most are small proteins (or shorter polypeptides). The reverse action of releasing stored glucose into the blood is controlled by glucagon, which is another small protein.

To perform its function, insulin must be stored in the pancreas (so that it can be released rapidly), and must maintain a specific shape as it is this that allows it to bind specifically to a cell surface protein (the insulin receptor), through which the message is relayed to the interior of the cell. As a result, insulin is a small compact molecule with packed secondary structure that is stabilized internally by an hydrophobic core and disulphide cross-links (between cysteine residues). The insulin molecule has two separate polypeptide chains (called A and B chains) and in its stored form, it is packed as a unit of six molecules (hexamer) stabilized by zinc ions. On release, the insulin molecule remains in the dimeric form shown in Figure B.8, which has two identical copies of the whole molecule (A-plus-B chains).

Despite its final compact form, insulin begins 'life' as a longer molecule. The gene itself typically has an intron and when the spliced RNA is translated, the nascent insulin molecule (called preproinsulin, pPI) has 110 residues, more than twice as long as its final form. The amino terminal part of pPI is an hydrophobic stretch that marks the protein chain as something to be exported from the cell (or with insulin, into a storage vesicle to await release). This signal sequence is then cleaved off, leaving proinsulin, which has a middle section called the C chain that is cut-out by enzymes leaving the two end parts called the A and B chains. Note that the B chain comes first in the sequence as the shorter A chain was the first to be sequenced by Fred Sanger in 1955, long before the gene sequence was known). This convoluted route, involving the excision of a part of the original gene product is probably a 'historical' relic from

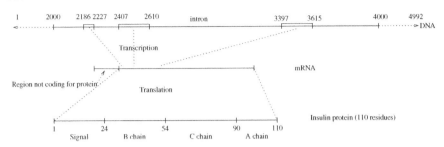

Figure B.9   Illustration of the generation of the insulin protein from its gene, and the location of the different chains of the protein.

before the time when insulin functioned as a stored molecule. The 'economic' route, to delete the part of the gene corresponding to the unwanted C chain, would leave a molecule that could not fold into an insulin-like molecule and consequently a dead individual who would not pass on that particular DNA variation. Figure B.9 illustrates the insulin protein.

As the structures of more hormones were determined, it was found that insulin was part of a large group of proteins that were similar in sequence and structure but not necessarily in function. Some of these are growth factors (so related to the function of insulin) while others, such as relaxin, have unrelated functions and act via a receptor that is completely unrelated to the insulin receptor. (Relaxin 'relaxes' tissue prior to child-birth.) There are also many insulin-like molecules in the genomes of flies and worms, the function of which can only be guessed. This rich variety of form and function makes for interesting results when comparing insulin sequences with large sequence databanks. Typically, the key cross-links (involving cysteine residues) will be found but between these may be large regions of sequence corresponding to C chains of differing lengths, making both the searches and the subsequent alignment of the sequences difficult. Moving to the DNA level, this is further compounded by the insertion of a variable length intron. So although insulin is a small molecule, it exhibits features both in structure and sequence that make it a challenging target for bioinformatics programs.

B.9   Bibliographic notes

An introduction to molecular biology can be found in Lodish et al. (1999), Lewin (1999), Nelson and Cox (2000) and Alberts et al. (2002). Protein structures are described in Branden and Tooze (1999) and Lesk (2001).

References

Alberts B, Johnson A, Lewis J, Raff M, Roberts K and Walter P 2002 *Molecular biology of the cell*. Garland Publishing, Inc.

Alesker V, Nussinov R and Wolfson HJ 1996 Detection of non-topological motifs in protein structures. *Prot. Eng.* **9**, 1103–1119.

Alexandrov N and Fischer D 1996 Analysis of topological and nontopological structural similarities in the PDB: new examples with old structures. *Proteins: Struct. Funct. Gen.* **25**, 354–365.

Alexandrov N, Takahashi K and Go N 1992 Common spatial arrantements of bacbone fragments in homologous and non-homologous proteins. *J. Mol. Biol.* **225**, 5–9.

Althaus E, Caprara A, Lenhof H and Reinert K 2002 Multiple sequence alignment with arbitrary gap costs: Computing an optimal solution using polyhedral combinatorics In *Proc. European Conference on Computational Biology 2002* (ed. Lengauer T and Lenhof H), pp. S4–S16.

Altschul S, and Gish G 1996 Local alignment statistics. *Methods Enzymol.* **266**, 460–480.

Altschul S, Boguski M, Gish W and Wootton J 1994 Issues in searching molecular sequence databases. *Nat. Gen.* **6**, 119–128.

Altschul SF 1991 Amino acid substitution matrices from an information theoretic perspective. *J. Mol. Biol.* **219**, 555–565.

Altschul SF, Bundschuh R, Olsen R and Hwa T 2001 The estimation of statistical parameters for local alignment score distribution. *Nucl. Acids Res.* **29**(2), 351–361.

Altschul SF, Carroll R and Lipman D 1989 Weights for data related by a tree. *J. Mol. Biol.* **207**, 647–653.

Altschul SF, Gish W, Miller W, Myers EW and Lipman DJ 1990 Basic local alignment search tool. *J. Mol. Biol.* **215**, 403–410.

Altschul SF, Madden TL, Schæffer AA, Zhang J, Zhang Z and Lipman DJ 1997 Gapped blast and psi-blast: a new generation of protein database search programs. *Nucl. Acids Res.* **25**(17), 3389–3402.

Argos P 1987 A sensitive procedure to compare amino acid sequences. *J. Mol. Biol.* **193**, 385–396.

Artymiuk PJ, Porrette AR, Grindley HM, Rice DW and Willett P 1994 A graph-theoretic approach to the identification of three-dimensional patterns of amino acid side-chains in protein structures. *J. Mol. Biol.* **243**, 327–344.

Protein bioinformatics: an algorithmic approach to sequence and structure analysis
I. Eidhammer, I. Jonassen and W. R. Taylor   © 2004 John Wiley & Sons, Ltd   ISBN: 0-470-84839-1

Aszódi A and Taylor WR 1994 Folding polypeptide α-carbon backbones by distance geometry methods. *Biopolymers* **34**, 489–506.

Bagley S and Altman R 1995 Characterizing the microenvironment surrounding protein sites. *Prot. Sci.* **4**, 622–635.

Bailey T and Gribskov M 2002 Estimating and evaluating the statistics of gapped local-alignment scores. *J. Comp. Biol.* **9**(3), 575–593.

Barton G 1993 An efficient algorithm to locate all locally optimal alignments between 2 sequences allowing for gaps. *CABIOS* **9**, 729–734.

Baxevanis A 2003 The molecular biology database collection:2003 update. *Nucl. Acids Res.* **31**(1), 1–12.

Benner S, Cohen M and Gonnet G 1993 Empirical and structural models for insertions and deletions in the divergent evolution of proteins. *J. Mol. Biol.* **229**, 1065–1082.

Benner S, Cohen M and Gonnet G 1994 Amino acid substitution during functionally constrained divergent evolution of protein sequences. *Prot. Eng.* **7**, 1323–1332.

Berman H, Westbrook J, Feng Z, Gilliland G, Bhat T, Weissig H, Shindyalov I and Bourne P 2000 The protein data bank. *Nucl. Acids. Res.* **28**, 235–242.

Betancourt M and Skolnick J 2001 Universal similarity measure for comparing protein structures. *Biopolymers* **59**, 305–309.

Blanchette M, Schwikowski B and Tompa M 2000 An exact algorithm to identify motifs in orthologous sequences from multiple species In *Proc. of Eighth Int. Conf. on Intelligent Systems for Molecular Biology* (ed. Altman R, Bailey T, Bourne P, Gribskov M, Lengauer T, Shindyalov I, Eyck L and Weissig H), pp. 37–45 AAAI Press.

Bowie JU, Clarke ND, Pabo CO and Sauer RT 1990 Identification of protein folds: matching hydrophobicity patterns of sequence sets with solvent accessibility patterns of known structures. *Proteins: Struct. Funct. Gen.* **7**, 257–264.

Bowie JU, Lüthy R and Eisenberg D 1991 A method to identify protein sequences that fold into a known three-dimensional structure. *Science* **253**, 164–170.

Branden C and Tooze J 1999 *Introduction to protein structure*. Garland Publishing, Inc.

Brazma A, Jonassen I, Eidhammer I and Gilbert D 1998 Approaches to the automatic discovery of patterns in biosequences. *J. Comp. Biol.* **2**, 279–305.

Brazma A, Jonassen I, Ukkonen E and Vilo J 1996 Discovering patterns and subfamilies in biosequences *Proc. of Fourth Int. Conf. on Intelligent Systems for Molecular Biology*, pp. 34–43. AAAI Press.

Brenner S, Chothia C and Hubbard T 1998 Assessing sequence comparison methods with reliable structurally identified distant evolutionary relationships. *Proc. Natl Acad. Sci* **95**(11), 6073–6078.

Brown NP, Orengo CA and Taylor WR 1996 A protein structure comparison methodology. *Comput. Chem.* **20**, 359–380.

Carrillo H and Lipman D 1988 The multiple sequence alignment problem in biology. *SIAM J. Appl. Math.* **48**, 1073–1082.

Chan S, Wong K and Chiu D 1992 A survey of multiple sequence comparison methods. *Bull. Math. Biol.* **54**(4), 563–598.

Chao K 1994 Computing all suboptimal alignments in linear space *Proc. 5th Symp. on Combinatorial Pattern Matching.*, pp. 1–14 Springer LNCS 807.

Chao K, Hardison R and Miller W 1994 Recent developments in linear-space alignment methods: a survey. *J. Comp. Biol.* **1**, 271–291.

Chew L, Huttenlocher D, Kedem K and Kleinberg J 1999 Fast detection of common geometric substructure in proteins. *Proc. 3rd Int. Conf. on Computational Molecular Biology, RECOMB'99.*

Cohen F and Sternberg M 1980 On the prediction of protein structure: the significance of the root-mean-square deviation. *J. Mol. Biol.* **138**, 321–333.

Cohen GH 1997 ALIGN: a program to superimpose protein coordinates, accounting for insertions and deletions. *J. Appl. Cryst.* **30**, 1160–1161.

Collins J, Coulsen A and Lyall A 1988 The significance of protein sequence similarities. *CABIOS* **4**(1), 61–71.

Collins JF 1993 A sequence alignment ranking function *Proceedings of the ISI 49th Session*, pp. 87–94.

Cook D, Holder L and Djoko S 1996 Scalable discovery of informative structural concepts using domain knowledge. *IEEE EXPERT* **11**(5), 59–68.

Dayhoff MO 1978 Atlas of protein sequence and structure, vol 5. Technical report, National Biomedical Research Foundation.

de Rinaldis M, Ausiello G, Cesareni G and Helmer-Citterich M 1998 Three-dimensional profiles: a new tool to identify protein surface similarities. *J. Mol. Biol.* **284**, 1211–1221.

Diamond R 1992 On the multiple simultaneous superposition of molecular structures by rigid body transformations. *Prot. Sci.* **1**, 1279–1287.

Diederichs K 1995 Structural superposition of proteins with unknown alignment and detection of topologically similarity using a six-dimensional search algorithm. *Proteins: Struct. Funct. Gen.* **23**, 187–195.

Dietmann S and Holm L 2001 Identification of homology in protein structure classification. *Nat. Struct. Biol.* **8**(11), 953–957.

Ding D, Qian J and Feng Z 1994 A differential geometric treatment of protein structure comparison. *Bull. Math. Biol.* **56**, 923–943.

Dror O, Benyamini H, Nussinov R and Wolfson H 2003 MASS: a method for multiple structural alignment by secondary structures. (Submitted for publication.)

Durbin R, Eddy S, Krogh A and Mitchison G 1998 *Biological Sequence Analysis, Probabilistic Models of Protein and Nucleic Acids.* Cambridge University Press.

Eddy S 1995 Multiple alignment using hidden Markov models. In *Proc. of Third Int. Conf. on Intelligent Systems for Molecular Biology* (ed. Rawlings C, Clark D, Altman R, Hunter L, Lengauer T and Wodak S), pp. 114–120.

Eddy S 1998 Profile hidden Markov models. *Bioinformatics* **14**, 755–763.

Eidhammer I, Jonassen I and Taylor W 2000 Structure comparison and structure patterns. *J. Comp. Biol.* **7**(5), 685–716.

Elofsson A and Sonnhammer E 1999 A comparison of sequence and structure protein domain families as a basis for structural genomics. *Bioinformatics* **15**(6), 480–500.

Escalier V, Pothier J, Soldano H and Viari A 1998 Pairwise and multiple identification of three-dimensional common substructures in proteins. *J. Comp. Biol.* **5**, 41–56.

Falicov A and Cohen F 1996 A surface of minimum area metric for the structural comparison of proteins. *J. Mol. Biol.* **258**, 871–892.

Fetrow JS and Skolnick J 1998 Method for prediction of protein function from sequence using the sequence-to-structure-to-function paradigm with application to glutaredoxins/thioredoxins and t_1 ribonucleases. *J. Mol. Biol.* **281**, 949–968.

Fischer D, Tsai CJ, Nussinov R and Wolfson H 1995 A 3D sequence-independent representation of the protein data bank. *Prot. Eng.* **8**, 981–997.

Fischer D, Wolfson H, Lin SL and Nussinov R 1994 Three-dimensional, sequence order-independent structural comparison of a serine protease against the crystallographic database reveals active site similarities: potential implications to evolution and to protein folding. *Prot. Sci.* **3**, 769–778.

Flockner H, Braxenthaler M, Lackner P, Jaritz M, Ortner M and Sippl MJ 1995 Progress in fold recognition. *Proteins: Struct. Funct. Gen.* **23**, 376–386.

Flores T, Moss D and Thornton J 1994 An algorithm for automatically generating protein topology cartoons. *Prot. Eng.* **7**(1), 31–37.

Friedrichs MS and Wolynes PG 1989 Toward protein tertiary structure recognition by means of associative memory hamiltonians. *Science* **246**, 371–373.

Friedrichs MS, Goldstein RA and Wolynes PG 1991 Generalised protein tertiary structure recognition using associative memory Hamiltonians. *J. Mol. Biol.* **222**, 1013–1034.

Gelbert W, Küstner H and Hellwich M 1977 *The VNR Concise Encyclopedia of Mathematics*. Von Nostrand Reinhold Company.

Gerstein M and Altman R 1995 Using a measure of structural variation to define a core for the globins. *CABIOS* **11**(6), 633–644.

Gerstein M and Levitt M 1998 Comprehensive assessment of automatic structural alignment against a manual standard, the scop classification of proteins. *Prot. Sci.* **7**, 445–456.

Gibbs A and McIntyre G 1970 The diagram, a method for comparing sequences. its use with amino acid and nucleotide sequences. *Eur. J. Biochem.* **16**, 1–11.

Gibrat J, Madej T and Bryant S 1996 Surprising similarities in structure comparison. *Curr. Op. Struct. Biol.* **6**, 377–385.

Gilbert D, Westhead D, Nagano N and Thornton J 1999a Motif-based searching in tops protein topology databases. *Bioinformatics* **15**, 317–326.

Gilbert D, Westhead D, Thornton J and Viksna J 1999b Tops cartoons: formalisation, searching and comparison Poster at RECOMB 99.

Godzik A 1996 The structural alignment between two proteins: is there a unique answer? *Prot. Sci.* **5**, 1325–1338.

Godzik A and Skolnick J 1994 Flexible algorithm for direct multiple alignment of protein structures and sequences. *CABIOS* **10**(6), 587–596.

Godzik A, Kolinski A and Skolnick J 1992 Topology fingerprint approach to the inverse protein folding problem. *J. Mol. Biol.* **227**, 227–238.

Gonnet G 1994 Analysis of amino acid substitution during divergent evolution: the 400 by 400 dipeptide substitution matrix. *Biochem. Biophys. Res. Commun.* **199**, 489–496.

Gonnet G, Cohen M and Brenner S 1992 Exhaust matching of the entire protein sequence database. *Science* **256**, 1443–1445.

Gotoh O 1982 An improved algorithm for matching biological sequences. *J. Mol. Biol.* **162**, 705–708.

Gotoh O 1996 Significant improvement in accuracy of multiple protein sequence alignments by iterative refinement as assessed by reference to structural alignments. *J. Mol. Biol.* **264**, 823–838.

Gribskov M 1990 Profile analysis. *Methods Enzymology*, vol. 183. Academic Press.

Gribskov M and Robinson N 1996 Use of receiver operating characteristic (roc) analysis to evaluate sequence matching. *Comput. Chem.* **20**, 25–33.

Gribskov M and Veretnik S 1996 Identification of sequence patterns with profile analysis. *Methods Enzymol.* **266**, 198–212.

Gribskov M, McLachland A and Eisenberg D 1987 Profile analysis: detection of distantly related proteins. *Proc. Natl Acad. Sci.* **84**, 4355–4358.

Grindley HM, Artymiuk PJ, Rice DW and Willett P 1993 Identification of tertiary structure resemblance in proteins using a maximal common subgraph isomorphism algorithm. *J. Mol. Biol.* **229**, 707–721.

Gupta S, Kececioglu J and Schaffer A 1995 Making the shortest-paths approach to sum-of-pairs multiple sequence alignment more space efficient in practice. In *Proc. 6th Symp. on Combinatorial Pattern Matching*, pp. 128–143.

Gusfield D 1993 Efficient methods for multiple sequence alignment with guaranteed error bounds. *Bull. Math. Biol.* **55**, 141–154.

Gusfield D 1997 *Algorithms on Strings, Trees, and Sequences.* Cambridge University Press.

Hadley C and Jones D 1999 A systematic comparison of protein structure classification: Scop, cath and fssp. *Structure* **7**, 1099–1112.

Henikoff JG and Henikoff S 1996 Using substitution probabilities to improve position-specific scoring matrices. *CABIOS* **12**, 135–143.

Henikoff S 1996 Scores for sequence searches and alignments. *Curr. Op. Str. Biol.* **6**, 353–360.

Henikoff S and Henikoff JG 1991 Automated assembly of protein blocks for database searching. *Nucl. Acids Res.* **19**, 6565–6572.

Henikoff S and Henikoff JG 1992 Amino acid substitution matrices from protein blocks. *Proc. Natl Acad. Sci.* **89**, 10915–10919.

Henikoff S and Henikoff JG 1994 Position-based sequence weights. *J. Mol. Biol.* **243**, 574–578.

Hirschberg D 1975 A linear space algorithm for computing maximal common subsequences. *Com. ACM* **18**, 341–343.

Hofmann K, Bucher P and Bairoch LFA 1999 The prosite database, its status in 1999. *Nucl. Acids Res.* **27**, 215–219.

Holliday J and Willet P 1997 Using a genetic algorithm to identify common structural features in sets of ligands. *J. Mol. Graphics and Modeling* **15**, 221–232.

Holm L and Sander C 1993a Parser for protein folding units. *Proteins: Struct. Funct. Gen.* **19**, 256–268.

Holm L and Sander C 1993b Protein structure comparison by alignment of distance matrices. *J. Mol. Biol.* **233**, 123–138.

Holm L and Sander C 1995 3-d lookup: Fast protein structure database searches at 90% reliability. In *Proc. 3rd Int. Conf. on Intelligent Systems for Molecular Biology* (ed. Rawlings C, Clark D, Altman R, Hunter L, Lengauer T and Wodak S), pp. 179–187 AAAI.

Holm L and Sander C 1996 Mapping the protein universe. *Science* **273**, 595–602.

Holm L and Sander C 1997 DALI/FSSP classification of three-dimensional protein folds. *Nucl. Acids Res.* **25**, 231–234.

Holm L and Sander C 1998a Dictionary of recurrent domains in protein structures. *Proteins: Struct. Funct. Gen.* **33**, 88–96.

Holm L and Sander C 1998b Touring protein fold space with DALI/FSSP. *Nucl. Acids Res.* **26**, 316–319.

Hooft R, Sander C and Vriend G 1997 Objectively judging the quality of a protein structure from a Ramachandran plot. *CABIOS* **13**, 425–430.

Islam SA, Luo J and Sternberg MJE 1995 Identification and analysis of domains in proteins. *Prot. Eng.* **8**, 513–525.

Johnson M and Overington J 1993 A structural basis for sequence comparison: an evolution of scoring methodologies. *J. Mol. Biol.* **233**, 716–738.

Jonassen I 1997 Efficient discovery of conserved patterns using a pattern graph. *CABIOS* **13**, 509–522.

Jonassen I, Collins JF and Higgins DG 1995 Finding flexible patterns in unaligned protein sequences. *Protein Science* **4**, 1587–1595.

Jonassen I, Eidhammer I and Taylor WR 1999 Discovery of local packing motifs in protein structures. *Proteins: Struct. Funct. Gen.* **34**, 206–219.

Jonassen I, Eidhammer I, Conklin D and Taylor WR 2000a Protein structure motif discovery and mining the PDB. In *German Conf. on Bioinformatics, GCB 2000* (ed. Bornberg-Bauer E, Rost U, Stoye J and Vingron M).

Jonassen I, Eidhammer I, Grindhaug S and Taylor W 2000b Searching the protein structure databank with weak sequence patterns and structural constraints. *J. Mol. Biol.* **304**, 599–619.

Jones D 1999a Protein secondary structure prediction based on position-specific scoring matrices. *J. Mol. Biol.* **292**, 195–202.

Jones D, Taylor W and Thornton J 1992a The rapid generation of mutation data matrices from protein sequences. *CABIOS* **8**(3), 275–282.

Jones DT 1999b GenTHREADER: an efficient and reliable protein fold recognition method for genomic sequences. *J. Mol. Biol.* **287**, 797–815.

Jones DT, Taylor WR and Thornton JM 1992b A new approach to protein fold recognition. *Nature* **358**, 86–89.

Jones S, Stewart M, Michie A, Swindells MB, Orengo C and Thornton JM 1998 Domain assignment for protein structures using a consensus approach: characterization and analysis. *Prot. Sci.* **7**, 233–242.

Just W 2001 Computational complexity of multiple sequence alignment with sp-score. *J. Comp. Biol.* **8**(6), 615–623.

Kabsch W 1978 A discussion of the solution for the best rotation to relate two sets of vectors. *Acta Cryst.* A **32**, 922–923.

Kabsch W and Sander C 1983 Dictionary of protein secondary structure: pattern recognition of hydrogen-bonded and geometrical features. *Biopolymers* **22**, 2577–2637.

Karlin S and Altschul SF 1990 Methods for assessing the statistical significance of molecular sequence features by using general scoring schemes. *Proc. Natl Acad. Sci.* **87**, 2264–2268.

Karlin S and Altschul SF 1993 Applications and statistics for multiple high-scoring segments in molecular sequences. *Proc. Natl Acad. Sci.* **90**, 5873–5877.

Karplus K and Hu B 2001 Evaluation of protein multiple alignments by sam-t99 using the balibase multiple alignment test set. *Bioinformatics* **17**, 713–720.

Karplus K, Barrett C and Hughey R 1998 Hidden Markov models for detecting remote protein homologies. *Bioinfomatics* **14**, 846–856.

Kastenmüller G, Kriegel HP and Seidl T 1998 Similarity search in 3D protein databases. In *German Conf. on Bioinformatics, GCB 98* (ed. Zimmermann O and Schomburg D). Cologne University.

Kasuya A and Thornton J 1999 Three-dimensional structure analysis of prosite patterns. *J. Mol. Biol.* **286**, 1673–1691.

Katoh K, Misawa K, Kuma K and Miyata T 2002 Mafft: a novel method for rapid multiple sequence alignment based on fast Fourier transform. *Nucl. Acid. Res.* **30**(14), 3059–3066.

Kearsley S 1990 An algorithm for the simultaneous superpostion of a structural series. *J. Comp. Chem.* **11**(10), 1187–1192.

Kim J and Pramanik S 1994 An efficient method for multiple sequence alignment. In *Proc. Second Int. Conf. on Intelligent Systems for Molecular Biology* (ed. Altman R, Brutlag D, Karp P, Lathrop R and Searls D), pp. 212–218.

Kimura M 1980 A simple method for estimating evolutionary rates of base substitutions through comparative sequences. *J. Mol. Evol.* **16**, 111–120.

Kimura M 1983 *The Neutral Theory of Molecular Evolution*. Cambridge University Press.

Kleywegt G and Jones T 1997 Detecting folding motifs and similarities in protein structures. *Methods Enzymology*, pp. 525–545. Academic Press.

Koch I, Lengauer T and Wanke E 1996 An algorithm for finding maximal common subtopologies in a set of protein structures. *J. Comp. Biol.* **3**(2), 289–306.

Lawrence C, Altschul S, Boguski M, Liu J, Neuwald A and Wootton J 1993 Detecting subtle sequence signals: a Gibbs sampling strategy for multiple alignment. *Science* **262**, 208–214.

Lee C, Grasso C and Sharlow M 2002 Multiple sequence alignment using partial order graphs. *Bioinformatics* **18**(3), 452–464.

Leibowitz N, Nussinov R and Wolfsen H 2001 Musta—a general, efficient, automated method for multiple structure alignment and detection of common motifs: application to proteins. *J. Comp. Biol.* **8**, 93–121.

Lesk A 2001 *Introduction to Protein Architecture*. Oxford University Press.

Levitt M and Gerstein M 1998 A unified statistical framework for sequence comparison and structure comparison. *Proc. Natl Acad. Sci. USA* **95**, 5913–5920.

Lewin B 1999 *Genes VII*. Oxford University Press.

Li W 1993 Unbiased estimation of the rates of synonymous and nonsynonymous substitution. *J. Mol. Evol.* **36**, 96–99.

Li W and Gu X 1996 Estimating evolutionary distances between DNA sequences. *Methods Enzymol.* **266**, 449–459.

Li WH 1997 *Molecular Evolution*. Sinauer Associates, Inc.

Lin K, May AC and Taylor WR 2002a Amino acid substitution matrices from an artificial neural network model. *J. Com. Biol.* **8**, 471–481.

Lin K, May AC and Taylor WR 2002b Threading using neural networks (TUNE): the measure of protein sequence-structure compatibility. *Bioinformatics* **18**(10), 1350–1357.

Lindahl E and Elofsson E 2000 Identification of related proteins on family, superfamily and fold level. *J. Mol. Biol.* **295**, 613–625.

Lipman D, Altscul S and Kececioglu J 1989 A tool for multiple sequence alignment. *Proc. Natl Acad. Sci.* **86**, 4412–4415.

Lipman D and Pearson W 1985 Rapid and sensative protein similarity searches. *Science* **227**, 1435–1441.

Liu J, Neuwald A and Lawrence C 1995 Bayesian models for multiple local sequence alignment and Gibbs sampling strategies. *J. Amer. Stat. Assoc.* **90**, 1156–1170.

Lodish H, Berk A, Zipursky SL, Matsudaira P, Baltimore D and Darnell J 1999 *Molecular Cell Biology*. W. H. Freeman.

Madej T, Gibrat J and Bryant S 1995 Threading a database of protein core. *Proteins: Struct. Funct. Gen.* **23**, 356–369.

Maiorov VN and Crippen GM 1994 Significance of root-mean-square deviation in comparing three-dimensional structures of globular proteins. *J. Mol. Biol.* **235**, 625–634.

Matsuda H, Taniguchi F and Hashimoto A 1997 An approach to detection of protein structural motifs using an encoding scheme of backbone conformation *PSB97*.

May ACW and Johnson MS 1995 Improved genetic algorithm-based protein structure comparisons: pairwise and multiple superpositions. *Prot. Eng.* **8**, 873–882.

McGuffin LJ, Bryson K and Jones DT 2001 What are the baselines for fold recognition? *Bioinformatics* **17**, 63–72.

McLachlan A 1972 A mathematical procedure for superimposing atomic coordinates of proteins. *Acta Cryst.* A **28**, 656–657.

McLachlan A 1979 Gene duplications in the structural evolution of chymotrypsin. *J. Mol. Biol.* **128**, 49–79.

Michie A, Orengo C and Thornton J 1996 Analysis of domain structural class using an automated class assignment protocol. *J. Mol. Biol.* **262**, 168–185.

Mizuguchi K and Go N 1995 Comparison of spatial arrangements of secondary structural elements in proteins. *Prot. Eng.* **8**, 353–362.

Mott R and Tribe R 1999 Approximate statistics of gapped alignments. *J. Comp. Biol.* **6**(1), 91–112.

Murzin AG, Brenner SE, Hubbard T and Chothia C 1995 SCOP: a structural classification of proteins database for the investigation of sequences and structures. *J. Mol. Biol.* **247**, 536–540.

Myers E and Miller W 1988 Optimal alignments in linear space. *CABIOS* **4**, 11–17.

Needleman S and Wunsch C 1970 A general method applicable to the search for similarities in the amino acid sequence of two proteins. *J. Mol. Biol.* **48**, 443–454.

Nelson D and Cox M 2000 *Lehninger Principles of Biochemistry*. Worth Publishers, Inc.

Neuwald A and Green P 1994 Detecting patterns in protein sequences. *J. Mol. Biol.* **239**, 698–712.

Notredame C and Higgins D 1996 Saga: sequence alignment by genetic algorithm. *Nucl. Acids Res.* **24**, 1515–1524.

Notredame C, Higgins D and Heringa J 2000 T-coffee: a novel method for fast and accurate multiple sequence alignment. *J. Mol. Biol.* **302**, 205–217.

Notredame C, Holm L and Higgins D 1998 Coffee: an objective function for multiple sequence alignments. *Bioinformatics* **14**(5), 407–422.

Nussinov R and Wolfson HJ 1991 Efficient detection of three-dimensional structural motifs in biological macromolecules by computer vision techniques. *Proc. Natl Acad. Sci. USA* **88**, 10495–10499.

Orengo CA and Taylor WR 1990 A rapid method for protein structure alignment. *J. Theor. Biol.* **147**, 517–551.

Orengo CA, Flores TP, Taylor WR and Thornton JM 1993 Identification and classification of protein fold families. *Prot. Eng.* **6**, 485–500.

Orengo CA, Michie AD, Jones S, Jones DT, Swindells MB and Thornton JM 1997 CATH—a hierarchic classification of protein domain structures. *Structure* **5**, 1093–1108.

Overington J, Donnelly D, Johnson MS, Šali A and Blundell TL 1992 Environment-specific amino-acid substitution tables—tertiary templates and prediction of protein folds. *Prot. Sci.* **1**, 216–226.

Pascarella S and Argos P 1992 Analysis of insertions/deletions in proteinstructures. *J. Mol. Biol.* **224**, 461–471.

Pavesi G, Mauni G and Pesole G 2001 An algorithm for finding signals of unknown length in DNA sequences. In *Proc. Ninth Int. Conf. on Intelligent Systems for Molecular Biology* (ed. Brunak S, Galisson F, Gribskov M, Krogh A, Pedersen A, Rouzé P, Stormo G and Tramontano A), pp. S207–S214.

Pearl F, Lee D, Bray J, Sillitoe I, Todd A, Harrison A, Thornton J and Orengo C 2000 Assigning genomic sequences to CATH. *Nucl. Acids Res.* **28**, 277–282.

Pearson W 1990 Rapid and sensitive sequence comparison with FASTP and FASTA. *Methods Enzymol.* **183**, 63–98.

Pearson W 1996 Effective protein sequence comparison. *Methods Enzymol.* **266**, 227–258.

Pearson W 1998 Empirical statistical estimates for sequence similarity searces. *J. Mol. Biol.* **276**, 71–84.

Pennec X and Ayache N 1998 A geometric algorithm to find small but highly similar 3D substructures in proteins. *Bioinformatics* **14**, 516–522.

Petersen K and Taylor WR 2003 Modelling zinc-binding proteins with GADGET: genetic algorithm and distance geometry for exploring topology. *J. Mol. Biol.* **325**, 1039–1059.

Petitjean M 1998 Interactive maximal common 3D substructure searching with the combined SDM/RMS algorithm. *Comput. Chem.* **22**, 463–465.

Pevzner P and Sze SH 2000 Combinatorial approaches to finding subtle signals in DNA sequences. In *Proc. of Eighth Int. Conf. on Intelligent Systems for Molecular Biology* (ed. Altman R, Bailey T, Bourne P, Gribskov M, Lengauer T, Shindyalov I, Eyck L and Weissig H), pp. 269–278. AAAI Press.

Qian N and Sejnowski T 1988 Predicting the secondary structure of globular proteins using neural network models. *J. Mol. Biol.* **202**, 865–884.

Ramachandran G, Ramakrishnan C and Sasisekharan V 1963 Stereochemistry of polypeptide chain conformations. *J. Mol. Biol.* **7**, 95–99.

Rao ST and Rossmann MG 1973 Comparison of super-secondary structures in proteins. *J. Mol. Biol* **76**, 241–256.

Reese J and Pearson W 2002 Empirical determination of effective gap penalties. *Bioinformatics* **18**, 1500–1507.

Reinert K, Stoye J and Will T 2000 An iterative method for faster sum-of-pairs multiple sequence alignment. *Bioinformatics* **16**, 808–814.

Rice DW and Eisenberg D 1997 A 3D–1D substitution matrix for protein fold recognition that includes predicted secondary structure of the sequence. *J. Mol. Biol.* **267**, 1026–1038.

Rossmann MG and Argos P 1975 A comparison of the heme binding pocket in globins and cytochrombe b5*. *J. Biol. Chem.* **250**, 7523–7532.

Rossmann MG and Argos P 1976 Exploring structural homology of proteins. *J. Mol. Biol.* **105**, 75–96.

Rost B and Sander C 1993 Prediction of protein secondary structure at better than 70% accuracy. *J. Mol. Biol.* **232**, 584–599.

Roytberg MA 1992 A search for common patterns in many sequences. *CABIOS* **8**(1), 57–64.

Rufino S and Blundell T 1994 Structure-based identification and clustering of protein families and superfamilies. *J. Computer-Aided Molecular Design* **8**, 5–27.

Russel R and Barton G 1992 Multiple protein sequence alignment from tertiary structure comparison: assignment of global and residue confidence levels. *Proteins: Struct. Funct. Gen.* **14**, 309–323.

Russel RB 1998 Detection of protein three-dimensional side-chain patterns: new examples of convergent evolution. *J. Mol. Biol.* **279**, 1211–1227.

Sadreyev R and Grishin N 2003 Compass: a tool for comparison of multiple protein alignments with assessment of statistical significance. *J. Mol. Biol.* **326**(1), 317–336.

Saitou N and Nei M 1987 The neighbour-joining method: a new method for reconstructing phylogenetic trees. *Mol. Biol. Evol.* **4**, 406–425.

Sali A and Blundell TL 1990 Definition of general topological equivalence in protein structures: a procedure involving comparison of properties and relationships through simulated annealing and dynamic programming. *J. Mol. Biol.* **212**, 403–428.

Sander C and Schneider R 1991 Homology-derived secondary structure of proteins and the structural meaning of sequence homology. *Proteins* **9**, 56–68.

Satow Y, Cohen GH, Padlan EA and Davies DR 1986. *J. Mol. Biol* **190**, 593–604.

Schæffer AA, Aravind L, Madden TL, Shavirin S, Spouge JL, Wolf YI, Koonin EV and Altschul SF 2001 Improving the accuracy of psi-blast protein database searches with composition-based statistics and other refinements. *Nucl. Acid. Res.* **29**, 2994–3005.

Schulz GE and Schirmer RH 1979 *Principles of Protein Structure*. Springer.

Sewell R and Durbin R 1995 Method for calculation of probability of matching a bounded regular expression in a random data string. *J. Comp. Biol.* **2**, 25–31.

Shapiro A, Botha JD, Pastore A and Lesk A 1992 A method for multiple superposition of structures. *Acta. Cryst.* A **48**, 11–14.

Shatsky M, Nussinov R and Wolfson H 2002a Flexible protein alignment and hinge detection. *Proteins: Struct. Funct. Gen.* **48**, 242–256.

Shatsky M, Nussinov R and Wolfson H 2002b Multiprot—a multiple protein structural alignment algorithm. In *Workshop on algorithms in bioinformatics* (ed. Guigo R and Gusfield D). Springer.

Shi J, Blundell TL and Mizuguchi K 2001 FUGUE: sequence-structure homology recognition using environment-specific substitution tables and structure dependent gap penalties. *J. Mol. Biol.* **310**, 243–257.

Siddiqui AS and Barton GJ 1995 continuous and discontinuous domains—an algorithm for the automatic generation of reliable protein domain definitions. *Prot. Sci.* **4**, 872–884.

Sippl MJ 1990 Calculation of conformational ensembles from potentials of mean force. an approach to the knowledge-based prediction of local structures in globular proteins. *J. Mol. Biol.* **213**, 859–883.

Sjølander K, Karplus K, Brown M, Hughey R, Krogh A, Mian I and Haussler D 1996 Dirichlet mixtures: a method for improving detection of weak but significant protein sequence homology. *CABIOS* **12**, 327–345.

Skolnick J and Kihara D 2001 Defrosting the frozen approximation: PROSPECTOR—a new approach to threading. *Proteins: Struct. Funct. Gen.* **42**, 319–331.

Skolnick J, Kolinski A, Kihara D, Betancourt MR, Rotkiewicz P and Boniecki M 2001 Ab initio protein structure prediction via a combination of threading, lattice folding, clustering, and structure refinement. *Proteins: Struct. Funct. Gen.* **5**, 149–156. Special CASP Issue.

Smith RF and Smith TF 1990 Automatic generation of primary sequence patterns from sets of related protein sequences. *Proc. Natl Acad. Sci.* **87**, 118–122.

Smith RF and Smith TF 1992 Pattern-induced multi-sequence alignment(pima).... *Prot. Eng.* **5**(1), 35–41.

Smith T and Waterman M 1981 Identification of common molecular subsequences. *J. Mol. Biol.* **147**, 195–197.

Sonnhammer E, Eddy S and Durbin R 1997 PFAM: a comprehensive database of protein domain families based on seed alignments. *Proteins: Struct. Funct. Gen.* **28**, 405–420.

Sonnhammer E, Eddy S, Bateman E and Durbin R 1998 PFAM: multiple sequence alignments and hmm-profiles of protein domains. *Nucl. Acids Res.* **26**, 320–322.

Stark A, Sunyaev S and Russell R 2003 A model for statistical significance of local similarities in structure. *J. Mol. Biol.* **326**, 1307–1316.

Stephen GA 1994 *String Searching Algorithms.* Lecture Notes Series on Computing, vol. 3. World Scientific.

Su S, Cook D and Holder L 1999 Applications of knowledge discovery to molecular biology: identifying structural regularities in proteins. *Pacific Symposium on Biocomputing '99.*

Sutcliffe M, Haneef I, Carney D and Blundell TL 1987 Knowledge based modeling of homologous proteins 1. 3-dimensional frameworks derived from the simultaneous superposition of multiple structures. *Prot. Eng.* **1**, 377.

Swindells MB 1995 A procedure for detecting structural domains in proteins. *Prot. Sci.* **4**, 103–112.

Swofford D, Olsen G, Waddel P and Hillis D 1996 Phylogenetic inference. In *Molecular Systematics* (ed. Hillis D), ch. 5. Sinauer Associates.

Tatusov R, Altschul S and Koonin E 1994 Detection of conserved segments in proteins: iterative scanning of sequence databases with alignment blocks. *Proc. Natl Acad. Sci.* **91**, 12091–12095.

Taylor W 1997a Random structural models for double dynamic programming score evaluation. *J. Mol. Evol.* **44**, 174–180.

Taylor W 1999a Protein structure comparison using iterated double dynamic programming. *Prot. Sci.* **8**, 654–665.

Taylor W 1999b Protein structure domain identification. *Prot. Eng.* **12**, 203–216.

Taylor W and Jones D 1993 Deriving an amino acid distance matrix. *J. Theor. Biol.* **164**, 65–83.

Taylor W, May A, Brown N and Aszódi A 2001 Protein structure: geometry, topology and classification. *Rep. Prog. Phys.* **64**, 517–590.

Taylor WR 1986 The classification of amino acid conservation. *J. Theor. Biol.* **119**, 205–218.

Taylor WR 1990 Hierarchical method to align large numbers of biological sequences. In *Methods Enzymol.*, vol 183 (ed. Doolittle R). Academic Press.

Taylor WR 1997b Multiple sequence threading: an analysis of alignment quality and stability. *J. Mol. Biol.* **269**, 902–943.

Taylor WR 2002 A periodic table for protein structure. *Nature* **416**, 657–660.

Taylor WR and Orengo CA 1989 Protein structure alignment. *J. Mol. Biol.* **208**, 1–22.

Taylor WR and Orengo CA 1997 Protein fold topology and structural families. In *Protein: A Comprehensive Treatise* (ed. Allen G), pp. 143–169. JAI Press Inc.

Taylor WR, Flores TP and Orengo CA 1994 Multiple protein structure alignment. *Prot. Sci.* **3**, 1858–1870.

Taylor WR, Thornton JM and Turnell WG 1983 A elipsoidal approximation of protein shape. *J. Mol. Graphics* **1**, 30–38.

Thomas D 1994 The graduation of secondary structure elements. *J. Mol. Graphics* **12**, 146–152.

Thompson J, Higgins D and Gibson T 1994a Improved sensitivity of profile searches through the use of sequence weights and gap excision. *CABIOS* **10**(1), 19–29.

Thompson J, Plewniak F and Poch 1999a Balibase: a benchmark alignment database for the evoluation of multiple alignment programs. *Bioinformatics* **15**, 87–88.

Thompson J, Plewniak F and Poch 1999b A comprehensive comparison of multiple sequence alignment program. *Nucl. Acids Res.* **27**, 2682–2690.

Thompson JE, Higgins D and Gibson T 1994b Clustal W: improving the sensitivity of progressive multiple sequence alignemnt through sequence weighting, position-specific gap penalties and weight matrix choice. *Nucl. Acids Res.* **22**, 4673–4680.

Verbitsky G, Nussinov R and Wolfson H 1999 Structural comparison allowing hinge bending, swiveling motions. *Proteins: Struct. Funct. Gen.* **34**(2), 232–254.

Vilo J 1998 Discovering frequent patterns from strings. Technical Report C-1998-9, Department of Computer Science, University of Helsinki.

Vingron M and Argos P 1991 Motif recognition and alignment for many sequences by comparison of dot-matrices. *J. Mol. Biol.* **218**, 33–43.

Vingron M and Sibbald P 1993 Weighting in sequence space: a comparison of methods in terms of generalized seqeunces. *Proc. Natl Acad. Sci.* **90**, 8777–8781.

Vogt G, Etzold T and Argos P 1995 An assessment of amino acid exchange matrices: the twilight zone re-visited. *J. Mol. Biol.* **249**, 816–831.

Vriend G and Sander C 1991 Detection of common three-dimensional substructures in proteins. *Proteins: Struct. Funct. Gen.* **11**, 52–58.

Wako H and Yamato T 1998 Novel method to detect a motif of local structures in different protein conformations. *Prot. Eng.* **11**, 981–990.

Wallace AC, Borkakoti N and Thornton JM 1997 TESS: a geometric hashing algorithm for deriving 3D coordinate templates for searching structural databases. Applications to enzyme active sites. *Prot. Sci.* **6**, 2308–2323.

Wallace AC, Laskowsi RA and Thornton JM 1996 Derivation of 3D coordinate templates for searching structural databases: applications to ser-his-asp catalytic triads in the serine proteinases and lipases. *Prot. Sci.* **5**, 1001–1013.

Waterman M and Eggert M 1987 A new algorithm for best subsequence alignment with application to TRNA–RRNA comparisons. *J. Mol. Biol.* **197**, 723–728.

Waterman MS 1995 *Introduction to Computational Biology*. Chapman and Hall.

Webber C and Barton G 2001 Estimation of *p*-values for global alignments of protein sequences. *Bioinformatics* **17**, 1158–1167.

Wolferstetter F, French K, Herrmann G and Werner T 1996 Identification of functional elements in aligned nucleic acid sequences by a novel tuple search algorithm. *CABIOS* **12**(1), 71–80.

Wolfson H 1997 Geometric hashing: an overview. *IEEE Comp. Science and Eng.* October–December, pp. 10–21.

Zhang Z, Schwartz S, Wagner L and Miller W 2000 A greedy algorithm for aligning DNA sequences. *J. Comp. Biol.* **7**, 203–214.

Zu-Kang F and Sippl MJ 1996 Optimum superimposition of protein structures: ambiguities and implications. *Folding and Design* **1**, 123–132.

Zuker M 1991 Suboptimal sequence alignment in molecular biology, alignment with error analysis. *J. Mol. Biol.* **221**, 403–420.

Index

3_{10}-helix, 176

AACH, 102
activation value, 298
active site, 333
algorithm, 319
  backtracking, 12
  BLAST, 42
  dynamic programming, 10
  forward-recursion with pruning,
    70
  PGMA, 82
  progressive multiple alignment, 87
alignment, 6, 187
  distribution, 119
  global, 3
  local, 26
α-helix, 176, 331
amino acid, 325
  class hierarchy, 102
  hydrophilic, 329
  hydrophobic, 329
analogous, 285
anchor, 273
architecture, 167
artificial neural network, 298
associative memory Hamiltonians,
  314
attractor, 287, 288

Protein bioinformatics: an algorithmic
approach to sequence and structure analysis
I. Eidhammer, I. Jonassen and W. R. Taylor
© 2004 John Wiley & Sons, Ltd
ISBN: 0-470-84839-1

back propagation, 299
backbone, 326
backtracking, 10
basis, 212
β-sheet, 331
β-strand, 331
bias matrix, 203
binding site, 333
binomial coefficient, 315
biological clock, 104
bit, 57, 146
blank, 4
block, 113
BLOSUM, 113
Boolean algebra, 316
bootstrapping, 85
bridge, 181
bucket, 214

CAFASP, 313
CASP, 313
CATH, 288
cell, 323
character-based reconstruction, 79
characteristic value, 52
chirality, 168
chromosome, 326
cis, 174
CLUSTAL, 94
clustering, 89
  average linkage method, 90
  by relation, 243
  by transformation, 237
  complete linkage method, 90

Printed and bound by CPI Group (UK) Ltd, Croydon, CR0 4YY

27/10/2024

14580157-0003